Transfer and Storage of Energy by Molecules

Volume 3

Rotational Energy

Transfer and Storage of Energy by Molecules

(A Multi-volume Treatise)

Volume 1 : Electronic Energy
Volume 2 : Vibrational Energy
Volume 3 : Rotational Energy
Volume 4 : Solid State Systems
Volume 5 : Biological Systems

Transfer and Storage of Energy by Molecules

Volume 3
Rotational Energy

Edited by

George M. Burnett *Professor of Physical Chemistry, University of Aberdeen*

Alastair M. North *Professor of Physical Chemistry, University of Strathclyde, Glasgow*

WILEY — INTERSCIENCE
A division of John Wiley & Sons Ltd
London — New York — Sydney — Toronto

Library of Congress catalog card number 77–78048

ISBN 0 471 12432 X

Printed in Great Britain by J. W. Arrowsmith Ltd., Bristol 3

Preface

This is the third volume in this series dealing with molecular energy transfer processes. In the articles comprising this text emphasis is placed on the significance of molecular rotation in a variety of apparently unrelated phenomena.

Energy transfer by molecular rotation is not an easy subject to review. This is because rotation appears briefly in innumerable theoretical and phenomenological discourses, but so often is specifically disregarded almost as soon as introduced. To establish at the outset the present position of rotational processes in our understanding of fluids we have asked Professors Dahler and Hoffman to summarize current theories of transport and relaxation processes in polyatomic fluids, and to show how these theories might apply to rotation-dependant relaxation phenomena.

Dielectric methods form the most widely used technique for the observation of molecular rotation, both in gases and in condensed phases. The second chapter of this volume summarizes the scope of this technique and the kind of information which can be obtained from it.

Although a wide range of physico-chemical phenomena have their origin in molecular rotation, there are two in particular where energy transfer is of paramount importance. These are viscoelasticity and magnetic spin relaxation. Full coverage of either subject could demand a volume in its own right. However in chapters three and four the ways in which energy transfer and rotation interact to give the observed phenomena are presented.

Contributing Authors

BULLOCK, A. T.	*Lecturer in Chemistry, University of Aberdeen, Scotland.*
DAVIES, MANSEL	*Professor in Chemistry, University College of Wales, Aberystwyth.*
DAHLER, J. S.	*Professor of Chemical Engineering and Chemistry, University of Minnesota, Minneapolis, U.S.A.*
EISENBERG, A.	*Professor of Chemistry, McGill University, Montreal, Canada.*
HOFFMAN, D. K.	*Institute for Atomic Research and Department of Chemistry, Iowa State University, Ames, Iowa, U.S.A.*

Glossary of Symbols

In Chapter 1 certain of the symbols below are given specific connotations by the addition or variation of indices. Where this occurs the significance is explained in the text. Particular reference may be made to Tables 1.1 and 1.2.

A	constant from Williams–Landel–Ferry equation.
A	Helmholtz free energy (Chapter 3): area of a closed loop (Chapter 1): local integral of motion in a fluid (Chapter 1).
$\hat{A}$	time independent quantum operator corresponding to classical probability density.
$\mathscr{A}$	electric permeability tensor.
α	polarizability tensor.
$a_n(t)$	scalar expansion coefficients.
$a_{(n)}$	nuclear spin operator functions.
B	constant from Williams–Landel–Ferry equation.
$\mathscr{B}$	dimensionless variable describing interaction of a molecule with a magnetic field.
b	macromolecular dimension parameter.
C	momentum flux.
$\mathbf{C}$	velocity function (Chapter 1): spin rotation interaction tensor (Chapter 4).
C_1, C_2	constants of Williams–Landel–Ferry equation.
c	concentration (Chapter 3): velocity of light (Chapter 2).
c_j	velocity of particle, j.
c_{int}	internal heat capacity.
$\hat{c}_\iota$	rotational heat capacity (per unit mass).
D	subsidiary function.
D_r	rotational diffusion coefficient.
$\mathbf{D}$	electric displacement vector.

d	molecular skeletal length.
E	elastic modulus: internal energy.
$\bar{E}$	effective electric field.
$\bar{E}_L$	local electric field.
$\bar{E}_0$	applied electric field (Chapter 2): Arrhenius activation energy (Chapter 4).
E_m	nuclear magnetic energy levels.
ΔE_{anis}	anisotropic intermolecular interaction energy.
e	molecular internal energy (energy density).
$\mathbf{e}$	unit vector.
eQ	nuclear electric quadrupole moment.
$\mathbf{F}, \mathbf{F}(x)$	time independent electric or magnetic field.
$\mathscr{F}$	functions derived from integration of fluxes (see Table 1.2).
f	stress; fractional free volume.
f_s	shear stress.
f_g	fractional free volume at glass transition temperature.
$f_0(t)$	correlation function for first order spherical harmonies.
f_r	microviscosity factor.
f_A, f_B	mole fractions of A, B in a binary mixture.
f_{JK}	fraction of molecules in appropriate state for energy absorption.
$f^{(N)}(t, \Omega)$	density of points in vicinity of Ω in phase space.
$f_{(n)}(t)$	lattice variable function in magnetic spin relaxation.
G	shear modulus.
G^*	complex shear modulus.
G'	shear storage modulus.
G''	shear loss modulus.
$\mathbf{G}$	probability density.
$G(\Omega)$	probability density.
G_j	local probability variable.
ΔG^*	activation free energy.
$\Delta G_\mu{}^*$	activation free energy for dipole orientation.
$\Delta G_\eta{}^*$	activation free energy for viscous flow.
$\mathscr{G}_i^{ik}$	collision integral (basis function).
$g_n(t)$	reduced correlation functions.
g	gravitational constant: general function: 'g-factor' for rotation charge.
$\mathbf{g}$	velocity of centre of mass of molecule normal to point of contact.

H	Hamiltonian.
$\hat{H}$	energy operator.
$H_1(\tau)$	distribution of relaxation times.
H_1	alternating magnetic field along x-axis.
H'	local magnetic field.
H''	magnetic field from dipole–dipole interactions.
$\mathbf{H}$	magnetic field.
$\mathbf{H}_{sr}$	spin rotation magnetic field.
H_0	magnitude of **H**.
$\mathscr{H}$	Hamiltonian under specific conditions such as nucleus in a magnetic field.
$\Delta\mathrm{H}^*$	activation enthalpy (Chapter 4).
$\Delta\mathrm{H}_\mathrm{E}^*$	Eyring activation enthalpy (Chapter 2).
$\Delta\mathrm{H}^*(\mathrm{T}\tau)$	dielectric activation enthalpy.
$\Delta\mathrm{H}^*(\eta)$	viscosity enthalpy.
h	Planck's constant : general function.
$\hbar$	Planck's constant.
I	electric current.
I_1	collected coordinates and conjugate momenta.
I^+, I^-	shift operators.
$\mathbf{I}$	spin angular momentum operator.
$\hat{\mathbf{I}}$	nuclear spin moment.
i	$\sqrt{-1}$.
J	principal rotational quantum number.
J_G	conductive contribution to flux.
$\mathbf{J}$	angular momentum index.
$J_\mathrm{n}(\omega)$	spectral density function.
$\mathbf{J}_\mathrm{s}$	flux of entropy.
$\hat{\mathbf{J}}$	total angular momentum of molecule.
$\mathscr{J}$	set of action variables.
$\mathbb{J}_G(t, \mathbf{x})$	flux of G at t and **x** in phase space.
J	spin flux.
J_P	polarization flux.
$\mathbf{j}$	spin flux.
j	$\sqrt{-1}$
K	kinetic energy : Fredholm kernel.
$\mathbf{K}$	collisional impulse.
$\hat{\mathrm{K}}_{12}$	pair kinetic energy.
k	friction factor : rate constant for orientation decay.

$\mathbf{k}$	Boltzmann's constant.
L	length of a rubber element.
L_{nm}	scalar phenomenological coefficients in expressions relating forces and fluxes.
$\mathbf{L}$	molecular rotational angular momentum.
L	Liouville operator.
l	flux tensor of rank zero.
l_x	projection length along x-axis.
M_i	molecular weight of molecule (or submolecule) i.
M_w	weight average molecular weight.
M_x, M_y, M_z	components of macroscopic magnetization.
$\mathbf{M}$	macroscopic magnetization: dipole moment per unit volume.
$\underset{\sim}{\mathscr{M}}$	excess magnetization.
m	magnetic quantum number (Chapter 4): inertial constant Chapter 2).
m_A	molecular mass of A.
$\mathbf{m}$	rotational angular momentum density.
$\|m_{JK}\|$	dipole moment matrix element determining transition intensity.
N	number of molecules in a system.
N_α, N_β	number of nuclei in spin state α, β.
N	Avogadro's number.
$\mathbf{N}$	torque produced by a field on a dipole.
n	number of bonds in a polymer chain.
n	refractive index.
n^*	complex refractive index.
$\hat{\mathbf{n}}$	unit vector (positional).
$\mathrm{P}(\theta, \phi, \mathrm{t})$	probability of a rotational position of time t.
$\bar{P}$	molar polarization.
$\mathbf{P}$	polarization vector: total momentum of a fluid.
$\hat{\mathrm{P}}$	parity operator.
$\mathscr{P}$	excess polarization.
P	pressure tensor: viscous stress.
P	momentum flux.
$\bar{p}$	specific polarization conjugate momentum in phase space.
p_n	linear momentum of particle j.
$\mathbf{p}$	dual of pressure tensor.

Q	work.
Qe	nuclear electric quadrupole moment.
$\mathbf{Q}$	energy flow vector.
q	electric charge (Chapter 2): normalization factor (Chapter 1).
q_n	generalized coordinate in phase space.
R	molecular radius; molecular separation (Chapters 1 and 4): tensor rank (Chapter 1).
$\hat{R}$	rotation operator.
$\mathbf{r}_{ij}$	intermolecular radius vector.
r_{ij}	magnitude of $\mathbf{r}_{ij}$.
S	molar entropy.
$\hat{S}$	total entropy.
$\hat{\mathbf{S}}$	electron spin moment.
$\Delta S_E{}^*$	Eyring activation entropy.
s	strain; entropy of a polymer chain (Chapter 3): number of molecules per cm^3 (Chapter 2).
s_s	shear strain
T	absolute temperature.
T_g	glass transition temperature.
T_1	spin–lattice relaxation time.
T_2	spin–spin relaxation time.
$\hat{T}$	time reversal operator.
T^k	thermodynamic state tensor.
$\hat{t}$	scattering operator.
U	molar internal energy.
$\hat{U}$	evolution operator.
u'	internal energy per unit volume.
$\mathbf{u}$	molecular velocity.
V	molar volume (Chapter 2): electric field (Chapter 4).
V_0	occupied volume.
V_f	free volume.
$\hat{V}_{1k}$	position dependent pair energy.
v	molecular volume.
$v(\mathbf{x})$	spherical volume centred at $\mathbf{x}$.
W	molar work (Chapter 3): lattice induced nuclear spin transition probability (Chapter 4): transport integral (Chapter 1).
$W(\mathrm{x})$	probability of encountering a chain end at x.

$\mathbf{W}$	molecular velocity function incorporating mass and temperature.
w	transition rate matrix.
X	transport integral.
X	isotropic state tensor.
Y	transport integral.
Y	thermodynamic state tensor.
Z	transport integral (Chapter 1): position in a gravitational field along z-axis (Chapter 1): number of adjacent sites involved in a reorientation process (Chapter 2).
α	molecular polarizability: radiation absorption coefficient: Cole–Davidson distribution parameter: extension ratio of rubber.
α_f	free volume expansion coefficient.
α_e	molecular electronic polarizability.
α_a	molecular atomic polarizability.
α_{dip}	dipolar orientation polarizability.
$\vert\alpha\rangle$	spin function.
β	damping constant for rotation: Fuoss–Kirkwood distribution parameter (Chapter 2): polymer chain dimension parameter (Chapter 3).
$\vert\beta\rangle$	spin function.
$\hat{\mathbf{\Gamma}}$	linear operator in absence of applied field.
γ	gyromagnetic ratio.
$\bar{\gamma}(L)$	gyromagnetic or gyroelectric ratio.
δ	loss angle.
δ_j	argument of three-dimensional Dirac delta function for molecule, j.
$\boldsymbol{\delta}$	flux tensor (rank two).
ε	flux tensor (rank three).
ε^*	complex permittivity.
ε'	real component of permittivity.
ε''	imaginary component of permittivity.
ε_0	permittivity at frequencies below dispersion.
ε_∞	permittivity at frequencies above dispersion.
$(\varepsilon_0-\varepsilon_\infty)_Q$	contribution of quadrupole moment to permittivity.
$\overset{\circ}{\underset{=}{\varepsilon}}$	anisotropic portion of electric permeability tensor.
ζ_i	quantum numbers of states involved in transition.
ζ_{ij}	function of value zero or unity.

η	viscosity coefficient: set of angle variables (Chapter 1).
$\hat{\Theta}$	operator for energy transfer in an applied field.
θ_{ij}	polar angle between two dipoles.
κ	dimensionless moment of inertia (Chapter 1): radius of gyration (Chapter 2).
$\Lambda_i, \Lambda_j^e, \Lambda_{jk}$	Poisson bracket operators.
$\Lambda\hbar$	component of electronic orbital angular momentum along internuclear axis.
$\mathbf{\Lambda}$	electronic orbital angular momentum.
λ	thermal conductivity coefficient (Chapter 1): wavelength of electromagnetic radiation.
μ	chemical potential.
μ_B	Bohr magneton.
μ_0	nuclear magneton.
$\bar{\mu}$	mean effective dipole moment.
$\boldsymbol{\mu}$	magnetic moment of nucleus.
$\boldsymbol{\mu}_j$	any magnetic or electric moment of particle j.
ν	frequency of electromagnetic radiation.
ν_c	correlation frequency.
ν_i	set of internal quantum numbers.
$\hat{\Xi}$	linear operator.
$\boldsymbol{\xi}$	displacement of force from centre of mass.
ξ	increment in time.
ρ	friction factor: density.
$\rho(t, \mathbf{x})$	density at t, $\mathbf{x}$ in phase space.
$\rho\hat{\mathbf{F}}^e$	body force on a molecule.
σ	effective molecular diameter: transition cross section.
σ_{ij}	mean intermolecular separation.
σ_s	rate of entropy production.
$\boldsymbol{\sigma}$	chemical shift tensor.
$\sigma_{\parallel}, \sigma_{\perp}$	components of the chemical shift tensor.
τ	relaxation time.
τ_c	correlation time.
τ_{pi}	relaxation time of pth mode of molecule i.
τ_{sr}	spin–rotational correlation time.
ϕ	bond angle to y axis: extent of displacement from Maxwell–Boltzmann distribution (Chapter 1).
ϕ_{ij}	azimuthal angle between two dipoles.
ϕ_n	orthonormal eigen functions of L.

ϕ_{ik}	intermolecular interaction energy (Chapter 1).
χ	scattering angle.
$\chi_G(t, \mathbf{x})$	rate of relaxation of G at t and $\mathbf{x}$ in phase space.
$\chi_G^e(t, \mathbf{x})$	rate of change of G at t and $\mathbf{x}$ in phase space.
ψ	single particle field (Chapter 1): bond angle to axis (Chapter 3).
$\boldsymbol{\psi}_k$	tensor function of $\mathbf{W}$ and $\boldsymbol{\Omega}$.
Ω	state of a system (point in phase space).
$\hat{\Omega}$	linear operator in presence of applied field.
$\boldsymbol{\Omega}$	molecular angular velocity.
ω	angular frequency.
ω_0	Larmor precession frequency (Chapter 4).
$\boldsymbol{\omega}_0$	mean rotational velocity.
$\boldsymbol{\omega}_L$	Larmor precession frequency (Chapter 1).

Contents

4 Nuclear Magnetic Relaxation in Fluids

1

Theory of Transport and Relaxation Processes in Polyatomic Fluids

John S. Dahler and David K. Hoffman

Our purpose in writing this lengthy chapter on transport and relaxation in polyatomic fluids is to provide a review of the theory which has been developed during the past few years. Some of the results presented here are new, some are abstracted from papers currently in press and many more are gleaned from the recent literature. The flavour of the presentation doubtlessly reflects the tastes and limitations of the authors, whose research experience with the classical theory of non-equilibrium processes is rather more extensive than with the quantum theory. Furthermore, we recognize that by chance or Freudian design the research contributions of some (notably D.K.H. and J.S.D.) may have been overemphasized while those of others may have been dealt with summarily, misinterpreted or completely overlooked.

Since it is unnatural and very nearly impossible to separate the theory of energy transfer and relaxation from that of other non-equilibrium processes, we really have not attempted to do so. But whenever a choice was presented, we did select methods and/or applications where the emphasis was on energy dissipation or transfer. We have refrained from considering mixtures simply to avoid complexity. The first section of the chapter is devoted to the presentation of general theoretical considerations of non-equilibrium processes in polyatomic fluids. In the second section we concentrate upon the rather fully developed theory for dilute gases. Finally, in the third section the emphasis is upon applications.

1.1 NON-EQUILIBRIUM STATISTICAL MECHANICS OF POLYATOMIC FLUIDS

1.1.1 The Classical Ensemble

Later some attention will be directed to the quantum theory of non-equilibrium processes in polyatomic fluids. However, our initial considerations will be based exclusively on classical mechanics. The mechanical systems which we shall examine consist of very many molecules, each of which will be represented by a collection of mass points or particles bound together by forces which to a first approximation are taken to be harmonic, i.e., forces which vary linearly with separation. To the same approximation this aggregate will behave as an almost-rigid three-dimensional array of particles upon which are superposed vibrational motions associated with the responses of the particles to the harmonic restoring forces. The potential of interaction between two of these molecules will be a complicated function dependent upon the relative orientations of the molecular skeletons as well as upon the separation of their centres of mass. Due to their very small masses the electrons of the molecules are of little direct consequence to the reckoning of the system's mass, momentum or energy, or to the fluxes of these quantities. Therefore, unless the system contains free electrons, or collision processes involving electronic excitation of molecules are of importance, the only rôles played by the electrons are indirect, that is, as the glue which binds the atoms of a molecule together and as contributors to the electric and magnetic multipolar structure of the molecules and to intermolecular forces.

Under standard conditions one cubic centimetre of a gas will contain of the order of 10^{19} molecules. Therefore, a characterization of the micro-state of such a system includes approximately 10^{20} bits of information. However, in place of this staggering amount of information it is likely that we have available only a few data such as the mean values of velocity, temperature and composition. Now, in general there is a many-to-one correspondence between the micro- and macroscopic states, that is, there will be a multitude of different microscopic states which are consistent with a single macroscopic one. In particular, even if the local values of fluid velocity, temperature and composition were known very precisely, the number of compatible microscopic states would be essentially unbounded.

It is the utter impossibility of determining the precise microstate of a many-body system which forces us from the very start to adopt a statistical point of view. There is no element of choice involved in the decision to adopt statistical methods: it is a necessity which should not be confused with the convenience which is sometimes afforded by representing the salient characteristics of large collections of data in terms of averages, deviations from the mean and the like. It is impossible to decide which one of the many microscopic states compatible with the available macroscopic data should be assigned to the actual physical system at hand. This is logically equivalent to the so-called 'assumption of equal *a priori* probabilities' according to which it is taken to be equally likely for the actual system to be in each of the microscopic states which is compatible with the available macroscopic data.

To proceed further we assume for the moment that the number of molecules in the system of interest is known to be N. The collection of microscopic dynamical states compatible with the initial macroscopic data then may be represented by a cloud of points in the phase space of this N-particle system. (It will be recalled that each point in the phase space of a system corresponds to one of its permitted dynamical states, Ω.) Next we observe that each of the points of this cloud corresponds to a dynamical system identical in constitution to the actual physical system and in a dynamical state which is consistent with the initial macroscopic data, that is, any one of these system points might describe the actual (but necessarily indeterminate) initial microscopic state. The density of these system points in the vicinity of the point Ω will be denoted by the symbol $f^{(N)}(t, \Omega)$. Since by their manner of definition the states compatible with the initial macroscopic data form a dense set in phase space, it is not unreasonable to assume that $f^{(N)}(0, \Omega)$ will be a rather decent function with as many derivatives as one might desire.

The totality of physical systems, each identical to that of interest and distributed over the microscopic states in accordance with $f^{(N)}(0, \Omega)$ is called the *ensemble* representative of the prescribed macroscopic state at the initial instant of time $t = 0$. It is convenient to normalize the function $f^{(N)}(0, \Omega)$ in such a way that its integral over all phase space is equal to unity, i.e., such that $\int f^{(N)}(0, \Omega)\, d\Omega = 1$. For then the integral of $f^{(N)}(0, \Omega)$ over a restricted region $\Delta\Omega$ may be interpreted as the probability that at time $t = 0$ the microscopic state of the real physical system lies within $\Delta\Omega$. Furthermore, with this convention it is obvious that $f^{(N)}(0, \Omega)$ itself has the meaning of the probability density in the immediate vicinity of the phase point Ω at time $t = 0$. It should be emphasized that it is only by an appeal to the ensemble theory which has just been outlined that one can justify the application of these statistical concepts to the study of mechanical systems. For us the probability concepts have very explicit interpretations in terms of the density of system points in the phase space of the ensemble representative of the physical system under consideration.

Each member of the ensemble has assigned to it a completely specified microscopic state at $t = 0$ and, in the course of time, will trace out a unique trajectory in the phase space of states Ω. This trajectory is fully determined by the system's initial state and by Hamilton's equations of motion,

$$\dot{q}_n = \partial H/\partial p_n, \qquad \dot{p}_n = -\partial H/\partial q_n \qquad (n = 1, 2, \ldots) \tag{1}$$

Here H denotes the Hamiltonian of the system, q_n a representative generalized coordinate, and p_n the corresponding conjugate momentum. Stated somewhat differently, in classical mechanics the temporal evolution of a dynamical system (the sequence of dynamical states which it traverses in the course of time) is completely determined by a specification of its Hamiltonian function and by an assignment of values to its coordinates and conjugate momenta at a single instant of time: from the former one calculates (by use of Hamilton's equations) the instantaneous velocities and accelerations; from the latter the integration constants needed to characterize a unique solution of the dynamical equations.

Taken collectively, the motion of the member systems of the ensemble can be compared with the flow of a fluid whose density, $f^{(N)}(t, \Omega)$, is proportional to the number of systems per unit extension in phase space. The population of the ensemble is constant so that $\int_{\Delta\Omega} f^{(N)}(t, \Omega)\, d\Omega$, which is proportional to the number of members within a fixed cell of phase space with volume $\Delta\Omega$, can be altered only by the flow of system points across the boundaries of $\Delta\Omega$. From arguments of continuity it then can be shown that the density function satisfies the Liouville equation

$$\partial_t f^{(N)} = -\sum_n \left(\dot{q}_n \frac{\partial f^{(N)}}{\partial p_n} + \dot{p}_n \frac{\partial f^{(N)}}{\partial q_n} \right) = [H, f^{(N)}] \tag{1.2}$$

where $[a, b]$ indicates the Poisson bracket of the functions a and b.

The Liouville equation also can be written in the form $\partial_t f^{(N)} = -iLf^{(N)}$ where

$$L = -i \sum_n \left(\frac{\partial H}{\partial p_n} \frac{\partial}{\partial q_n} - \frac{\partial H}{\partial q_n} \frac{\partial}{\partial p_n} \right) \tag{1.3}$$

is called the Liouville operator. Since the energy and momentum of the system must be finite, $f^{(N)}$ tends to zero for very large values of the molecular momentum. Furthermore, we may without loss of generality assume that the system is bounded in space (this can be done explicitly by including wall forces in the Hamiltonian function). With respect to this class of distribution functions, L is clearly an Hermitian operator, i.e.

$$(g, Lh) \equiv \int g^*(Lh)\,\mathrm{d}\Omega = \int L(g^*h)\,\mathrm{d}\Omega - \int (Lg^*)h\,\mathrm{d}\Omega$$

$$= \int (L^*g^*)h\,\mathrm{d}\Omega \equiv (Lg, h) \tag{1.4}$$

for two functions g and h belonging to this class.

Now when various special cases are examined it is invariably found that the eigenfunctions of L form a complete set. Consequently it is reasonable to postulate that this will always be so. From this it follows immediately that the probability density function may be written in the form

$$f^{(N)}(t, \Omega) = \sum_n \mathrm{a}_n(t)\phi_n(\Omega) \tag{1.5}$$

where the scalar expansion coefficients $\mathrm{a}_n(t)$ are independent of the dynamical variables $\{q_n p_n\}$ and where the ϕ_n's are the orthonormal eigenfunctions of L, i.e., $L\phi_n = \lambda_n\phi_n$, $\lambda_n^* = \lambda_n$, $(\phi_n, \phi_m) = \delta_{nm}$. Furthermore, it is readily established that $\dot{\mathrm{a}}_n + \mathrm{i}\lambda_n\mathrm{a}_n = 0$ for all n so that

$$f^{(N)}(t, \Omega) = \sum_n \mathrm{a}_n(t_0)\,\mathrm{e}^{-\mathrm{i}\lambda_n(t-t_0)}\phi_n(\Omega) = \sum_n \mathrm{a}_n(t_0)\,\mathrm{e}^{-\mathrm{i}L(t-t_0)}\phi_n(\Omega)$$

$$= \mathrm{e}^{-\mathrm{i}L(t-t_0)}\sum_n \mathrm{a}_n(t_0)\phi_n(\Omega) = \mathrm{e}^{-\mathrm{i}L(t-t_0)}f^{(N)}(t_0, \Omega) \tag{1.6}$$

It is clear that equation (1.6) is the formal solution of the Liouville equation which one would obtain by treating the operator L as though it were a complex number. The eigenfunction analysis which we have just performed provides a 'justification' for this formal manipulation.

An analogous procedure can be used to generate formal solutions of Hamilton's equations of motion. Thus, by direct calculation one can verify that $\mathrm{i}\dot{q}_n = \mathrm{i}\partial H/\partial p_n = -Lq_n$ and $\mathrm{i}\dot{p}_n = -\mathrm{i}\partial H/\partial q_n = -Lp_n$ and from this conclude that

$$q_n(t) = \mathrm{e}^{\mathrm{i}L_0(t-t_0)}q_{n0} \qquad p_n(t) = \mathrm{e}^{\mathrm{i}L_0(t-t_0)}p_{n0} \tag{1.7}$$

The meaning of these results can be couched in the language of the Lagrange coordinates used in hydrodynamics. Thus, $q_n(t)$ and $p_n(t)$ are the Lagrange coordinates along the trajectory in phase space, given parametrically in terms of the time $(t-t_0)$, which passes through the point $\{q_{n0}, p_{n0}\}$ at $t = t_0$. Since the linear transformation $L = L(\{q_n, p_n, \partial/\partial q_n, \partial/\partial p_n\})$ is a function of all the dynamical coordinates and associated derivative operators, we have denoted by $L_0 = L(\{q_{n0}, p_{n0}, \partial/\partial q_{n0}, \partial/\partial p_{n0}\})$ the Liouville operator 'at the point $\Omega_{t_0} = \{q_{n0}, p_{n0}\}$'.

One of our objectives is to derive the familiar equations of continuum mechanics directly from the Liouville equation. As a first step in this direction let us examine

$$(G, f^{(N)}(t)) = \int G(\Omega) f^{(N)}(t, \Omega)\, d\Omega, \tag{1.8}$$

the inner product of the probability density at time t with a real function $G = G(\Omega)$ which depends upon the dynamical coordinates of the system but not explicitly upon time. For purposes of illustration we take G to be the function $\Delta(\mathbf{x}; \mathbf{x}_i)$ defined as,

$$\Delta(\mathbf{x}; \mathbf{x}_i) = \begin{cases} v^{-1}; & \text{if } \mathbf{x}_i \text{ is in } v(\mathbf{x}) \\ 0; & \text{otherwise} \end{cases} \tag{1.9}$$

Here $v(\mathbf{x})$ denotes a spherical region of volume v in ordinary three-space with its centre at the point $\mathbf{x}$. It is clear that with this choice for G the only non-vanishing contributions to $(G, f^{(N)}(t))$ come from those members of the representative ensemble with dynamical states at time t that place particle i within the region $v(\mathbf{x})$. From this and the statistical interpretation of $f^{(N)}$ we conclude that $(G, f^{(N)}(t))$, the 'ensemble average' of $G = \Delta(\mathbf{x}; \mathbf{x}_i)$, is to be identified as the probability per unit volume at time t that particle i lies within $v(\mathbf{x})$, i.e., the 'probability' of this event which we compute on the basis of our limited knowledge of the actual microscopic state of the physical system. It follows that the ensemble averages of $G_1 = \sum_i \Delta(\mathbf{x}; \mathbf{x}_i)$ and $G_2 = \sum_i m_i \Delta(\mathbf{x}; \mathbf{x}_i)$, respectively, are to be interpreted as the instantaneous values for the particle concentration and mass density within $v(\mathbf{x})$. To define the values of these densities at a point $\mathbf{x}$ we replace $\Delta(\mathbf{x}; \mathbf{x}_i)$ with a sequence $\Delta_k(\mathbf{x}-\mathbf{x}_i) = \Delta_k(x-x_i)\Delta_k(y-y_i)$ $\Delta_k(z-z_i)$ which is equivalent to the three-dimensional Dirac delta function.

Thus

$$n(\mathbf{x}, t) = \sum_i \int \delta_3(\mathbf{x}-\mathbf{x}_i) f^{(N)}(t, \Omega)\, d\Omega$$

and

$$\rho(\mathbf{x}, t) = \sum_i m_i \int \delta_3(\mathbf{x}-\mathbf{x}_i) f^{(N)}(t, \Omega)\, d\Omega \tag{1.10}$$

are the statistical analogues to the particle concentration and mass density of continuum mechanics.

From the preceding results we see that the ensemble average of a function which has no explicit dependence upon time may be written as:

$$\begin{aligned}(G, f^{(N)}(t)) &= \int G(\Omega_t) f^{(N)}(t, \Omega_t)\, d\Omega_t = \int G(\Omega_t) f^{(N)}(t_0, \Omega_0)\, d\Omega_0 \\ &= \int G(t, \Omega_0) f^{(N)}(t_0, \Omega_0)\, d\Omega_0 = (G(t), f^{(N)}(t_0))\end{aligned} \tag{1.11}$$

where the meaning of the relationship,

$$\begin{aligned}G(t, \Omega_0) = G(\Omega_t) &= G(\{\exp[iL_0(t-t_0)]q_{n0}, \exp[iL_0(t-t_0)]p_{n0}\}) \\ &= \exp[iL_0(t-t_0)]G(\Omega_0)\end{aligned} \tag{1.12}$$

(by which the function $G(t, \Omega_0)$ has been introduced) is that a description of the dynamical evolution of the system can be phrased either in terms of the sequence of instantaneous states Ω_t which lie along its trajectory or in terms of a family of trajectories each of which is parametrized by the time t and labelled according to the point Ω_0 through which it passes at the parameter value $t = t_0$. In fluid dynamics the corresponding alternative modes of description are called the Euler and Lagrange representations, respectively. In quantum mechanics they are referred to as the Schroedinger and Heisenberg representations.

A practical consequence of this dichotomy of representation is that we may choose to express the ensemble average of $G = G(\Omega)$ either: (i) in terms of the instantaneous state of the ensemble $f^{(N)}(t, \Omega_t)$ and the corresponding values of $G(\Omega_t)$ or; (ii) in terms of the ensemble probability density $f^{(N)}(t_0, \Omega_0)$ at a fixed time t_0 and the values $G(t, \Omega_0) = \exp[iL_0(t-t_0)]G(\Omega_0)$ of G which are associated with states occupied by the member systems of the ensemble at the later instant of time $t > t_0$.

The function $G(t, \Omega)$, which may be defined more precisely as the solution of the equation $i\partial_t G(t, \Omega) = -LG(t, \Omega)$ with boundary condition $G(0, \Omega) = G(\Omega)$, will be called the 'Heisenberg representative of $G(\Omega)$'. From these considerations it follows that,

$$\partial_t(G, f^{(N)}(t)) = (G, \partial_t f^{(N)}(t)) = (\partial_t G(t), f^{(N)}(t_0)) = (iLG(t), f^{(N)}(t_0)) \quad (1.13)$$

and so, to derive the equation of change for the ensemble average of $G(\Omega)$ it is first necessary to determine the effect of the Liouville operator upon this function.

1.1.2 Local Variables

In this section we examine the equations of change, $\partial_t G(t;\Omega) - iLG(t;\Omega) = 0$, for the (very large) class of functions called LOCAL VARIABLES. These variables, of which the previously considered $\sum_j \delta_3(\mathbf{x}-\mathbf{x}_i)$ is an example, are defined so that their ensemble averages correspond to the macroscopic fields of continuum mechanics. In particular, a local variable is one which can be expressed in the manner

$$G(t\mathbf{x}) = \sum_j G_j \delta_j \quad (1.14)$$

where the argument of the three-dimensional Dirac Delta function $\delta_j \equiv \delta_3(\mathbf{x}-\mathbf{x}_j)$ is the difference between the vector-valued parameter $\mathbf{x}$ and the vector $\mathbf{x}_j$ which locates the centre of mass of molecule j. The range of definition of $\mathbf{x}$ (which is measured from the same origin as the $\mathbf{x}_j$s) and hence of $G(t\mathbf{x})$ extends throughout the region of ordinary 3-space to which the physical system is confined. The parameters t and $\mathbf{x}$ are the 'time' and 'place' coordinates of the macroscopic fields associated with the system. Although we shall often write $G(t\mathbf{x})$ in place of the more cumbersome symbol $G(t\mathbf{x};\Omega)$, it should not be forgotten that this function depends upon the phase point Ω as well as upon the parameters t and $\mathbf{x}$.

The function G_j which appears in the defining formula (1.14) may depend only upon the dynamical coordinates of molecule j, upon those of two molecules or even upon those of three and more molecules. The corresponding ensemble averages are called single-particle fields, two-particle fields and multi-particle fields. For convenience we occasionally refer to the variables $G(t\mathbf{x})$ by these same terms.

Let us now assume that the Hamiltonian of the system is given by

$$H = \sum_j (p_j^2/2m_j + H_j^e + H_j) + \sum_j \sum_{k>j} \phi_{jk} \quad (1.15)$$

where $\mathbf{p}_j \equiv m_j\mathbf{c}_j$ is the linear momentum associated with the centre of mass of molecule j and where m_j is the molecular mass. The function $H_j = H_j(q_jp_j)$ is the Hamiltonian specific to the internal coordinates and conjugate momenta of molecule j. We denote these *collections* of variables by the symbols q_j and p_j, respectively. If the molecule is 'structureless' then H_j is zero.

The function $H_j^{\mathrm{e}} = H_j^{\mathrm{e}}(q_jp_j, \mathbf{x}_j)$ is the interaction energy of molecule j with static external fields. Although it is reasonable to assume that this energy will be independent of the molecular momentum, we cannot dismiss the possibility that it depends upon the momenta p_j which are conjugate to the internal coordinates q_j. For example, let us suppose that the system is composed of electrically neutral molecules and that it is immersed in an electric or magnetic field with a time-independent amplitude $\mathbf{F} = \mathbf{F}(\mathbf{x})$. Then, provided that this field is not too intense we can express H_j^{e} as the sum of a 'boundary interaction' which represents the confining influence of the immediate surroundings of the system and a term $-\boldsymbol{\mu}_j \,.\, \mathbf{F}(\mathbf{x}_j)$. In general the (electric or magnetic) dipole moment $\boldsymbol{\mu}_j$ will depend upon molecular orientation and/or rotational angular momentum and spin. We shall discuss these moments more fully in a later section. For present purposes it may be assumed that $\boldsymbol{\mu}_j = \bar{\gamma}(L_j)\mathbf{L}_j$ where $\mathbf{L}_j$ is the molecular rotational angular momentum and $\bar{\gamma}(L)$ the gyromagnetic or gyroelectric ratio.

The function $\phi_{jk} = \phi_{jk}(\mathbf{x}_{jk}, q_jp_j, q_kp_k)$, with $\mathbf{x}_{jk} = \mathbf{x}_j - \mathbf{x}_k$, is the energy of interaction of molecules j and k, i.e. the potential of the intermolecular force. Although we have assumed that these forces are independent of the particle momenta $\mathbf{p}_j$ and $\mathbf{p}_k$, we do not exclude the possibility of dependence upon the momenta conjugate to the internal coordinates q_j and q_k. This permits more flexibility in the choice of the variables q_jp_j. For example, let us suppose that the interaction ϕ_{jk} appropriate to two rigid rotors can be expressed fully in terms of $\mathbf{x}_{jk}$ and of the vectors $\mathbf{e}_j$ and $\mathbf{e}_k$ which coincide with the molecular symmetry axes. One can then choose for q_j the polar spherical angles (θ_j and ϕ_j) of $\mathbf{e}_j$ and so identify p_j with the associated pair of conjugate momenta $p_{\theta j}$ and $p_{\phi j}$. The function ϕ_{jk} clearly is independent of these momenta. However, one might prefer a description in terms of the action-angle variables of the rotor. The generalized momenta are then L_j and L_{jz}, the magnitude and z-projection of the rotational angular momentum $\mathbf{L}_j$. The coordinate which is conjugate to L_{jz} is the angle which locates the position of the line of nodes in the plane perpendicular to the space-fixed z-axis.

The conjugate of $L_j = |\mathbf{L}_j|$ is the angle between the line of nodes and $\mathbf{e}_j$. With this choice of coordinates $\mathbf{e}_j$—and hence ϕ_{jk}—is a function of L_{jz}/L_j as well as of both angular variables.

The Liouville operator can be expressed most conveniently in the form

$$\mathrm{i}L = \sum_j \left[\frac{\mathbf{p}_j}{m_j} \cdot \frac{\partial}{\partial \mathbf{x}_j} + \Lambda_j + \Lambda_j^{\mathrm{e}} + \sum_{k>j} \Lambda_{jk} \right] \tag{1.16}$$

where Λ_j, Λ_j^{e} and Λ_{jk} are the Poisson bracket operators associated with H_j, H_j^{e} and ϕ_{jk}, respectively, i.e. $\Lambda_j = [\ , H_j]$, $\Lambda_j^{\mathrm{e}} = [\ , H_j^{\mathrm{e}}]$ and $\Lambda_{jk} = \Lambda_{kj} = [\ , \phi_{jk}]$. Then, since

$$\dot{\delta}_j = \mathrm{i}L\delta_j = \frac{\mathbf{p}_j}{m_j} \cdot \frac{\partial}{\partial \mathbf{x}_j} \delta_j = -\nabla \cdot \frac{\mathbf{p}_j}{m_j} \delta_j \tag{1.17}$$

with $\nabla = \partial/\partial \mathbf{x}$, it follows that

$$\dot{G}(t\mathbf{x}) + \nabla \cdot \left[\sum_i \frac{\mathbf{p}_i}{m_i} G_i \delta_i \right] = \sum_i \delta_i \left[\sum_j \left(\frac{\mathbf{p}_j}{m_j} \cdot \frac{\partial}{\partial \mathbf{x}_j} + \Lambda_j + \Lambda_j^{\mathrm{e}} \right) + \sum_j \sum_{k>j} \Lambda_{jk} \right] G_i \tag{1.18}$$

There are two cases of particular interest to us here.

1.1.3 Single-Particle Fields

Let us first suppose that G_j depends only upon the dynamical variables $\mathbf{x}_j$, $\mathbf{p}_j$ and $q_j p_j$ of molecule j. Then,

$$\sum_i \sum_j \delta_i \left(\sum_{k>j} \Lambda_{jk} \right) G_i = \sum_i \sum_j \delta_i (\dot{G}_i)_j$$

where $(\dot{G}_i)_j \equiv \Lambda_{ij} G_i$ is the rate of change of G_i due to the interaction of molecule i with molecule j. We next write this result in the more symmetrical form

$$\tfrac{1}{2} \sum_i \sum_{j \neq i} \{ \delta_i (\dot{G}_i)_j + \delta_j (\dot{G}_j)_i \} = \tfrac{1}{2} \sum_i \sum_{j \neq i} \delta_i \{ (\dot{G}_i)_j + (\dot{G}_j)_i \} + \tfrac{1}{2} \sum_i \sum_j (\delta_j - \delta_i)(\dot{G}_j)_i$$

The factor $(\dot{G}_i)_j + (\dot{G}_j)_i$ which appears in the first term is the rate of change of $G_i + G_j$ due to the mutual interaction of molecules i and j. Functions G_i for which this rate vanishes are called BINARY INVARIANTS.

The physical significance of the second sum* can be made more obvious by introducing the Taylors series

$$\delta_j - \delta_i = \sum_{n\geq 1} \frac{1}{n!}(\mathbf{x}_{ij} \cdot \boldsymbol{\nabla})^n \delta_i = \boldsymbol{\nabla} \cdot \{\mathbf{x}_{ij} 0_{ij} \delta_i\} \tag{1.19}$$

with

$$0_{ij} \equiv \sum_{n\geq 0} \frac{1}{(n+1)!}(\mathbf{x}_{ij} \cdot \boldsymbol{\nabla})^n \tag{1.20}$$

Thus, we find that

$$\dot{G}(t\mathbf{x}) + \boldsymbol{\nabla} \cdot \mathbb{J}_G = \chi_G + \chi_G{}^{\mathrm{e}} \tag{1.21}$$

where

$$\mathbb{J}_G(t\mathbf{x}) = \sum_j \left[\frac{\mathbf{p}_j}{m_j} G_j + \sum_{k>j} \mathbf{x}_{jk}(\Lambda_{kj} G_k) 0_{jk} \right] \delta_j \tag{1.22}$$

is to be interpreted as the 'flux' of G,

$$\chi_G(t\mathbf{x}) = \sum_j \left[\left\{ \frac{\mathbf{p}_j}{m_j} \cdot \frac{\partial}{\partial \mathbf{x}_j} + \Lambda_j \right\} G_j + \sum_{k>j} \Lambda_{jk}(G_j + G_k) \right] \delta_j \tag{1.23}$$

as the corresponding 'rate of relaxation', and

$$\chi_G{}^{\mathrm{e}}(t\mathbf{x}) = \sum_j (\Lambda_j{}^{\mathrm{e}} G_j) \delta_j \tag{1.24}$$

as the rate of change of $G(t\mathbf{x})$ due to the action of the external field.

1.1.4 Two-Particle Fields

We next examine the case where $G_i = \sum_{j\neq i} G_{ij}$ is the sum of functions G_{ij}, each dependent upon the dynamical states of two molecules. In this situation,

$$\sum_i \sum_j \frac{\mathbf{p}_j}{m_j} \cdot \frac{\partial G_i}{\partial \mathbf{x}_j} \delta_i = \sum_i \sum_j \delta_i \left[\frac{\mathbf{p}_j}{m_j} \cdot \frac{\partial}{\partial \mathbf{x}_j} + \frac{\mathbf{p}_i}{m_i} \cdot \frac{\partial}{\partial \mathbf{x}_i} \right] G_{ij}$$

$$\sum_i \sum_j \delta_i \Lambda_j G_i = \sum_i \sum_j \delta_i [\Lambda_j + \Lambda_i] G_{ij}$$

* The second sum,

$$\tfrac{1}{2} \sum_i \sum_j (\delta_j - \delta_i)(\dot{G}_j)_i = \tfrac{1}{2} \sum_i \sum_j \delta_i [(\dot{G}_i)_j - (\dot{G}_j)_i]$$

vanishes when G_i is an ANTI-INVARIANT, a function for which $(\dot{G}_i)_j = (\dot{G}_j)_i$

and

$$\sum_i \sum_j \delta_i \left(\sum_{k>j} \Lambda_{jk} \right) G_i = \sum_i \sum_j \sum_k \delta_i (\dot{G}_{ij})_k$$

where $(\dot{G}_{ij})_k \equiv (\Lambda_{ik} + \Lambda_{jk}) G_{ij}$ is the rate of change of G_{ij} caused by the interaction of the pair (ij) with a third molecule k. By writing this last in the form

$$\sum_i \sum_j \sum_k \delta_i (\dot{G}_{ij})_k = \sum_i \sum_j \sum_k [(\delta_i - \delta_k)(\dot{G}_{ij})_k + \delta_k \{(\dot{G}_{ij})_k + (\dot{G}_{ki})_j + (\dot{G}_{kj})_i\} - \delta_k (\dot{G}_{ki})_j - \delta_k (\dot{G}_{kj})_i]$$

we see that

$$\sum_i \sum_j \delta_i \left(\sum_{k \neq j} \Lambda_{jk} \right) G_i = \tfrac{1}{3} \sum_i \sum_j \sum_k [\delta_k \{(\dot{G}_{ij})_k + (\dot{G}_{ki})_j + (\dot{G}_{kj})_i\} + (\delta_i - \delta_k)(\dot{G}_{ij})_k] \tag{1.25}$$

If G_{ij} is symmetric in its indices then the factor $(\dot{G}_{ij})_k + (\dot{G}_{ki})_j + (\dot{G}_{kj})_i$ is the rate of change of $G_{ij} + G_{ik} + G_{jk}$ due to the mutual interactions of molecules i, j and k.

From these results one can verify that the equation of change for the two-particle field is of the form of equation (1.21) with the flux and rate of relaxation defined according to:

$$\mathbb{J}_G(t\mathbf{x}) = \sum_i \sum_{j \neq i} \left[\frac{\mathbf{p}_i}{m_i} G_{ij} - \tfrac{1}{3} \sum_{k \neq i,j} \mathbf{x}_{ij} (\dot{G}_{jk})_i 0_{ij} \right] \delta_i \tag{1.26}$$

$$\chi_G(t\mathbf{x}) = \sum_i \sum_{j \neq i} \left[\left\{ \frac{\mathbf{p}_j}{m_j} \cdot \frac{\partial}{\partial \mathbf{x}_j} + \frac{\mathbf{p}_i}{m_i} \cdot \frac{\partial}{\partial \mathbf{x}_i} + \Lambda_j + \Lambda_i \right\} G_{ij} + \tfrac{1}{3} \sum_{k \neq i,j} \left\{ (\dot{G}_{ij})_k + (\dot{G}_{ik})_j + (\dot{G}_{jk})_i \right\} \right] \delta_i \tag{1.27}$$

and

$$\chi_G^{\mathrm{e}}(t\mathbf{x}) = \sum_i \sum_{j \neq i} (\Lambda_i^{\mathrm{e}} + \Lambda_j^{\mathrm{e}}) G_{ij} \delta_i \tag{1.28}$$

1.1.5 Transport-Relaxation Equations

With G_j set equal to m_j we obtain from equation (1.21) the 'equation of continuity',

$$\partial_t \sum_j m_j \delta_j + \boldsymbol{\nabla} \cdot \sum_j \mathbf{p}_j \delta_j = 0 \tag{1.29}$$

wherein $\sum_j m_j\delta_j$ and $\sum_j \mathbf{p}_j\delta_j$ are the generalized functions whose ensemble averages are the densities of mass and of linear momentum, respectively.

Let us now factor each $G(t\mathbf{x})$ into the product of the density $\rho(t\mathbf{x})$ and a function $g(t\mathbf{x}) \equiv g(t\mathbf{x}, \Omega)$. Then, from the identity

$$\dot{G} = \dot{\rho}g + \dot{\rho}g = -g\nabla.(\rho\mathbf{u}) + \dot{g}\rho = \rho(\partial_t + \mathbf{u}.\nabla)g - \nabla.(\mathbf{u}G) \quad (1.30)$$

it follows that equation (1.21) can be written in the form

$$\rho d_t g = -\nabla . J_G + \chi_G + \chi_G^{\mathrm{e}} \quad (1.31)$$

with $\mathrm{d}_t \equiv \partial_t + \mathbf{u}.\nabla$ and where $J_G = \mathbb{J}_G - \mathbf{u}G$ is the generalized function whose ensemble average is the CONDUCTIVE CONTRIBUTION TO THE FLUX OF G.

The prescriptions of the previous section provide us with the means to construct equations of change for whatever single- and two-particle fields may be of interest. Some of these fields will be familiar from continuum mechanics while others may not. For example, it is obvious that we should interpret the ensemble average of $\rho^{-1}\sum_i \mathbf{p}_i\delta_i$ to be the fluid velocity. From equation (1.31) we then obtain the equation of motion

$$\rho \mathrm{d}_t\mathbf{u} = -\nabla . \mathsf{p} + \rho\hat{\mathbf{F}}^{\mathrm{e}} \quad (1.32)$$

and so naturally are inclined to identify

$$\mathsf{p} = \left\langle \sum_j \left[m_j\mathbf{C}_j\mathbf{C}_j - \sum_{k>j} \mathbf{x}_{jk}\frac{\partial\phi_{jk}}{\partial\mathbf{x}_{jk}}0_{jk} \right] \delta_j; t \right\rangle; \qquad \mathbf{C}_j = \mathbf{p}_j/m_j - \mathbf{u} \quad (1.33)$$

and

$$\rho\hat{\mathbf{F}}^{\mathrm{e}} = \left\langle \sum_j (\Lambda_j^{\mathrm{e}}\mathbf{p}_j)\delta_j; t \right\rangle \quad (1.34)$$

with the pressure tensor and body force, respectively*. Furthermore, since $\Lambda_i^{\mathrm{e}}\mathbf{p}_i \equiv [\mathbf{p}_i, H_i^{\mathrm{e}}]$ is indeed the external force on molecule i the correctness of the latter of these identifications is beyond dispute. Finally, it is a straightforward but rather lengthy task to prove that equation (1.33) does equal the non-convective rate of transfer of momentum and so really qualifies as the statistical counterpart of the macroscopic pressure tensor.

* Here we have introduced the abbreviation $\langle A; t\rangle$ for the ensemble average, $(A, f^{(N)}(t))$, of a dynamical variable A. When there is little danger of ambiguity we shall often write $\langle A\rangle$ in place of $\langle A; t\rangle$.

The ensemble average of the kinetic energy $K\left(\equiv \sum_j (p_j^2/2m_j)\delta_j\right)$ can be separated into the sum of $\frac{1}{2}\rho u^2$, the translational energy associated with the local state of motion of the fluid, and a 'thermal' contribution

$$U_K(t\mathbf{x}) \equiv \rho e_K(t\mathbf{x}) = \left\langle \sum_j \frac{m_j}{2} C_j^2 \delta_j \right\rangle \tag{1.35}$$

which arises from the dispersion of the molecular translational velocities about the local mean value of $\mathbf{u}(t\mathbf{x})$. The equation of change for this single-particle field is

$$\rho \mathrm{d}_t e_K = -\nabla . [\mathbf{Q}_K + \mathbf{Q}'_\phi] - \mathsf{p}^\dagger : \nabla \mathbf{u} + \chi_K + \chi_K^{\mathrm{e}} \tag{1.36}$$

where

$$\mathbf{Q}_K = \left\langle \sum_j C_j(\tfrac{1}{2} m_j C_j^2)\delta_j \right\rangle \tag{1.37}$$

and

$$\mathbf{Q}'_\phi = -\left\langle \sum_j \sum_{k>j} \mathbf{x}_{jk} \left(\mathbf{C}_k . \frac{\partial \phi_{jk}}{\partial \mathbf{x}_{jk}} \right) 0_{jk} \delta_j \right\rangle \tag{1.38}$$

are contributions to the conductive flow of energy which have been discussed more fully elsewhere[4,20]. The physical interpretations of the two functions

$$\chi_K = \left\langle \sum_j \sum_{k>j} (\mathbf{C}_k - \mathbf{C}_j) . \frac{\partial \phi_{jk}}{\partial \mathbf{x}_{jk}} \delta_j \right\rangle \tag{1.39}$$

and

$$\chi_K^{\mathrm{e}} = \left\langle \sum_j [\tfrac{1}{2} m_j C_j^2, H_j^{\mathrm{e}}] \delta_j \right\rangle \tag{1.40}$$

will be more obvious after we have examined the equations of change for $e_\iota = \rho^{-1}\left\langle \sum_j H_j \delta_j \right\rangle$ and $e_\phi = \rho^{-1}\left\langle \sum_j \sum_{k>j} \phi_{jk}\delta_j \right\rangle$, the densities of internal and of interactional energy, respectively. These equations are found to be

$$\rho \mathrm{d}_t e_\iota = -\nabla . \mathbf{Q}_\iota + \chi_\iota + \chi_\iota^{\mathrm{e}} \tag{1.41}$$

and

$$\rho \mathrm{d}_t e_\phi = -\nabla . \mathbf{Q}_\phi'' + \chi_\phi + \chi_\phi^{\mathrm{e}} \tag{1.42}$$

where

$$\mathbf{Q}_\iota = \left\langle \sum_j \left(\mathbf{C}_j H_j - \sum_{k>j} \mathbf{x}_{jk}[H_j, \phi_{jk}]0_{jk} \right) \delta_j \right\rangle \tag{1.43}$$

$$\mathbf{Q}''_\phi = \left\langle \sum_j \mathbf{C}_j \left(\sum_{k>j} \phi_{jk} \right) \delta_j \right\rangle \tag{1.44}$$

$$\chi_\iota^{\mathrm{e}} = \left\langle \sum_j [H_j, H_j^{\mathrm{e}}] \delta_j \right\rangle \tag{1.45}$$

$$\chi_\phi^{\mathrm{e}} = \left\langle \sum_j \sum_{k>j} [\phi_{jk}, H_j^{\mathrm{e}} + H_k^{\mathrm{e}}] \delta_j \right\rangle \tag{1.46}$$

$$\chi_\iota = \left\langle \sum_j \sum_{k>j} [H_j + H_k, \phi_{jk}] \delta_j \right\rangle \tag{1.47}$$

and $\chi_\phi = -\chi_K - \chi_\iota$. In most of what follows we shall ignore the possibility of molecular vibrations and so identify H_j with the Hamiltonian of a non-deformable rotor. However, if one wished to be less restrictive he might express H_j as the sum $H_{rj} + H_{vj}$ of a rotational and of a vibrational energy. Then, each of e_ι, $\mathbf{Q}_\iota$, χ_ι and χ_ι^{e} would separate into the sum of two contributions and in place of equation (1.41) there would be two distinct equations.

The structure of this set of energy equations discloses some facts which, although trivial in retrospect, are perhaps not quite so obvious *a priori*. No terms providing for direct collisional coupling among the various degrees of freedom appear. Instead, the translational, rotational and vibrational energies are all coupled through e_ϕ, which acts as a sort of clearing house or transducer for the channelling of energy among the different mechanical degrees of freedom. Now the autonomy of the energy densities e_K, e_{rot} and e_{vib} will depend upon the relative magnitudes of the inter- and intramolecular forces and also upon the level of thermal excitation of the fluid. Thus, at moderate densities it may be possible to associate separate temperatures with each of these energy contributions and to observe the associated thermal relaxation processes. On the other hand, in solids, liquids and dense gases the rotational degrees of freedom may—by virtue of the non-central interactions—be so strongly coupled to the translational motions that resolution of the rotational and translational energies is experimentally impossible and so without physical significance. The energy density e_ϕ is, of course, dependent upon structural parameters related to the local state of aggregation or spatial organization of the molecular system rather than upon the degree of thermal excitation. While one easily can cite examples of systems for which these structural parameters are coupled only weakly to the rotational and vibrational excitations, a quantity such as the radial distribution function invariably will exhibit a strong dependence upon the translational temperature. Therefore, a rather convincing argument can be presented for combining equations (1.36) and (1.42) into the single equation

$$\rho \mathrm{d}_t e_\kappa = -\nabla . [\mathbf{Q}_K + \mathbf{Q}_\phi] - \mathsf{p}^\dagger : \nabla \mathbf{u} - \chi_\iota + \chi_K{}^{\mathrm{e}} + \chi_\phi{}^{\mathrm{e}} \tag{1.48}$$

with $\mathbf{Q}_\phi = \mathbf{Q}'_\phi + \mathbf{Q}''_\phi$ and where $e_\kappa = e_K + e_\phi$ is the energy density exclusive of contributions arising from the molecular internal degrees of freedom. Furthermore, the energy couplings which appear in the set of equations (1.41) and (1.48) are less obscure than before: the function χ_ι obviously equals the rate of energy exchange between e_κ and e_ι.

As a last example let us consider the equation of change

$$\rho \mathrm{d}_t \mathbf{m} = -\nabla . \mathsf{C} + \chi_L + \chi_L{}^{\mathrm{e}} \tag{1.49}$$

for the density $\mathbf{m} = \rho^{-1}\left\langle \sum_i \mathbf{L}_i \delta_i \right\rangle$ of rotational angular momentum. Here

$$\mathsf{C} = \left\langle \sum_i \left(\mathbf{C}_i \mathbf{L}_i - \sum_{j>i} \chi_j [\mathbf{L}_j, \phi_{ij}] 0_{ij} \right) \delta_i \right\rangle \tag{1.50}$$

is the flux of this momentum and

$$\chi_L{}^{\mathrm{e}} = \left\langle \sum_i [\mathbf{L}_i, H_i{}^{\mathrm{e}}] \delta_i \right\rangle \equiv \rho \hat{\mathbf{G}}^{\mathrm{e}} \tag{1.51}$$

the body torque per unit volume. Finally, since $\mathbf{J}_i = \mathbf{x}_i \times \mathbf{p}_i + \mathbf{L}_i$ is a binary invariant we see that

$$\chi_L = \left\langle \sum_i \sum_{j>i} [\mathbf{L}_i + \mathbf{L}_j, \phi_{ij}] \delta_i \right\rangle = -\left\langle \sum_i \sum_{j>i} [\mathbf{x}_i \times \mathbf{p}_i + \mathbf{x}_j \times \mathbf{p}_j, \phi_{ij}] \delta_i \right\rangle$$

$$= -\boldsymbol{\varepsilon} : \left\langle \sum_i \sum_{j>i} \mathbf{x}_{ij} \frac{\partial \phi_{ij}}{\partial \mathbf{x}_{ij}} \delta_i \right\rangle \doteq \boldsymbol{\varepsilon} : \mathsf{p} \tag{1.52}$$

where $\mathbf{p}^{\mathrm{a}} = -\frac{1}{2}\boldsymbol{\varepsilon} : \mathsf{p}$ is the dual of the tensor p. (The only thing which prevents this last from being an identity is the absence of the differential operator 0_{ij}.)

Near equilibrium we expect for e_K+e_ϕ to depend only upon an appropriate entropy density $\hat{S}_\kappa$ and upon the specific volume ρ^{-1}. The rotational energy density $e_\iota = e_{\text{rot}}$ (we here ignore vibrational degrees of freedom) should be a function of the rotational entropy density $\hat{S}_\iota$ and of $\mathbf{m}$ as well. Consequently, we postulate the Gibbs relations

$$de_\kappa = T_\kappa \, d\hat{S}_\kappa - p \, d\rho^{-1} \tag{1.53}$$

and

$$de_\iota = de'_\iota + \boldsymbol{\omega}_0 \cdot d\mathbf{m} = T_\iota \, dS_\iota + \boldsymbol{\omega}_0 \cdot d\mathbf{m} \tag{1.54}$$

where $T_\kappa = (\partial e_\kappa/\partial \hat{S}_\kappa)_\rho$, $p = -(\partial e_\kappa/\partial \rho^{-1})_{\hat{S}}$, and $T_\iota = (\partial e_\iota/\partial \hat{S}_\iota)_{\mathbf{m}}$ have the obvious interpretations and where $\boldsymbol{\omega}_0 = (\partial e_\iota/\partial \mathbf{m})_S$ is the mean rotational velocity of the tumbling molecules. For the total entropy $\hat{S} = \hat{S}_\kappa + \hat{S}_\iota$ we then obtain the equation of change $\rho \, d_t\hat{S} = -\boldsymbol{\nabla} \cdot \mathbf{J}_S + \sigma_S$ where

$$\mathbf{J}_S = (\mathbf{Q}_K + \mathbf{Q}_\phi)T_\kappa + \mathbf{Q}'_\iota/T_\iota \tag{1.55}$$

is the reversible flux of entropy and where

$$\sigma_S = (\mathbf{Q}_K + \mathbf{Q}_\phi) \cdot \boldsymbol{\nabla}\frac{1}{T_\kappa} + \mathbf{Q}'_\iota \cdot \boldsymbol{\nabla}\frac{1}{T_\iota} - \frac{\mathsf{P}^\dagger}{T_\kappa} : \left(\boldsymbol{\nabla}\mathbf{u} - \frac{T_\kappa}{T_\iota}\boldsymbol{\varepsilon} \cdot \boldsymbol{\omega}_0\right)$$
$$- \frac{\mathsf{C}^\dagger}{T_\iota} : \boldsymbol{\nabla}\boldsymbol{\omega}_0 + \left(\frac{1}{T_\iota} - \frac{1}{T_\kappa}\right)\chi_\iota + \frac{1}{T_\kappa}(\chi_K{}^{\mathrm{e}} + \chi_\phi{}^{\mathrm{e}}) + \frac{1}{T_\iota}(\chi_\iota{}^{\mathrm{e}} - \boldsymbol{\omega}_0 \cdot \rho\hat{\mathbf{G}}^{\mathrm{e}}) \tag{1.56}$$

is the rate of entropy production. In these formulas $\mathbf{Q}'_\iota = \mathbf{Q}_\iota - \mathsf{C} \cdot \boldsymbol{\omega}_0$* and $\mathsf{P} = \mathsf{p} - \boldsymbol{\delta} p$.

If there is no external field then this rate obviously reduces to the sum of five terms, each the product of a flux, $\mathbf{Q}_K + \mathbf{Q}_\phi$, $\mathbf{Q}'_\iota$, P, C, χ, and the associated conjugate force, $\boldsymbol{\nabla}T_\kappa^{-1}$, $\boldsymbol{\nabla}T_\iota^{-1}$, $-T_\kappa^{-1}\boldsymbol{\nabla}\mathbf{u} + T_\iota^{-1}\boldsymbol{\varepsilon} \cdot \boldsymbol{\omega}_0$, $-T_\iota^{-1}\boldsymbol{\nabla}\boldsymbol{\omega}_0$, $T_\iota^{-1} - T_\kappa^{-1}$, respectively. In the presence of a field the situation is more interesting and only slightly more complicated. To illustrate this we suppose that $H_j{}^{\mathrm{e}} = -\boldsymbol{\mu}_j \cdot \mathbf{F}(\mathbf{x}_j)$, with $\boldsymbol{\mu}_j = \bar{\gamma}(L_j)\mathbf{L}_j$, and that $H_j = L_j{}^2/2\Gamma$. It then can be shown that $\chi_\iota{}^{\mathrm{e}} = 0$ and that

$$\rho\hat{\mathbf{F}}^{\mathrm{e}} = \boldsymbol{\nabla}\mathbf{F} \cdot \mathbf{M} \tag{1.57}$$

$$\rho\hat{\mathbf{G}}^{\mathrm{e}} = \mathbf{M} \times \mathbf{F} \tag{1.58}$$

$$\chi_K{}^{\mathrm{e}} = \left\langle \sum_i \delta_i \mathbf{C}_i \boldsymbol{\mu}_i \right\rangle^\dagger : \boldsymbol{\nabla}\mathbf{F} \tag{1.59}$$

* If $H_i = L_i{}^2/2\Gamma$ then to obtain $\mathbf{Q}'_\iota$ and e'_ι one has only to replace H_i with $H'_\iota = (\mathbf{L}_i - \Gamma\boldsymbol{\omega}_0)^2/2\Gamma$ in the defining formulas for $\mathbf{Q}_\iota$ and e_ι, respectively.

$$\chi_\phi{}^{\mathrm{e}} = \left\langle \sum_i \sum_{j>i} \delta_i[\boldsymbol{\mu}_i + \boldsymbol{\mu}_j, \phi_{ij}] \right\rangle . \mathbf{F} - \left\langle \sum_i \sum_{j>i} \delta_i[\boldsymbol{\mu}_j, \phi_{ij}] . (\mathbf{F}_i - \mathbf{F}_j) \right\rangle \quad (1.60)$$

$$\doteq \left\langle \sum_i \sum_{j>i} \delta_i[\boldsymbol{\mu}_i + \boldsymbol{\mu}_j, \phi_{ij}] \right\rangle . \mathbf{F} - \left\langle \sum_i \sum_{j>i} \mathbf{x}_{ij}[\boldsymbol{\mu}_j, \phi_{ij}]\delta_i \right\rangle^{\dagger} : \nabla\mathbf{F} \quad (1.61)$$

where $\mathbf{M} = \left\langle \sum_i \delta_i \boldsymbol{\mu}_i \right\rangle$ is the dipole moment per unit volume. If $\bar{\gamma}(L_j)$ is a constant we immediately conclude that

$$\chi_K{}^{\mathrm{e}} + \chi_\phi{}^{\mathrm{e}} = \mathsf{C}^\dagger : \nabla\boldsymbol{\omega}_L - \mathsf{P}^\dagger : (\boldsymbol{\varepsilon} . \boldsymbol{\omega}_L) \quad (1.62)$$

and

$$\sigma_S = (\mathbf{Q}_K + \mathbf{Q}_\phi) . \nabla\frac{1}{T_\kappa} + \mathbf{Q}'_\iota . \nabla\frac{1}{T_\iota} - (\mathsf{P}/T_\kappa)^\dagger : \left[\nabla\mathbf{u} + \boldsymbol{\varepsilon} . \left(\boldsymbol{\omega}_L - \frac{T_\kappa}{T_\iota}\boldsymbol{\omega}_0 \right) \right]$$

$$- (\mathsf{C}/T_\iota)^\dagger : \left[\frac{T_\kappa}{T_\iota}\nabla\boldsymbol{\omega}_0 - \nabla\boldsymbol{\omega}_L \right] + \left(\frac{1}{T_\iota} - \frac{1}{T_\kappa} \right)\chi_\iota - (\rho/T_\iota)\boldsymbol{\omega}_0 \times \mathbf{m} . \mathbf{F} \quad (1.63)$$

where $\boldsymbol{\omega}_L \equiv \bar{\gamma}\mathbf{F}$ is the 'Larmor' angular velocity. Since it is to be expected that $\mathbf{m}$ will be parallel to $\boldsymbol{\omega}_0$, the last term of equation (1.63) vanishes.

From this result it follows that—insofar as the rate of entropy production is concerned—the net effect of the field is simply to replace the rotational velocity $\boldsymbol{\omega}_0$ with $\boldsymbol{\omega}_F \equiv \boldsymbol{\omega}_0 - (T_\iota/T_\kappa)\bar{\gamma}\mathbf{F} \approx \boldsymbol{\omega}_0 - \boldsymbol{\omega}_L$. (Although this conclusion is strictly justified only if $\bar{\gamma}(L_j)$ is constant, the formalism is not altered significantly when this restriction is relaxed.) However, the constitutive equations which relate the fluxes to the forces are profoundly influenced by the presence of the field. For not only is it to be expected that the phenomenological coefficients will be dependent upon the strength of this field, but one also must contend with the fluid anisotropies which are induced by the field. Later we shall examine some of these ef`ʻ`ts in more detail but for the present our attention will be confined to the field-free situation.

1.1.6 Linear Phenomenology

Our specific objective is the construction of the linear constitutive relations which are appropriate to the fluid under consideration. A first step in that direction is the decomposition of the forces and fluxes into linearly independent, irreducible components. The scalars χ_ι and $T_\iota^{-1} - T_\kappa^{-1}$ and the vectors $\mathbf{Q}_K + \mathbf{Q}_\phi$, $\mathbf{Q}'_\iota$, ∇T_κ^{-1} and ∇T_ι^{-1} are already in irreducible form. A second rank tensor such as P can be decomposed uniquely into the sum of three parts, a traceless symmetric tensor $\overset{\circ}{\overline{\overline{\mathsf{P}}}} = \frac{1}{2}(\mathsf{P}+\mathsf{P}^{\dagger})-\frac{1}{3}\boldsymbol{\delta}\ \mathrm{tr}\ \mathsf{P}$, a diagonal tensor $\boldsymbol{\delta}\frac{1}{3}\ \mathrm{tr}\ \mathsf{P}$, and an antisymmetrical tensor $\mathsf{P}^{\mathrm{a}} = \frac{1}{2}(\mathsf{P}-\mathsf{P}^{\dagger}) = \boldsymbol{\varepsilon}\,.\,\mathbf{P}^{\mathrm{a}}$. Here the vector $\mathbf{P}^{\mathrm{a}} = -\frac{1}{2}\boldsymbol{\varepsilon}:\mathsf{P}$ denotes the dual of P. Since P is a 'proper' or polar dyadic, tr P is a true scalar and $\mathbf{P}^{\mathrm{a}}$ is an axial vector. However, C is an axial dyadic (it transforms as **pL**) and so its trace is a pseudoscalar and its dual is a polar vector.

The next step in the construction is to express each flux as a linear combination of the forces. The coefficients in these expansions are properties of the fluid and as such can depend only upon its local thermodynamical state. If the medium is isotropic then these phenomenological coefficients must be proportional to one of the isotropic tensors 1 (rank zero), $\boldsymbol{\delta}$ (rank two) or $\boldsymbol{\varepsilon}$ (rank three). On the basis of these rules we are led directly to the constitutive equations

$$\mathbf{Q}_K+\mathbf{Q}_\phi = L_{11}\nabla\frac{1}{T_\kappa}+L_{12}\nabla\frac{1}{T_\iota}+L_{19}\left(\frac{1}{T_\iota}\,\mathrm{curl}\,\boldsymbol{\omega}_0\right) \tag{1.64}$$

$$\mathbf{Q}'_\iota = L_{21}\nabla\frac{1}{T_\kappa}+L_{22}\nabla\frac{1}{T_\iota}+L_{29}\left(\frac{1}{T_\iota}\,\mathrm{curl}\,\boldsymbol{\omega}_0\right) \tag{1.65}$$

$$\chi_\iota = L_{33}\left(\frac{1}{T_\iota}-\frac{1}{T_\kappa}\right)+L_{35}\left(-\frac{1}{T_\kappa}\nabla\,.\,\mathbf{u}\right) \tag{1.66}$$

$$\mathsf{P} = L_{44}\left(-\frac{1}{T_\kappa}\overset{\circ}{\overline{\overline{\nabla\mathbf{u}}}}\right)+\boldsymbol{\delta}\left[L_{53}\left(\frac{1}{T_\iota}-\frac{1}{T_\kappa}\right)+L_{55}\left(-\frac{1}{T_\kappa}\nabla\,.\,\mathbf{u}\right)\right]$$
$$+L_{66}\boldsymbol{\varepsilon}\,.\,\frac{1}{T_\kappa}\left(\mathrm{curl}\,\mathbf{u}-\frac{T_\kappa}{T_\iota}2\boldsymbol{\omega}_0\right) \tag{1.67}$$

and

$$\mathsf{C} = L_{77}\left(-\frac{1}{T_\iota}\overset{\circ}{\overline{\overline{\nabla\boldsymbol{\omega}_0}}}\right)+\boldsymbol{\delta}\left[L_{88}\left(-\frac{1}{T_\iota}\nabla\,.\,\boldsymbol{\omega}_0\right)\right]+L_{91}\boldsymbol{\varepsilon}\,.\,\nabla\frac{1}{T_\kappa}$$
$$+L_{92}\boldsymbol{\varepsilon}\,.\,\nabla\frac{1}{T_\iota}+L_{99}\boldsymbol{\varepsilon}\,.\,\frac{1}{T_\iota}\,\mathrm{curl}\,\boldsymbol{\omega}_0 \tag{1.68}$$

where the coefficients L_{mn} are all true scalars. These coefficients must conform to the conditions of Onsager and Casimir and to the second law of thermodynamics. According to the first of these, $L_{mn} = T_n T_m L_{nm}$ where T_n indicates the eigenvalue of the time-reversal operator which is appropriate to the nth flux. For example, time reversal transforms the dynamical variable of which $\mathbf{Q}'$ is the ensemble average, into its negative.

The same is true for $\mathbf{Q}_K + \mathbf{Q}_\phi$. Therefore $L_{12} = L_{21}$. A similar argument leads one to conclude that $L_{35} = L_{53}$. On the other hand

$$\sum_j \left[\mathbf{C}_j \times \mathbf{L}_j - \sum_{k>j} \mathbf{x}_{jk} \times [\mathbf{L}_k, \phi_{jk}] 0_{jk} \right] \delta_j$$

is unaltered by time reversal and so $L_{19} = -L_{91}$ and $L_{29} = -L_{92}$. Thus, there are four non-trivial 'reciprocal relations' to which the matrix of coefficients L_{nm} must conform.

In order to invoke the second law of thermodynamics we now insert the constitutive equations into equation (1.63):

$$\begin{aligned} \sigma_S = {} & L_{11}(\nabla T_\kappa^{-1})^2 + L_{22}(\nabla T_\iota^{-1})^2 + L_{33}(T_\iota^{-1} - T_\kappa^{-1})^2 \\ & + L_{44}(-T_\kappa^{-1} \overset{\circ}{\overline{\overline{\nabla \mathbf{u}}}})^2 + L_{55}(-T_\kappa^{-1} \nabla . \mathbf{u})^2 \\ & + L_{66}[T_\kappa^{-1}\{\text{curl}\, \mathbf{u} - (T_\kappa/T_\iota) 2\boldsymbol{\omega}_0\}]^2 + L_{77}(-T_\iota^{-1} \overset{\circ}{\overline{\overline{\nabla \boldsymbol{\omega}_0}}})^2 \\ & + L_{88}(-T_\iota^{-1} \nabla . \boldsymbol{\omega}_0)^2 + L_{99}(T_\iota^{-1}\, \text{curl}\, \boldsymbol{\omega}_0)^2 \\ & + 2L_{12}(\nabla T_\kappa^{-1}).(\nabla T_\iota^{-1}) + 2L_{35}(T_\iota^{-1} - T_\kappa^{-1})(T_\kappa^{-1} \nabla . \mathbf{u}) \end{aligned} \quad (1.69)$$

Since the fluxes are linearly independent, the second law $\sigma_S \geqslant 0$ demands that each of the diagonal elements L_{ii}, $i = 1 - 9$, be non-negative. The algebraic signs of the off-diagonal elements L_{12} and L_{35} are not prescribed by the second law but their magnitudes are restricted by the conditions $L_{12}^2 \leqslant L_{11}L_{22}$ and $L_{35}^2 \leqslant L_{33}L_{55}$, respectively. The second law provides no restrictions whatsoever upon the two phenomenological coefficients L_{19} and L_{29}.

Most of these quantities can be identified with familiar phenomenological coefficients. Thus, when we ignore the difference between the two temperatures T_κ and T it is easy to see that $\lambda = T^{-2}(L_{11} + 2L_{12} + L_{22})$ corresponds to the coefficient of thermal conductivity and that $\eta = \frac{1}{2}T^{-1}L_{44}$, $\eta_b = T^{-1}L_{55}$ and $\zeta = T^{-1}L_{66}$ are the coefficients of shear, bulk and vortex viscosity, respectively. Similarly, $\frac{1}{2}T^{-1}L_{77}$, $T^{-1}L_{88}$ and $T^{-1}L_{99}$ are the spin flux coefficients ν_i of Reference (1), $(\rho T^2)[(\hat{c}_{v,k}^{-1} + \hat{c}_{v,\text{int}}^{-1})L_{33}]^{-1}$ the rotational relaxation time, and $T^{-1}(L_{91} + L_{92})$ the coefficient θ of Reference (1).

In the absence of external fields the balance equations (1.29), (1.32), (1.49), (1.48) and (1.41) can be written

$$\mathrm{d}_t\rho = -\rho\boldsymbol{\nabla}\cdot\mathbf{u} \tag{1.70}$$

$$\rho\mathrm{d}_t\mathbf{u} = -\boldsymbol{\nabla}p - \boldsymbol{\nabla}\cdot\mathsf{P} \tag{1.71}$$

$$\rho\mathrm{d}_t\mathbf{m} = -\boldsymbol{\nabla}\cdot\mathsf{C} - 2\mathbf{P}^{\mathrm{a}} \tag{1.72}$$

$$\mathrm{d}_t u_\kappa = -h_\kappa\boldsymbol{\nabla}\cdot\mathbf{u} - \boldsymbol{\nabla}\cdot\mathbf{Q}_\kappa - \mathsf{P}^\dagger:\boldsymbol{\nabla}\mathbf{u} - \chi_\iota \tag{1.73}$$

$$\mathrm{d}_t u'_\iota = -u'_\iota\boldsymbol{\nabla}\cdot\mathbf{u} - \boldsymbol{\nabla}\cdot\mathbf{Q}'_\iota - \mathsf{C}^\dagger:\boldsymbol{\nabla}\boldsymbol{\omega}_0 + 2\boldsymbol{\omega}_0\cdot\mathbf{P}^{\mathrm{a}} + \chi_\iota \tag{1.74}$$

with $\mathbf{Q}_\chi = \mathbf{Q}_K + \mathbf{Q}_\phi$ and where $u_K = \rho e_K$, $u'_\iota = \rho e'_\iota$ and $h_K = u_K + p$ are the energies and enthalpy per unit volume. Furthermore, it is reasonable to assume that the inertial tensor density is very nearly isotropic so that $\mathbf{m} \doteq (\Gamma/m)\boldsymbol{\omega}_0$ and $\rho\mathrm{d}_t\mathbf{m} \doteq n\Gamma\mathrm{d}_t\boldsymbol{\omega}_0$. When the constitutive relations (1.64–1.68) are inserted into these equations of balance one obtains a set of coupled differential equations for the macroscopic fields. To complete the programme solutions of these equations should then be constructed which conform to suitable boundary and/or initial conditions. One of the simplest, non-trivial applications of this sort is to ignore the temperature fields, e.g., to assume that T_κ and T_ι are both constants, and focus exclusively upon the two coupled equations

$$\rho\mathrm{d}_t\mathbf{u} = -\nabla p + 2\zeta\,\mathrm{curl}\,\boldsymbol{\omega}_0 + (\tfrac{1}{3}\eta + \eta_{\mathrm{b}} - \zeta)\boldsymbol{\nabla}(\boldsymbol{\nabla}\cdot\mathbf{u}) + (\eta + \zeta)\nabla^2\mathbf{u} \tag{1.75}$$

$$n\Gamma\mathrm{d}_t\boldsymbol{\omega}_0 = -4\zeta(\boldsymbol{\omega}_0 - \tfrac{1}{2}\,\mathrm{curl}\,\mathbf{u}) + (\tfrac{1}{3}\nu_2 + \nu_1 - \nu_3)\boldsymbol{\nabla}(\boldsymbol{\nabla}\cdot\boldsymbol{\omega}_0) + (\nu_2 + \nu_3)\nabla^2\boldsymbol{\omega}_0 \tag{1.76}$$

which govern the velocity and spin fields. These dynamical equations characterize what has come to be called the 'micropolar fluid'.[2] They have been solved in a variety of cases with particular attention devoted to flows induced by the body torque $\mathbf{X}_L{}^{\mathrm{e}} = \rho\hat{\mathbf{G}}^{\mathrm{e}} \doteq \bar{\gamma}\mathbf{m}\times\mathbf{F}$.[3]

1.1.7 Formal Solution of Liouville Equation

These and other hydrodynamical applications of the theory are of considerable intrinsic interest. However, they are not our primary concern here. The principal task remaining before us is the derivation of formulas which relate the phenomenological coefficients L_{mn} to molecular dynamics. In an earlier publication[4] we accomplished this by adapting a theory of H. S. Green's[5] to the problems of transport and relaxation in polyatomic fluids. Actually, there are several similar but distinct approaches[6,7,8,9,10] which would have served our purposes almost equally well. Among these the theories due to McLennan[11] and Zubarev[12] are our favourites. Therefore, we shall pattern our development for polyatomic fluids very closely upon the work of these two authors. We could, of course, simply state the results which are obtained by applying these theories to the problems at hand. At the opposite extreme would lie a development and discussion as thorough as those of the original authors. We have chosen a middle course which includes enough of the mathematical details to permit easy verification of the final results but which presumes that the reader will turn to McLennan's papers for a truly complete description and interpretation of the theory.

What we want is a solution of the Liouville equation which will be characteristic of a non-equilibrium state. McLennan's procedure for constructing such a solution is to begin (at $t = -\infty$) with the system at equilibrium and then to turn on 'boundary interactions' which displace it from that state. After a time the effects of these perturbing influences will have spread throughout the system, establishing gradients of the macroscopic fields, stimulating various relaxation processes and causing flows of heat, mass, momentum and the like. Once these processes have begun their origins should be of little consequence to the description of the system's behaviour—provided that one concerns oneself only with regions which are reasonably distant from the boundaries. Under these circumstances the predictions of McLennan's theory do indeed become independent of the forces which originally displaced the system from equilibrium. This suggests that, unless one is specifically concerned with the boundary interactions themselves, one need never introduce them at all. Consistent with this is the fact that Zubarev obtains the very same 'asymptotic' results without explicitly introducing the perturbing forces. Since Zubarev's treatment of this aspect of the theory is so commendably brief we adopt it as our starting point.

Now there is no difficulty in principle of treating a system which is coupled to a non-reactive electromagnetic field. However, the presence of such a field does introduce complexities—largely notational—which we prefer to avoid. Therefore, we neglect the influence of electromagnetic fields and similarly ignore the effects of gravity.

Let it be supposed that we have selected a set of local field variables, $G_\alpha(t\mathbf{x};\Omega) \equiv \sum_k \delta_k G_k^{(\alpha)}(t\Omega)$, each of which then satisfies an equation of motion

$$\partial_t G_\alpha = iLG_\alpha = -\nabla \cdot \mathbb{J}_\alpha + \chi_\alpha \tag{1.77}$$

wherein $\mathbb{J}_\alpha$ indicates the gross flux and χ_α the rate of relaxation of G_α. [Here, lower case Greek indices are used to distinguish between the various local field variables.] The choice of variables G_α depends only upon the nature of the *macro*scopic fields with which one wishes to characterize the local state of the system. For example, if one intends that this characterization coincide with the traditional description of a fluid—in terms of its density, velocity and internal energy—then one must limit the set of G_α's to $\hat{n} = \sum_j \delta_j$, $\hat{\mathbf{p}} = \sum_j \mathbf{p}_j \delta_j$ and

$$\hat{H} = \sum \delta_j \left[\frac{1}{2m} p_j^2 + H_j + \sum_{k>j} \phi_{jk} \right]$$

The macrofields associated with these three local variables are $-v+\beta\frac{1}{2}mu^2$, $-\beta\mathbf{u}$ and $\beta = 1/\mathbf{k}T$, respectively. [Here $v = \beta\mu$ and μ is the chemical potential.] In the present situation we wish to treat as distinct the rotational and translational temperatures $T_\iota = 1/\mathbf{k}\beta_\iota$ and $T_\kappa = 1/\mathbf{k}\beta_\kappa$. We also want to assign to each element of the fluid a spin velocity $\boldsymbol{\omega}_0 = \boldsymbol{\omega}_0(t\mathbf{x})$. Therefore, we select the set of five G_α's which appear, together with their associated macrofields $a_\alpha(t\mathbf{x})$, in the first two columns of Table 1.1. The symbols $\hat{H}_\kappa$, $\hat{H}_\iota$ and $\hat{\mathbf{L}}$ refer to the two energy functions

$$\sum_j \delta_j \left[\frac{1}{2m} p_j^2 + \sum_{k>j} \phi_{jk} \right] \text{ and } \sum_j \delta_j H_j$$

and to the angular momentum $\sum_j \delta_j \mathbf{L}_j$, respectively.

Table 1.1

Local variables G_α, their macroscopic 'conjugates' a_α, associated fluxes J_α, and the averages of G_α and J_α in the local equilibrium ensemble.

G_α	a_α	J_α	$\langle G_\alpha \rangle_L$	$\langle J_\alpha \rangle_L$
$\hat{n}$	$-v_\kappa + \frac{1}{2}\beta_\kappa mu^2$ $-v_\iota + \frac{1}{2}\beta_\iota \Gamma \omega_0^2$	$\underset{\sim}{\mathbb{J}} \equiv \hat{\mathbf{p}}$	n	ρu
$\hat{\mathbf{p}}$	$-\beta_\kappa \mathbf{u}$	$\underset{\approx}{\mathbb{J}}_p$	$\rho\mathbf{u}$	$p\boldsymbol{\delta} + \rho\mathbf{u}\mathbf{u}$
$\hat{\mathbf{L}}$	$-\beta_\iota \boldsymbol{\omega}_0$	$\underset{\approx}{\mathbb{J}}_L$	$n\Gamma\boldsymbol{\omega}_0$	$n\Gamma\mathbf{u}\boldsymbol{\omega}_0$
$\hat{H}_\kappa$	β_κ	$\underset{\sim}{\mathbb{J}}_\kappa$	$u_\kappa + \frac{1}{2}\rho u^2$	$(h_\kappa + \frac{1}{2}\rho u^2)\mathbf{u}$
$\hat{H}_\iota$	β_ι	$\underset{\sim}{\mathbb{J}}_\iota$	$u'_\iota + \frac{1}{2}n\Gamma\omega_0^2$	$(u'_\iota + \frac{1}{2}n\Gamma\omega_0^2)\mathbf{u}$

It is the *sine qua non* of the Zubarev theory that to each field variable G_α [and associated macrofield $a_\alpha(t\mathbf{x})$] there corresponds a local integral of the motion

$$A_\alpha(t\mathbf{x}\Omega) = a_\alpha(t\mathbf{x})G_\alpha(\mathbf{x}\Omega) - \int_{-\infty}^{0} \mathrm{d}s\, \mathrm{e}^{\varepsilon s}\partial_s\{a_\alpha(t+s\mathbf{x})G_\alpha(s\mathbf{x}\Omega)\}$$

$$= \varepsilon \int_{-\infty}^{0} \mathrm{d}s\, \mathrm{e}^{\varepsilon s} a_\alpha(t+s\mathbf{x})G_\alpha(s\mathbf{x}\Omega) \qquad (1.78)$$

that is, a function A_α which itself is a solution of the Liouville equation $(\mathrm{d}/\mathrm{d}t)A_\alpha \equiv (\partial_t + \mathrm{i}L)A_\alpha = 0$. It is to be understood that one proceeds to the limit $\varepsilon \to 0^+$ only after one has permitted the volume of the system to tend to infinity.

The local variables A_α play the same rôle in the theory of non-equilibrium ensembles as do the gross integrals of the motion—such as the energy, the total momentum and the particle number—in the case of equilibrium. In particular, since each of the A_α's is an integral of the motion, any functional of these quantities will satisfy the Liouville equation. Although the choice of this functional is arbitrary we know from experience which choices correspond to physically realizable situations and/or are compatible with experimental means for measuring the macrofields with which we wish to characterize the local state of the system. Therefore, it is altogether natural to select the pseudo-Gibbsian

$$f(t\Omega) = \exp\left\{-q - \sum_\alpha \int \mathrm{d}^3x A_\alpha(t\mathbf{x}\Omega)\right\} \qquad (1.79)$$

and to define the normalization factor q so that $\sum_N \int \mathrm{d}\Omega f(t\Omega) = 1$. According to the Liouville theorem this normalization should be independent of time. This is obviously so since each of the A_α's is separately invariant. Furthermore, by imposing suitable initial conditions upon the functions $a_\alpha(t\mathbf{x})$ it can be arranged* so that as $t \to -\infty$ the function

* From equations (1.77) and (1.78) we see that

$$\int \mathrm{d}^3x A_\alpha(t\mathbf{x}\Omega) = \int \mathrm{d}^3x a_\alpha(t\mathbf{x})G_\alpha(\mathbf{x}\Omega) - \int \mathrm{d}^3x \int_{\infty}^{0} \mathrm{d}s\, \mathrm{e}^{\varepsilon s}\{G_\alpha(s\mathbf{x}\Omega)\partial_s a_\alpha(t+s\mathbf{x})$$

$$+ \mathbb{J}_\alpha^\dagger(s\mathbf{x}\Omega) . \nabla_\mathbf{x} a_\alpha(t+s\mathbf{x}) + \chi_\alpha(s\mathbf{x}\Omega)a_\alpha(t+s\mathbf{x})\}$$

and, since it is understood that $\varepsilon \to 0^+$,

$$\lim_{(t\to-\infty)} \int_{-\infty}^{0} \mathrm{d}s\, \mathrm{e}^{\varepsilon s}\chi_\alpha(s\mathbf{x})a_\alpha(t+s\mathbf{x}) = \lim_{(t\to-\infty)} \int_{-1/\varepsilon}^{0} \mathrm{d}s\, \mathrm{e}^{\varepsilon s}\chi_\alpha(s\mathbf{x})a_\alpha(t+s\mathbf{x})$$

$$= \lim_{(t\to-\infty)} a_\alpha(t\mathbf{x}) \int_{-\infty}^{0} \mathrm{d}s\, \mathrm{e}^{\varepsilon s}\chi_\alpha(s\mathbf{x})$$

defined by equation (1.79) tends to the grand ensemble density

$$f_{\mathrm{eq}} = \exp\{-q-\beta\mathbf{H}+\nu N\} \tag{1.80}$$

where $\mathbf{H} = \int d^3x(\hat{H}_\kappa+\hat{H}_\iota)$ is the Hamiltonian and

$$\begin{aligned}\exp q &= \sum_{N\geq 0}\frac{1}{N!}\int d\Omega \exp\left\{-\sum_\alpha\int d^3x a_\alpha(-\infty\mathbf{x})G_\alpha(\mathbf{x}\Omega)\right\}\\ &= \exp \beta pV\end{aligned} \tag{1.81}$$

the grand partition function.

It is reasonable to expect that in first approximation $f(t\Omega)$ should reduce to the function

$$f_L(t\Omega) = \exp\left\{-q(t)-\sum_\alpha\int d^3x a_\alpha(t\mathbf{x})G_\alpha(\mathbf{x}\Omega)\right\} \tag{1.82}$$

These same manipulations apply to the other terms of $\int d^3xA_\alpha(t\mathbf{x}\Omega)$ so that

$$\lim_{t\to-\infty}\int d^3xA_\alpha(t\mathbf{x}\Omega) = \lim_{t\to-\infty}\int d^3x\Bigg[a_\alpha(t\mathbf{x})\left\{G_\alpha(x\Omega)-\int_{-\infty}^0 ds\, e^{\varepsilon s}\chi_\alpha(s\mathbf{x})\right\}$$
$$+\left\{\int_{-\infty}^0 ds\, e^{\varepsilon s}\mathbf{J}_\alpha{}^\dagger(s\mathbf{x}\Omega)\right\}.\nabla_{\mathbf{x}}a_\alpha(t\mathbf{x})+\left\{\int_{-\infty}^0 ds\, e^{\varepsilon s}G_\alpha(s\mathbf{x}\Omega)\right\}\partial_t a_\alpha(t\mathbf{x})\Bigg]$$

Therefore, if we assume $a_\alpha(t\mathbf{x})$ to be independent of time and position in the limit $t\to-\infty$,

$$\lim_{t\to-\infty}\int d^3xA_\alpha(t\mathbf{x}\Omega) = a_\alpha\int d^3x\left\{G_\alpha(\mathbf{x}\Omega)-\int_{-\infty}^0 ds\, e^{\varepsilon s}\chi_\alpha(s\mathbf{x})\right\}$$

For the set of variables listed in Table 1.1 this becomes

$$\lim_{t\to-\infty}\sum_\alpha\int d^3xA_\alpha(t\mathbf{x}\Omega) = \int d^3x\Bigg[\beta_\kappa(H'_\kappa-n\mu_\kappa)+\beta_\iota(H'_\iota-n\mu_\iota)-\beta_\kappa\int_{-\infty}^0 ds\, e^{\varepsilon s}(-\chi_\iota)$$
$$-\beta_\iota\int_{-\infty}^0 ds\, e^{\varepsilon s}(+\chi_\iota)-(-\beta_\iota\boldsymbol{\omega}_0).\int_{-\infty}^0 ds\, e^{\varepsilon s}\boldsymbol{\varepsilon}:\mathbf{p}\Bigg]_{t=-\infty}$$

and so, if we insist that $\beta_\kappa(t=-\infty) = \beta_\iota(t=-\infty)$, that $\mu_\kappa+\mu_\iota\to\mu$ and $\boldsymbol{\omega}_0(t=-\infty) = 0$, it follows that $H'_\iota\to H_\iota$ and

$$\begin{aligned}\lim_{t\to-\infty}\sum_\alpha\int d^3xA_\alpha(t\mathbf{x}\Omega) &= \int d^3x\beta[H'_\kappa+H_\iota-n\mu]\\ &= \beta\sum_j\left(\frac{m}{2}C_j{}^2+\frac{L_j{}^2}{2\Gamma}+\sum_{k>j}\phi_{kj}\right)-\nu N\end{aligned}$$

where $\nu = \beta\mu$.

which describes a state of local equilibrium. For the particular set of variables listed in Table 1.1

$$f_L(t\Omega) = \exp\left\{-q(t) - \int d^3x[\beta_\kappa(\hat{H}'_\kappa - n\mu_\kappa) + \beta(\hat{H}'_\iota - n\mu_\iota)\right\} \quad (1.83)$$

where $\hat{H}'_\iota = \sum_j \delta_j H'_j$ and

$$\hat{H}'_\kappa = \sum_j \delta_j\left[\tfrac{1}{2}mC_j^2 + \sum_{k>j} \phi_{jk}\right]$$

It is obvious that averages computed with the distribution function f_L will have the same dependence upon the local variables β_κ, β_ι, n, v_κ and v_ι as they would upon the corresponding variables associated with a state of thermodynamic equilibrium. Thus, we are provided with a very natural way of establishing connections between the variables of thermodynamics and the quantities which appear in the statistical theory. (Indeed, many of these connections have been implicit throughout the preceding discussion.) For example, we surely must identify $\langle\hat{H}'_\kappa; t\rangle_L$ and $\langle\hat{H}'_\iota; t\rangle_L$ with the values of the thermodynamic energy densities u_κ and u'_ι specific to the local and instantaneous thermodynamic state. By way of summary, we have listed in the fourth column of Table 1.1 values of the functions $\langle G_\alpha; t\rangle_L$ which one obtains from direct computations based upon the definition of f_L.

Now although we have been successful in identifying the functions $\langle G_\alpha; t\rangle_L$ with quantities which are familiar from thermostatics and fluid dynamics, the basic elements of the statistical theory are the averages $\langle G_\alpha; t\rangle$ computed in the full ensemble $f(t\Omega)$ and not those associated with local equilibrium. We resolve this situation and eliminate all ambiguities concerning the identification of the functions $\langle G_\alpha; t\rangle$ by *assigning* to each the value of the corresponding local average $\langle G_\alpha; t\rangle_L$. It is this step which distinguishes the G_α's from other dynamical variables such as the $\mathbb{J}_\alpha$'s and χ_α's. An immediate consequence of this is that the ensemble averages of the equations of motion $\partial_t G_\alpha = -\nabla . \mathbb{J}_\alpha + \chi_\alpha$ become equations of change

$$\partial_t\langle G_\alpha;\rangle_L = -\nabla . \langle \mathbb{J}_\alpha;\rangle + \langle \chi_\alpha;\rangle \quad (1.84)$$

for the local macroscopic variables $\langle G_\alpha;\rangle_L$. To further emphasize and clarify the distinction which we have introduced between the G_α's and the J_α's and χ_α's we decompose the ensemble averages of the latter into the sums

$$\langle \mathrm{J}_\alpha; t\rangle = \langle \mathrm{J}_\alpha; t\rangle_\mathrm{L} + \langle \mathrm{J}_\alpha^*; t\rangle \tag{1.85}$$

$$\langle \chi_\alpha; t\rangle = \langle \chi_\alpha; t\rangle_\mathrm{L} + \langle \chi_\alpha^*; t\rangle \tag{1.86}$$

with $\mathrm{J}_\alpha^* = \mathrm{J}_\alpha - \langle \mathrm{J}_\alpha;\rangle_\mathrm{L}$ and $\chi_\alpha^* = \chi_\alpha - \langle \chi_\alpha;\rangle_\mathrm{L}$. The computed values of the functions $\langle \mathrm{J}_\alpha; t\rangle_\mathrm{L}$ are listed in the fifth column of Table 1.1. Inspection of these entries reveals that they are the convective contributions to the fluxes of the quantities $\langle G_\alpha;\rangle$. Therefore, the conductive, dissipative portions of the fluxes are to be identified with the functions $\langle \mathrm{J}_\alpha^*;\rangle$.

The situation with regard to the relaxation rates is somewhat different because the first-order estimates of the functions $\langle \chi_\alpha;\rangle_\mathrm{L}$ are zero. To prove that this is so one has only to expand f_L about the equilibrium function f_eq and then invoke the fact that $\int d^3x\chi_\alpha(t\mathbf{x}\Omega)$ is the rate of change of a gross dynamical variable. For example, in the case of rotational temperature relaxation

$$\langle \chi_\iota;\rangle_\mathrm{L} \doteq \langle \chi_\iota[1-(\beta_\iota-\beta_\kappa)(H_\iota - N\mu)]\rangle_\mathrm{eq} = -\langle \chi_\iota H_\iota\rangle_\mathrm{eq}(\beta_\iota - \beta_\kappa)$$

$$= -V^{-1}\langle X_\iota H_\iota\rangle_\mathrm{eq}(\beta_\iota-\beta_\kappa) = -V^{-1}\left\langle \frac{d}{dt}\left(\frac{1}{2}H_\iota^2\right)\right\rangle_\mathrm{eq}(\beta_\iota-\beta_\kappa) \tag{1.87}$$

where $X_\iota = \int d^3x\chi_\iota = \dot{H}_\iota$. No first-order terms in the field gradients appear since the parities of the corresponding contributions to the integrand are odd and so contribute nothing to the average constructed in the equilibrium ensemble. Finally, since the equilibrium average of a rate of change must vanish it follows that, to the stated approximation, $\langle \chi_\iota;\rangle_\mathrm{L}$ is indeed zero. At first glance it might seem that the spin–relaxation rate $\langle \chi_\mathrm{L}\rangle_\mathrm{L}$ does not conform to this rule. For since $\langle \mathsf{P};\rangle_\mathrm{L}$ is symmetric (indeed it is zero), $\langle \chi_\mathrm{L};\rangle_\mathrm{L} = \langle \boldsymbol{\varepsilon}:\mathsf{P};\rangle_\mathrm{L}$ certainly vanishes identically. However, it must be remembered that our identification of χ_L with $\boldsymbol{\varepsilon}:\mathsf{P}$ was correct only through terms of first order in the gradients.

We turn now to the task of manipulating the expression for the non-equilibrium distribution function into a more manageable and interpretable form. In terms of f_L this function can be written as the product $f = f_\mathrm{L}\exp D$ with

$$D = -q + q(t) + \sum_\alpha \int d^3x \int_{-\infty}^{0} ds\, e^{\varepsilon s}\partial_s\{a_\alpha(t+s\mathbf{x})G_\alpha(s\mathbf{x}\Omega)\} \tag{1.88}$$

To evaluate the normalization factor $q(t)$ we differentiate the defining formula

$$\exp q(t) = \sum_{N \geqslant 0} \frac{1}{N!} \int \mathrm{d}\Omega \exp \left\{ -\sum_{\alpha} \int \mathrm{d}^3 x a_\alpha(t\mathbf{x}) G_\alpha(\mathbf{x}\Omega) \right\} \tag{1.89}$$

with respect to time:

$$\dot{q}(t) = -\sum_{\alpha} \int \mathrm{d}^3 x \dot{a}_\alpha(t\mathbf{x}) \langle G_\alpha(\mathbf{x}); t \rangle \tag{1.90}$$

Then, since $q(-\infty) = q$ it follows that

$$q(t) - q = -\sum_{\alpha} \int \mathrm{d}^3 x \int_{-\infty}^{0} \mathrm{d}s \dot{a}_\alpha(t+sx) \langle G_\alpha(x); s \rangle \tag{1.91}$$

and

$$D = \sum_{\alpha} \int \mathrm{d}^3 x \int_{-\infty}^{0} \mathrm{d}s\, \mathrm{e}^{\varepsilon s} [\{ G_\alpha(s\mathbf{x}\Omega) - \langle G_\alpha(x); s \rangle \} \dot{a}_\alpha(t+s\mathbf{x})$$
$$+ \mathbf{J}_\alpha^{\dagger}(s\mathbf{x}\Omega) \cdot \nabla a_\alpha(t+s\mathbf{x}) + \chi_\alpha(s\mathbf{x}\Omega) a_\alpha(t+s\mathbf{x})] \tag{1.92}$$

To obtain the last of these formulas it has been assumed that the fluxes vanish at the boundaries of the system.

Although we now are in a position to examine the general structure of the constitutive equations we shall confine our attention to the linear relationships

$$\langle \mathbf{J}_\alpha^{*}(\mathbf{x}); t \rangle \doteq \sum_{\beta} \int \mathrm{d}^3 x' \int_{-\infty}^{0} \mathrm{d}s\, \mathrm{e}^{\varepsilon s}$$
$$\times [\langle \mathbf{J}_\alpha(\mathbf{x}) \{ G_\beta(s\mathbf{x}') - \langle G_\beta(\mathbf{x}'); s \rangle \}; t \rangle_{\mathrm{L}} \dot{a}_\beta(t+s\mathbf{x}')$$
$$+ \langle \mathbf{J}_\alpha(\mathbf{x}) \mathbf{J}_\beta^{\dagger}(s\mathbf{x}'); t \rangle_{\mathrm{L}} \cdot \nabla a_\beta(t+s\mathbf{x}')$$
$$+ \langle \mathbf{J}_\alpha(\mathbf{x}) \chi_\beta(s\mathbf{x}'); t \rangle_{\mathrm{L}} a_\beta(t+s\mathbf{x}')] \tag{1.93}$$

based upon the approximation $f = f_{\mathrm{L}}\, \mathrm{e}^{D} \approx f_{\mathrm{L}} + f_{\mathrm{L}} D$. According to these formulas the (conductive) flux associated with each local variable depends upon the amplitudes and upon the space and time derivatives of all the macrofields. This dependence is non-local but it is certainly to be expected that correlation functions such as $\langle \mathbf{J}_\alpha(\mathbf{x}) \mathbf{J}_\beta^{\dagger}(s\mathbf{x}'); t \rangle_{\mathrm{L}}$ will sensibly differ from zero only when the separation $|\mathbf{x} - \mathbf{x}'|$ is quite small. Finally, we observe that the fluxes at one instant do, in general, depend (casually) upon the values of the field variables at all prior times. However, the distances over which there are significant correlations are usually so very small that the lag between stimulus and response must be correspondingly brief. In the event that the field variables do change significantly during the lifetime of these dynamical correlations it may be necessary to introduce frequency-dependent transport and relaxation coefficients. To allow for this contingency we Fourier analyse the variables $a_\alpha(t\mathbf{x})$ in the manner $a_\alpha(t\mathbf{x}) = \int d\omega \tilde{a}_\alpha(\omega x) \exp i\omega t$ and so obtain from equation (1.93) the formula

$$\begin{aligned}\langle \mathbb{J}_\alpha^*(\mathbf{x}); t\rangle = \sum_\beta d^3x' \int d\omega\, e^{i\omega t} \\ \times[\langle \mathbb{J}_\alpha(x)\{G_\beta(\mathbf{x}') - \langle G_\beta(\mathbf{x}');\rangle\}; t\rangle_{\omega L} i\omega \tilde{a}_\beta(\omega \mathbf{x}') \\ + \langle \mathbb{J}_\alpha(\mathbf{x})\mathbb{J}_\beta{}^\dagger(\mathbf{x}'); t\rangle_{\omega L} \cdot \nabla \tilde{a}_\alpha(\omega \mathbf{x}') \\ + \langle \mathbb{J}_\alpha(\mathbf{x})\chi_\beta(\mathbf{x}'); t\rangle_{\omega L} \tilde{a}_\alpha(\omega \mathbf{x}')]\end{aligned} \tag{1.94}$$

where

$$\langle A(\mathbf{x})B(\mathbf{x}'); t\rangle_{\omega L} = \int_{-\infty}^{0} ds\, e^{(\varepsilon + i\omega)s} \langle A(\mathbf{x})B(s\mathbf{x}'); t\rangle_L \tag{1.95}$$

1.1.8 Correlation Integrals and Linear Phenomenological Coefficients

Let us turn again from the general to the specific. From the thermodynamic relations $n dv_\kappa = u_\kappa\, d\beta_\kappa + d\beta_\kappa p$ and $n\, dv_\iota = u'_\iota\, d\beta_\iota$ we see that

$$\begin{aligned} a(t) - q &= -\int d^3x \int_{-\infty}^{0} ds[u_\kappa \partial_s \beta_\kappa + u'_\iota \partial_s \beta_\iota - n\partial_s(v_\kappa + v_\iota)] \\ &= \int d^3x \int_{-\infty}^{0} ds \partial_s(\beta_\kappa p)\end{aligned} \tag{1.96}$$

and so $q(t) = \int d^3x \beta_\kappa p$. This is precisely what one expects since at equilibrium $q = \beta p V$.

For the system of interest to us here the function D is of the form

$$
\begin{aligned}
D = \int \mathrm{d}^3x \int_{-\infty}^{0} & \mathrm{d}s\, \mathrm{e}^{\varepsilon s}[(\hat{n}-n)\partial_s(-\nu_\kappa-\nu_\iota+\tfrac{1}{2}\beta_\kappa m u^2+\tfrac{1}{2}\beta_\iota\Gamma\omega_0{}^2) \\
& +\hat{\mathbf{p}}^* \cdot \partial_s(-\beta_\kappa \mathbf{u})+\hat{\mathbf{L}}^* \cdot \partial_s(-\beta_\iota \boldsymbol{\omega}_0) \\
& +\hat{H}_\kappa{}^*\partial_s\beta_\kappa+\hat{H}_\iota{}^*\partial_s\beta+\mathbb{J}_n{}^* \cdot \\
& \nabla(-\nu_\kappa-\nu_\iota+\tfrac{1}{2}\beta_\kappa m u^2+\tfrac{1}{2}\beta_\iota\Gamma\omega_0{}^2) \\
& +\mathbb{J}_p{}^*:(-\nabla\beta_\kappa\mathbf{u}+\beta_\iota\boldsymbol{\varepsilon} \cdot \boldsymbol{\omega}_0)^\dagger+\underset{\approx}{\mathbb{J}}_{\mathrm{L}}{}^*:(-\nabla\beta\boldsymbol{\omega}_0) \\
& +\underset{\approx}{\mathbb{J}}_\kappa{}^* \cdot \nabla\beta_\kappa+\underset{\approx}{\mathbb{J}}_\iota{}^* \cdot \nabla\beta_\iota \\
& +\chi_\iota(\beta_\iota-\beta_\kappa)-\nabla \cdot (\mathbf{u}\beta_\kappa p)]
\end{aligned} \tag{1.97}
$$

[The last term of this expression can be discarded since we require the velocity field to be tangential to the boundaries of the system.] To obtain this result one first substitutes quantities from Table 1.1 directly into the formula (1.92). This leads to an expression which differs from (1.97) in only two ways: the final term $-\nabla \cdot (\mathbf{u}\beta_\kappa p)$ is missing and in place of each $\mathbb{J}_\alpha{}^*$ there appears the corresponding $\mathbb{J}_\alpha$. One then invokes the identity

$$
\begin{aligned}
n\mathbf{u} \cdot \nabla(-\nu_\kappa-\nu_\iota+\tfrac{1}{2}\beta_\kappa m u^2+\tfrac{1}{2}\beta_\iota\Gamma\omega_0{}^2)+\nabla \cdot (\mathbf{u}\beta_\kappa p)+\langle\underset{\approx}{\mathbb{J}}_{\mathrm{p}}\rangle_{\mathrm{L}}{}^\dagger:\nabla(-\beta_\kappa\mathbf{u}) \\
+\langle\underset{\approx}{\mathbb{J}}_{\mathrm{L}}\rangle_{\mathrm{L}}{}^\dagger:\nabla(-\beta_\iota\boldsymbol{\omega}_0)+\langle\mathbb{J}_\kappa\rangle_{\mathrm{L}} \cdot \nabla\beta_\kappa+\langle\mathbb{J}_\iota\rangle_{\mathrm{L}} \cdot \nabla\beta_\iota = 0
\end{aligned} \tag{1.98}
$$

to arrive at the final formula (1.97).

Our objective now is to derive expressions for the fluxes and relaxation rates which will be correct to first order in the driving forces. It is assumed that $\mathbf{u}$ and $\boldsymbol{\omega}_0$ are so small that we can replace the integrand of equation (1.97) with

$$
\begin{aligned}
-(\hat{n}-n)\partial_s(\nu_\kappa+\nu_\iota)-\beta_\kappa\hat{\mathbf{p}} \cdot \partial_s\mathbf{u}-\beta_\iota\hat{\mathbf{L}} \cdot \partial_s\boldsymbol{\omega}_0+(\hat{H}_\kappa-u_\kappa)\partial_s\beta_\kappa+(\hat{H}_\iota-u'_\iota)\partial_s\beta_\iota \\
-\mathbb{J}_n \cdot \nabla(\nu_\kappa+\nu_\iota)+(\underset{\approx}{\mathbb{J}}_p-p\boldsymbol{\delta})^\dagger:(-\beta_\kappa\nabla\mathbf{u}+\beta\boldsymbol{\varepsilon} \cdot \boldsymbol{\omega}_0)+\underset{\approx}{\mathbb{J}}_{\mathrm{L}}{}^\dagger:(-\beta_\iota\nabla\boldsymbol{\omega}_0) \\
+\mathbb{J}_\kappa \cdot \nabla\beta_\kappa+\underset{\approx}{\mathbb{J}}_\iota \cdot \nabla\beta_\iota+\chi_\iota(\beta_\iota-\beta_\kappa)
\end{aligned} \tag{1.99}
$$

To the stated approximation the time derivatives $\partial_s(\nu_\kappa+\nu), \ldots, \partial_s\boldsymbol{\omega}_0$ need be evaluated only to first order in the gradients. Thus, $\partial_t\boldsymbol{\omega}_0 \doteq 0$,

$$
\begin{aligned}
\partial_t\beta_\kappa &= (\partial\beta_\kappa/\partial u_\kappa)_n\partial_t u_\kappa+(\partial\beta_\kappa/\partial n)_{u_\kappa}\partial_t n \doteq -[h_\kappa(\partial\beta_\kappa/\partial u_\kappa)_n+n(\partial\beta_\kappa/\partial n)_{u_\kappa}]\nabla \cdot \mathbf{u} \\
&= \beta_\kappa(\partial p/\partial u_\kappa)_n\nabla \cdot \mathbf{u}
\end{aligned} \tag{1.100}
$$

and similarly, $\partial_t \beta_\iota \doteq -u'_\iota(\mathrm{d}\beta_\iota/\mathrm{d}u'_\iota)\nabla\cdot\mathbf{u} = \beta_\iota \nabla\cdot\mathbf{u}$, $\partial_t v_\kappa \doteq -\beta_\kappa(\partial p/\partial n)_{u_\kappa} \times \nabla\cdot\mathbf{u}$, and $\partial_t v_\iota \doteq -(u_\iota'^2/n)(\mathrm{d}\beta_\iota/\mathrm{d}u_\iota')\nabla\cdot\mathbf{u} = (\hat{c}_\iota m/k)\nabla\cdot\mathbf{u}$. To obtain these formulas we have made use of the identities $h_\kappa = -\beta_\kappa(\partial p/\partial\beta_\kappa)_{v_\kappa}$, $n = \beta_\kappa(\partial p/\partial v_\kappa)_{\beta_\kappa}$, $(\partial\beta_\kappa/\partial n) = -(\partial v_\kappa/\partial u_\kappa)_n$ and introduced the rotational heat capacity per unit mass, $\hat{c}_\iota = \rho^{-1}\,\mathrm{d}u'_\iota/\mathrm{d}T_\iota$. Finally, $\partial_\iota\mathbf{u} \doteq -\rho^{-1}\nabla p$ and $n\nabla(v_K+v_\iota) = h_K\nabla\beta_K + u'\nabla\beta + \beta_K\nabla p$.

When these approximations are substituted into equation (1.99) it is found that

$$D = \int \mathrm{d}^3x \int_{-\infty}^{0} \mathrm{d}s e^{\varepsilon s}\Bigg[\left(\underset{\approx}{\mathrm{J}}_\kappa - \hat{\mathbf{p}}\frac{h_\kappa}{m}\right)\cdot\nabla\beta_\kappa + \left(\underset{\approx}{\mathrm{J}}_\iota - \hat{\mathbf{p}}\frac{u'_\iota}{\rho}\right)\cdot\nabla\beta_\iota$$

$$+(\underset{\approx}{\mathrm{J}}_\mathrm{p} - \mathrm{J}_\mathrm{p}\boldsymbol{\delta})^\dagger : (-\beta_\kappa\nabla\mathbf{u} + \beta_\iota\boldsymbol{\varepsilon}\cdot\boldsymbol{\omega}_0) + \underset{\approx}{\mathrm{J}}_\mathrm{L}{}^\dagger : (-\beta_\iota\nabla\boldsymbol{\omega}_0) + (\beta_\iota - \beta_\kappa)\chi_\iota\Bigg] \tag{1.101}$$

where

$$\mathrm{J}_p = p + (\hat{H}_\kappa - u_\kappa)(\partial p/\partial u_\kappa)_n + (\hat{H}_\iota - u'_\iota)(\beta_\iota/\beta_\kappa) + (\hat{n} - n)[(\partial p/\partial n)_{u_\kappa} + \hat{c}_\iota m/k\beta_\kappa] \tag{1.102}$$

Now to the linear approximation we can neglect the position dependence of the driving forces and so replace equation (1.101) with

$$D = \int_{-\infty}^{0} \mathrm{d}s\, e^{\varepsilon s}\Bigg[\left(\mathscr{F}_\kappa - \mathbf{P}\frac{h_\kappa}{m}\right)\cdot\nabla\beta_\kappa + \left(\mathscr{F}_\iota - \mathbf{P}\frac{u'_\iota}{m}\right)\cdot\nabla\beta$$

$$+(\underset{\approx}{\mathscr{F}}_p - \mathscr{F}_p\boldsymbol{\delta})^\dagger : (-\beta_\kappa\nabla\mathbf{u} + \beta_\iota\boldsymbol{\varepsilon}\cdot\boldsymbol{\omega}_0) + \underset{\approx}{\mathscr{F}}_\mathrm{L}{}^\dagger : (-\beta_\iota\nabla\boldsymbol{\omega}_0) + X_\iota(\beta_\iota - \beta_\chi)\Bigg] \tag{1.103}$$

Here $\mathbf{P} = \int \mathrm{d}^3x\hat{\mathbf{p}} = \sum_j \mathbf{p}_j$ is the total momentum of the fluid and $\mathscr{F}_\alpha = \int \mathrm{d}^3x\mathrm{J}_\alpha$. In particular,

$$\mathscr{F}_p = pV + (N - \overline{N})[(\partial p/\partial n)_{u_\kappa} + \hat{c}_\iota m/k\beta_\kappa] + (H_\kappa - \overline{H}_\kappa)(\partial p/\partial u_\kappa)_n$$
$$+ (H_\iota - \overline{H}_\iota)(\beta_\iota/\beta_\kappa) \tag{1.104}$$

where $\overline{N} = nV$, $\overline{H}_\kappa = u_\kappa V$ and $\overline{H} = u'_\iota V$ are the average values of the particle number, and of translational and rotational energies, respectively.

We now rewrite equation (1.104) as the sum of inner products

$$D = \int_{-\infty}^{0} \mathrm{d}s\, \mathrm{e}^{\varepsilon s} \sum_{\beta} j_{\beta}(s\mathbf{x}\Omega) \odot x_{\beta}(t+s\mathbf{x}) \tag{1.105}$$

of the irreducible fluxes j_β and forces x_β which are listed in Table 1.2. The column headings P, T, R and K refer to the parity, time-reversal eigenvalue, tensor rank and tensor character (K = p for a polar tensor, K = a for an axial or pseudotensor) of the corresponding force and flux.

To the present approximation we conclude that

$$\langle \mathbb{J}_\alpha{}^*;\rangle \doteq \langle \mathbb{J}_\alpha D;\rangle_{\mathrm{L}} \doteq \langle \mathbb{J}_\alpha D\rangle_0 = \sum_{\beta} \int_{-\infty}^{0} \mathrm{d}s\, \mathrm{e}^{\varepsilon s} \langle \mathbb{J}_\alpha j_\beta(s)\rangle_0 \odot x_\beta(t+s) \tag{1.106}$$

where $\langle\ \rangle_0$ denotes an average computed in the equilibrium, grand canonical ensemble. Then, provided that the forces x_β change very little throughout the extremely short intervals when the correlations $\langle \mathbb{J}_\alpha j_\beta(s)\rangle_0$ differ from zero, we can replace equation (1.106) with

$$\langle \mathbb{J}_\alpha{}^*;\rangle \doteq \sum_{\beta} [\mathscr{F}_\alpha j_\beta] \odot x_\beta \tag{1.107}$$

where

$$[\mathscr{F}_\alpha j_\beta] = V^{-1} \int_{-\infty}^{0} \mathrm{d}s\, \mathrm{e}^{\varepsilon s} \langle \mathscr{F}_\alpha j_\beta(s)\rangle_0 = V^{-1} \int_{-\infty}^{0} \mathrm{d}s\, \mathrm{e}^{\varepsilon s} \langle j_\alpha j_\beta(s)\rangle_0 \tag{1.108}$$

To obtain the second of these expressions for $[\mathscr{F}_\alpha j_\beta]$ from the first we have made use of the fact that $\mathscr{F}_\alpha = j_\alpha - \bar{j}_\alpha$ where $\bar{j}_\alpha$ is the time average of the variable j_α. In particular, it can be proved that $\mathscr{F}_P$, $\mathbf{P}h_\kappa/m$ and $\mathbf{P}u'_\iota/m$ are the time averages of $\frac{1}{3}\,\mathrm{tr}\,\underset{\approx}{\mathscr{F}}_P$, $\mathscr{F}_\kappa$ and $\mathscr{F}_\iota$, respectively.

The integrals $\langle j_\alpha j_\beta(s)\rangle_0$ satisfy the reciprocal relations $\langle j_\alpha j_\beta(s)\rangle_0 = T_\alpha T_\beta \langle j_\alpha(s) j_\beta\rangle_0$ wherein T_α denotes the time-reversal eigenvalue of the dynamical variable j_α. Let us introduce the notation $\Omega_t = \mathrm{e}^{\mathrm{i}Lt}\Omega$ and denote by Ω_T the image of Ω under time reversal. Notice that $L(\Omega_T) = -L(\Omega)$. The proof of the reciprocal relations then proceeds as follows:

$$\begin{aligned}\langle j_\alpha j_\beta(s)\rangle_0 &= \int \mathrm{d}\Omega f_{\mathrm{eq}}(\Omega) j_\alpha(\Omega) j_\beta(\Omega_s) = \int \mathrm{d}\Omega_s f_{\mathrm{eq}}(\Omega_s) j_\alpha(\mathrm{e}^{-\mathrm{i}Ls}\Omega_s) j_\beta(\Omega_s) \\ &= \int \mathrm{d}\Omega_T f_{\mathrm{eq}}(\Omega_T) j_\alpha(\mathrm{e}^{\mathrm{i}Ls}\Omega_T) j_\beta(\Omega_T) = T_\alpha T_\beta \int \mathrm{d}\Omega f_{\mathrm{eq}}(\Omega) j_\alpha(\Omega_s) j_\beta(\Omega)\end{aligned} \tag{1.109}$$

Since f_{eq} is an even function of the momenta, $\langle j_\alpha j_\beta(s)\rangle_0$ can be different from zero only if the parities of j_α and j_β are the same. Furthermore, f_{eq} is the equilibrium distribution characteristic of an isotropic fluid and so these correlation integrals must reduce to multiples of the isotropic polar tensors 1, $\boldsymbol{\delta}$, etc. This means that the only non-zero correlation integrals will be those involving irreducible fluxes j_α and j_β which have the same tensor rank, i.e., for which $R_\alpha = R_\beta$. By observing these restrictions we obtain from equation (1.107) the same set of linear constitutive equations (1.55–1.59) as before and are able to identify the phenomenological coefficients $L_{\alpha\beta} = T_\alpha T_\beta L_{\beta\alpha}$ with the multiples

$$L_{\alpha\beta} = \frac{\delta(R_\alpha, R_\beta)\delta(P_\alpha, P_\beta)}{kV\sqrt{(2R_\alpha+1)(2R_\beta+1)}} \int_{-\infty}^{0} \mathrm{d}s\, \mathrm{e}^{\varepsilon s} \langle j_\alpha \odot j_\beta(s)\rangle_0 \tag{1.110}$$

of the time-averaged correlation integrals.

Table 1.2

Force and flux conjugate pairs. The symbols P, T, R and K denote the parity, time-reversal eigenvalue, tensor rank and tensor 'character', respectively.

α	χ_α	j_α	P	T	R	K
1	$\nabla\beta_\kappa$	$\mathscr{F}_\kappa - \mathbf{P}(h_\kappa/m)$	$-$	$-$	1	p
2	$\nabla\beta_\iota$	$\mathscr{F}_\iota - \mathbf{P}(u'_\iota/m)$	$-$	$-$	1	p
3	$\beta_\iota - \beta_\kappa$	X_ι	$+$	$+$	0	p
4	$-\beta_\kappa \overline{\overline{\nabla \mathbf{u}}}^{\circ}$	$\overline{\overline{\underset{\approx}{\mathscr{F}}_p}}^{\circ}$	$+$	$+$	2	p
5	$-\beta_\kappa \nabla . \mathbf{u}$	$\frac{1}{3}\,\mathrm{tr}\, \underset{\approx}{\mathscr{F}}_p - \mathscr{F}_p$	$+$	$+$	0	p
6	$-\beta_\kappa\, \mathrm{curl}\, \mathbf{u}$ $+2\beta_\iota \boldsymbol{\omega}_0$	$\frac{1}{2}\boldsymbol{\varepsilon} : \underset{\approx}{\mathscr{F}}_p$	$+$	$+$	1	a
7	$-\beta_\iota \overline{\overline{\nabla \boldsymbol{\omega}_0}}^{\circ}$	$\overline{\overline{\underset{\approx}{\mathscr{F}}_{\mathrm{L}}}}^{\circ}$	$-$	$+$	2	a
8	$-\beta_\iota \nabla . \boldsymbol{\omega}_0$	$\frac{1}{3}\,\mathrm{tr}\, \underset{\approx}{\mathscr{F}}_{\mathrm{L}}$	$-$	$+$	0	a
9	$-\beta_\iota\, \mathrm{curl}\, \boldsymbol{\omega}_0$	$\frac{1}{2}\boldsymbol{\varepsilon} . \underset{\approx}{\mathscr{F}}_{\mathrm{L}}$	$-$	$+$	1	p

As a final consideration we examine the entropy which, in accordance with the general philosophy of our theory, is to be identified with the same functional,

$$S = -k \int \mathrm{d}\Omega f_{\mathrm{L}} \log f_{\mathrm{L}} = k\left[q_{\mathrm{L}}(t) + \sum_k \int \mathrm{d}^3 x \alpha_k \langle G_k ; t\rangle_{\mathrm{L}} \right] \tag{1.111}$$

of the local quasi-equilibrium distribution function, f_L, as that which defines the entropy associated with a system in a state of thermodynamic equilibrium. Then, since we require that $\langle G_k;t\rangle = \langle G_k;t\rangle_L$, for all k, it follows that

$$S = -k\int \mathrm{d}\Omega f \log f_L \tag{1.112}$$

and

$$\begin{aligned}\dot{S} &= -k\left[\dot{q}_L + \sum_k \int \mathrm{d}^3x \langle(\dot{\alpha}_k G_k + \alpha_k \dot{G}_k);\rangle\right] \\ &= k\int \mathrm{d}^3x \sum_k \alpha_k \langle \dot{G}_k;t\rangle\end{aligned} \tag{1.113}$$

Therefore, the entropy per unit volume, s, and its rate of change, $\dot{s}$, are given by

$$s = k\left[\beta_\kappa p + \sum_k \alpha_k \langle G_k;\rangle\right] = k[\beta_\kappa h_\kappa + \beta_\iota u'_\iota - n(v_\kappa + v_\iota)] \tag{1.114}$$

and

$$\dot{s} = k\sum_k \alpha_k[-\boldsymbol{\nabla}.\langle \mathbb{J}_k;\rangle + \langle \chi_k;\rangle] \tag{1.115}$$

respectively. It follows from this that

$$\begin{aligned}\frac{1}{k}(\partial_t s + \nabla.\mathbf{u}s) &= -\beta_\kappa \boldsymbol{\nabla}.\langle \underset{\sim}{\mathbb{J}}_\kappa{}^*\rangle - \beta_\iota \boldsymbol{\nabla}.\langle \underset{\sim}{\mathbb{J}}_\iota{}^*\rangle + (\beta_\iota - \beta_\kappa)\langle \chi_\mathrm{i}\rangle \\ &\quad + (\beta_\kappa \mathbf{u}\boldsymbol{\nabla} - \beta_\iota \boldsymbol{\omega}_0.\boldsymbol{\varepsilon}):\langle \underset{\approx}{\mathbb{J}}_p{}^*\rangle + \beta_\iota \boldsymbol{\omega}_0 \boldsymbol{\nabla}:\langle \underset{\sim}{\mathbb{J}}_L{}^*\rangle \\ &= -\boldsymbol{\nabla}.\mathbf{J}_s + \sigma_s\end{aligned} \tag{1.116}$$

where

$$\mathbf{J}_s = \beta_\kappa\lfloor\langle \underset{\sim}{\mathbb{J}}_\kappa{}^*\rangle - \langle \underset{\approx}{\mathbb{J}}_p{}^*\rangle.\mathbf{u}] + \beta_\iota[\langle \underset{\sim}{\mathbb{J}}_\iota{}^*\rangle - \langle \underset{\sim}{\mathbb{J}}_L{}^*\rangle.\boldsymbol{\omega}_0] \tag{1.117}$$

and

$$\begin{aligned}\sigma_s = {} & [\langle \underset{\sim}{\mathbb{J}}_\kappa{}^*\rangle - \langle \underset{\sim}{\mathbb{J}}_p{}^*\rangle.\mathbf{u}].\boldsymbol{\nabla}\beta_\kappa + [\langle \underset{\sim}{\mathbb{J}}_\iota{}^*\rangle - \langle \underset{\approx}{\mathbb{J}}_L{}^*\rangle.\boldsymbol{\omega}_0].\boldsymbol{\nabla}\beta_\iota \\ & - \beta_\kappa\langle \underset{\approx}{\mathbb{J}}_p{}^*\rangle^\dagger:[\boldsymbol{\nabla}\mathbf{u} - (\beta_\iota/\beta_\kappa)\boldsymbol{\varepsilon}.\boldsymbol{\omega}_0] - \beta_\iota\langle \underset{\approx}{\mathbb{J}}_L{}^*\rangle^\dagger:\boldsymbol{\nabla}\boldsymbol{\omega}_0 \\ & + (\beta_\iota - \beta_\kappa)\langle \chi_\iota\rangle\end{aligned} \tag{1.118}$$

This confirms (what we already know to be true) that it is $\langle \mathbb{J}_k^*;\rangle = \langle \mathbb{J}_k;\rangle - \langle \mathbb{J}_k;\rangle_L$, the excess of the flux of G_k over that calculated in the local equilibrium ensemble, which determines the irreversible rate of entropy production.

1.1.9 The Quantum Ensemble

Up to now our considerations have been based upon the assumption that the molecular kinematics and dynamics are adequately described by classical mechanics. And indeed, there is a large range of physical conditions and a broad category of processes for which this assumption is fully justified. However, there also exist situations where it is not and so we must be prepared to shift from a classical to a quantum theory when circumstances demand. In classical theory the ensemble representative of a particular dynamical system contains one replica of that system for each point in the phase space, that is, for each dynamical state of the system, which is consistent with the limited amount of information available about its state. In quantum theory a dynamical state is not characterized by a phase point but by a state vector belonging to the Hilbert space which is appropriate to the system. These state vectors, $|t\rangle$, and their duals, $\langle t|$, are solutions of the Schroedinger equations $i\hbar\partial_t|t\rangle = \hat{H}|t\rangle$ and $-i\hbar\partial_t\langle t| = \langle t|\hat{H}$ where $\hat{H} = \hat{H}^\dagger$ is the self-adjoint Hamiltonian operator descriptive of the particular physical system under consideration.

Associated with the state $|t\rangle$ is the expectation value of the (time-independent) operator $\hat{A}$,

$$\langle t|\hat{A}|t\rangle = \iint \mathrm{d}\alpha\, \mathrm{d}\alpha' \langle t|\alpha\rangle\langle\alpha|\hat{A}|\alpha'\rangle\langle\alpha'|t\rangle = \iint \mathrm{d}\alpha\, \mathrm{d}\alpha' \langle\alpha|A|\alpha'\rangle\langle\alpha'|t\rangle\langle t|\alpha\rangle$$

$$= \mathrm{Tr}\,(\hat{A}\,.\,\hat{\rho}(t))$$

Here the vectors $|\alpha\rangle$ denote a complete set or basis for the space, $\langle\alpha|t\rangle$ is the wave function in this representation, Tr C indicates the trace of the operator matrix C, and $\hat{\rho}(t) = |t\rangle\langle t|$ is the 'density matrix' associated with the system in the pure quantum state $|t\rangle$. Since it is customary to deal with state vectors which are normalized to unity, we can assume without loss of generality that Tr $\hat{\rho} = 1$. Finally, it is easily proved that the equation of motion for this density matrix is $i\hbar\partial_t\hat{\rho} = [\hat{H}, \hat{\rho}]_-$ with $[\hat{a}, \hat{b}] = \hat{a}\hat{b} - \hat{b}\hat{a}$.

Although the state of a particular many-body system can always be described by a specific vector $|tm\rangle$, the available data generally will not be sufficient to allow a complete experimental identification of the system's quantum state. Therefore, all predictions concerning the future of this system must reflect our incomplete knowledge of its initial state. This is done by basing all predictions upon an ensemble which contains one member in each state that is consistent with the available data. If there are *a priori* reasons for weighting some states more heavily than others, then the composition of the ensemble can be altered accordingly. With all this in mind, let us suppose that from the information available about a particular system we are able to assign the value p_m to the relative probability that at time t the state of this system is described by the vector $|tm\rangle$. Therefore, the expectation value of $\hat{A}$ will be given by

$$\begin{aligned}\langle \hat{A}\,;t\rangle &= \sum_m p_m \langle tm|\hat{A}|tm\rangle = \sum_m \iint \mathrm{d}\alpha\,\mathrm{d}\alpha' \langle\alpha|\hat{A}|\alpha'\rangle\langle\alpha'|tm\rangle p_m \langle tm|\alpha\rangle \\ &= \mathrm{Tr}\,(\hat{A}\,.\,\hat{\rho}(t)) \end{aligned} \tag{1.119}$$

where $\hat{\rho}(t) = \sum_m |tm\rangle p_m \langle tm|$, the density matrix for the ensemble representative of this physical system, is the quantum analogue of the classical distribution function f. Since the vectors $|tm\rangle$ are normalized to unity and the probabilities p_m such that $\sum_m p_m = 1$, we see that $\mathrm{Tr}\,\hat{\rho}(t) = 1$. Furthermore, it is readily verified that the equation of change for this 'mixed state' ensemble density matrix is the same as that for a pure state.

Since $|tm\rangle = \mathrm{e}^{-\mathrm{i}Ht/\hbar}|0m\rangle$ it follows that $\hat{\rho}(t) = \mathrm{e}^{-\mathrm{i}\hat{H}t/\hbar}\hat{\rho}(0)\,\mathrm{e}^{\mathrm{i}\hat{H}t/\hbar}$ and

$$\langle \hat{A}\,;t\rangle = \mathrm{Tr}\,(\hat{A}\,.\,\hat{\rho}(t)) = \mathrm{Tr}\,(\hat{A}(t)\,.\,\hat{\rho}(0)) \tag{1.120}$$

with $\hat{A}(t) = \mathrm{e}^{\mathrm{i}\hat{H}t/\hbar}\hat{A}\,\mathrm{e}^{-\mathrm{i}\hat{H}t/\hbar}$. Thus, just as in the classical theory one may choose to work either in the Schroedinger picture with the time-dependent density $\hat{\rho}(t)$ and the time-independent operators $\hat{A}$, or in the Heisenberg picture with operators $\hat{A}(t)$ whose variations with time are governed by

$$\partial_t \hat{A}(t) = -\frac{1}{\mathrm{i}\hbar}[\hat{H}, \hat{A}(t)]_- \tag{1.121}$$

the quantum analogue of the classical equation of motion, $\partial_t G(t, \Omega) = -\mathrm{i}LG(t, \Omega) = -[H, G]_{\mathrm{PB}}$.

From this point on the quantum and classical formalisms differ scarcely at all. To each classical local variable $G(t\Omega)$ there corresponds a unique quantum operator. The equations of change for the ensemble averages of these two are precisely the same but because the quantum operator is Hermitean its equation of motion will usually be a bit more complicated than that of the classical counterpart. For example, the mass flux operator which appears in the equation of motion $\partial_t\hat{\rho}+\nabla\,.\,\hat{\mathbf{J}} = 0$ for the quantum operator $\hat{\rho} = \sum_k m\hat{\delta}_k = \sum_k m\delta_3(\hat{\mathbf{x}}_k-\mathbf{x})$ is given by $\hat{\mathbf{J}} = \frac{1}{2}\sum_k (\hat{\mathbf{p}}_k\hat{\delta}_k+\hat{\delta}_k\hat{\mathbf{p}}_k)$. Furthermore, the Hermitean (local) velocity operator $\hat{\mathbf{u}} = \frac{1}{2}(\hat{\rho}^{-1}\hat{\mathbf{J}}+\hat{\mathbf{J}}\rho^{-1})$ satisfies the equation of change

$$\tfrac{1}{2}[\hat{\rho}(\partial_t\hat{\mathbf{u}})+\tfrac{1}{2}(\hat{\rho}\hat{\mathbf{u}}+\hat{\mathbf{u}}\hat{\rho})\,.\,\nabla\hat{\mathbf{u}}]+\tfrac{1}{2}[(\partial_t\hat{\mathbf{u}})\hat{\rho}+(\nabla\hat{\mathbf{u}})^{\dagger}\,.\,\tfrac{1}{2}(\hat{\boldsymbol{\rho}}\hat{\mathbf{u}}+\hat{\mathbf{u}}\hat{\boldsymbol{\rho}})] = -\nabla\,.\,\hat{\mathbf{p}}+\hat{\rho}\hat{\mathbf{F}}^{e}$$

which is very similar but somewhat more complicated than the familiar equation $\rho(\partial_t\mathbf{u}+\mathbf{u}\,.\,\nabla\mathbf{u}) = -\nabla\,.\,\mathbf{p}+\rho\hat{\mathbf{F}}^{e}$ satisfied by the ensemble average of this operator. (For more examples and details concerning the quantum equations of motion for local variables the reader should consult reference 13.)

Generally it is possible to express ensemble averages of the densities, fluxes and relaxation rates of single- and two-particle fields in terms of the singlet and pair density matrices

$$\hat{\rho}^{(1)}(1) = [N!/(N-1)!]\,\mathrm{Tr}_{(2,3,\ldots,N)}\hat{\rho} \tag{1.122}$$

and

$$\hat{\rho}^{(2)}(12) = [N!/(N-2)!]\,\mathrm{Tr}_{(3,4,\ldots,N)}\hat{\rho} \tag{1.123}$$

where the symbol $\mathrm{Tr}_{(i,j,\ldots,N)}$ indicate a trace over the states of molecules $i, j, \ldots$ and N. For example, if $\hat{a}_j$ denotes a single-particle operator, then $\langle\sum_j \hat{a}_j; t\rangle = \sum_j \mathrm{Tr}_{(1,2,\ldots,N)}(\hat{a}_j\,.\,\hat{\rho}) = \mathrm{Tr}_{(1)}(\hat{a}_1\,.\,\hat{\rho}^{(1)}(1))$

$$= \sum_{\beta}\int \mathrm{d}^3p\langle\mathbf{p}\beta|\hat{a}\hat{\rho}^{(1)}|\mathbf{p}\beta\rangle$$

$$= \sum_{\beta}\sum_{\beta'}\iint \mathrm{d}^3p\,\mathrm{d}^3p'\,\mathrm{d}^3x\,\mathrm{e}^{-(i/\hbar)(\mathbf{p}'-\mathbf{p}).\mathbf{x}}\langle\mathbf{p}\beta|\hat{a}|\mathbf{p}'\beta'\rangle$$

$$\times f^{(1)}_{\beta'\beta}(t\mathbf{x}, \tfrac{1}{2}(\mathbf{p}+\mathbf{p}')) \tag{1.124}$$

where $\mathbf{p}$ and β, respectively, are the momentum and quantum numbers descriptive of the internal state of an individual molecule. Finally, the Wigner function

$$f^{(1)}_{\beta\beta'}(t\mathbf{x}\mathbf{p}) = (2\pi\hbar)^{-3}\int \mathrm{d}^3q\,\mathrm{e}^{(i/\hbar)\mathbf{q}.\mathbf{x}}\langle\beta, \mathbf{p}+\tfrac{1}{2}\mathbf{q}|\hat{\rho}^{(1)}|\boldsymbol{\beta}', \mathbf{p}-\tfrac{1}{2}\mathbf{q}\rangle \tag{1.125}$$

is the quantum analogue of the classical singlet distribution function. In terms of these functions the ensemble averages of several of the fields we have considered previously are given by

$$\begin{Bmatrix} \rho \\ \rho\mathbf{u} \\ \rho e_K \\ \rho e_\iota \\ \mathsf{p}_K \\ \mathbf{Q}_K \\ \mathbf{Q}_\iota \end{Bmatrix} = \sum_\beta \int \mathrm{d}^3p f^{(1)}_{\beta,\beta}(t\mathbf{x}\mathbf{p}) \begin{Bmatrix} m \\ \mathbf{p} \\ \frac{1}{2}mC^2 \\ \varepsilon_\beta \\ m\mathbf{C}\mathbf{C} \\ \mathbf{C}(\frac{1}{2}mC^2) \\ \mathbf{C}\varepsilon_\beta \end{Bmatrix} \tag{1.126}$$

with $\mathbf{C} = m^{-1}(\mathbf{p} - m\mathbf{u})$ and where ε_β is an eigenvalue of the internal state energy operator. Thus, it is clear that the diagonal elements of the Wigner function enter into the expressions for these particular densities and fluxes in precisely the same way as do the distribution functions for the separate chemical species (labelled by the index β) in a mixture of structureless molecules. However, other single-particle fields such as

$$\rho\mathbf{m} = \sum_\beta \sum_{\beta'} \int \mathrm{d}^3p f^{(1)}_{\beta,\beta'}(t\mathbf{x}\mathbf{p}) \langle \beta' | \mathbf{L} | \beta \rangle \tag{1.127}$$

and

$$\mathsf{C}_K = \sum_\beta \sum_{\beta'} \int \mathrm{d}^3p f^{(1)}_{\beta,\beta'}(t\mathbf{x}\mathbf{p}) \mathbf{C} \langle \beta' | \mathbf{L} | \beta \rangle \tag{1.128}$$

depend upon the off-diagonal elements as well and so cannot be interpreted in such a direct manner. Later we shall consider the Wigner functions in more detail and remark further about the conditions under which the quantum and classical theories are applicable.

As yet we have not addressed ourselves to the task of actually evaluating the characteristic relaxation times and transport coefficients of a polyatomic fluid. In principle the problem is that of computing numerical values for the correlation integrals derived in the previous section. However, this approach has been used in only a few situations and so in the next section we turn to the kinetic theory method which has been the source of most theoretical predictions concerning transport and relaxation in polyatomic gases and liquids. But before proceeding to that let us list briefly some items that could have but were not included in our discussion and developments which can be anticipated but have not yet materialized:

(i) It is straightforward (but a rather unrewarding exercise) to construct the quantum analogue of the classical ensemble theory given above.

(ii) The generalization of the present theory to include the effects of external fields is an item of current research. It is also of importance to allow for variations with time of these external fields.

(iii) There appears to be no difficulty in redoing our Zubarev–McLennan treatment of the non-equilibrium ensemble for polyatomic fluids from the slightly different point of view of Robertson[9] and Piccirelli[10]. Either of these formulations of the theory can serve as a natural starting point for the investigation of non-linear transport phenomena and of memory effects in polyatomic fluids. Investigations of this sort are already in progress.

(iv) The quantum version of the non-equilibrium theory has already been applied to the investigation of spin relaxation and diffusion in liquids and dense gases[14].

(v) As more and more experience is amassed about how to construct useful approximations to correlation integrals it will become possible to proceed directly from theories of the sort presented here to numerical predictions about the outcome of transport and relaxation experiments.

(vi) Our mechanical and statistical considerations can be extended to include systems containing chemically reactive species[15].

1.2 KINETIC THEORY OF POLYATOMIC GASES

1.2.1 The Boltzmann Equation

As has just been indicated we now intend to shift focus from the general to the specific. At the present time we understand best how to deal with the theoretical problems of fluid transport and relaxation in the limit of low densities and, more specifically, at pressures ranging from a few millimetres to a few hundred centimetres of mercury. Under these conditions the effects of gas imperfections (as evidenced by 'virial corrections' to gas transport coefficients and thermodynamic properties) and of Knudsen corrections due to free molecule flow are both negligible. It is generally accepted that under these circumstances a very satisfactory theory of gas transport and relaxation phenomena can be based upon

Boltzmann integrodifferential equations for the singlet distribution functions. The variety of proofs or rationalés for these Boltzmann equations is almost as great as the number of investigators and, it has been said that the arguments used are really far more revealing of the researchers' mentalities than of nature's workings. However, since these methods of derivation have seldom been applied to polyatomic species and since several novel features arise when they are, it is worthwhile to examine in detail at least one of the ways of proceeding from the firm foundation of the Liouville equation to the approximate Boltzmann equation.

To minimize notational complexity we confine our considerations to a single component gas. As before, we denote by $f^{(N)}$ the generic distribution function on the space of the conjugate variables appropriate to a gas of N identical polyatomic molecules. Since there are no sources or sinks of members of the ensemble representative of this system, $f^{(N)}$ satisfies the continuity (Liouville) equation $\partial_t f^{(N)} + [f^{(N)}, H] = 0$ where $[\phi, H]$ is the Poisson bracket of ϕ and the Hamiltonian function,

$$H = \sum_{j}^{N} \left(p_j^2/2m + H_j + H_j^{\mathrm{e}} + \sum_{k>j} \phi_{jk} \right)$$

for the N-particle system.

At the low densities with which we are currently concerned the macroscopic state of the fluid can be characterized wholly in terms of single-particle fields and the microstate can be described fully by the singlet distribution function

$$f^{(1)}(1) \equiv f^{(1)}(t\mathbf{x}\mathbf{c}_1 I_1) = N \int \mathrm{d}X_2 \ldots \int \mathrm{d}X_N f^{(N)} \tag{1.129}$$

Here, t and $\mathbf{x}$ refer to time and to the location of the molecular centre of mass, $\mathbf{c}_1$ is the molecular velocity and the symbol I_1 denotes a collection of coordinates and conjugate momenta descriptive of a molecule's internal degrees of freedom. Finally, $\mathrm{d}X_i = \mathrm{d}^3x_i \mathrm{d}^3c_i \mathrm{d}I_i$ is the differential extension in the space of a single molecule.

We now assume that there exist no molecular aggregates with lifetimes which are comparable in duration to the interval between the successive collisions suffered by a molecule. (This restriction upon the mechanical properties of the system can be replaced with the assumption that the concentrations of long-lived aggregates are negligible in comparison with that of monomers.) Under these circumstances there is a natural tendency to equate the singlet distribution function to the density of molecules which are, in some vaguely defined sense, isolated from all others. We can use this to advantage by shifting our attention from $f^{(1)}$ to a more conveniently defined function such as

$$\tilde{f}^{(1)}(1) = N \int \mathrm{d}X_2 \ldots \int \mathrm{d}X_N \left(\prod_{j=2}^{N} \zeta_{j,1} \right) f^{(N)} \tag{1.130}$$

where $\zeta_{j,i}$ assumes the value zero whenever the centre of mass of j lies within a convex region $\sigma_{i,j} = \sigma(\mathbf{x}_i I_i, \mathbf{x}_j I_j)$ surrounding molecule i and is otherwise equal to unity. This new function can be expressed in terms of the set $\{f^{(n)}(1 \ldots n) \equiv [N!/(N-n)!] \int \mathrm{d}X_{n+1} \ldots \int \mathrm{d}X_N f^{(N)}\}$ as follows:

$$\tilde{f}^{(1)}(1) = f^{(1)}(1) - \int_{\sigma_{1,2}} \mathrm{d}X_2 f^{(2)}(12)$$

$$+\tfrac{1}{2} \int_{\sigma_{1,2}} \mathrm{d}X_2 \int_{\sigma_{1,3}} \mathrm{d}X_3 f^{(3)}(123) + \cdots \tag{1.131}$$

Although this would appear to be an infinite series, it does in fact truncate after a finite number of terms. Thus, for any realistic choice of the intermolecular forces only a limited number of molecules can be fitted within the region σ. The physical interpretation of the series is clear: the first term is just the molecule density, the second term subtracts from this the number of paired molecules, the third corrects for the counting of pairs which are embedded within molecular trios. As the density tends to zero any difference between $\tilde{f}^{(1)}$ and $f^{(1)}$ can be attributed exclusively to the formation of bound pairs, trimers and the like, that is, to aggregates which we have assumed either to have very short lifetimes or to exist only in very small concentrations.

To obtain the equation of change for $\tilde{f}^{(1)}$ we multiply the Liouville equation with $N\Xi_1 = \prod_{j \geq 2} \zeta_{j,1}$ and integrate to obtain

$$\partial_t \tilde{f}^{(1)}(1) + [\tilde{f}^{(1)}(1), \mathscr{H}_1] = \int \mathrm{d}X_2 \tilde{f}^{(2)}(12)[\zeta_{2,1}, \mathscr{H}_1 + \mathscr{H}_2]$$

$$- \int \mathrm{d}X_2 \ldots \int \mathrm{d}X_N N\Xi_1 \left[f^{(N)}, \sum_{k>j} \sum \phi_{jk} \right] \tag{1.132}$$

with $\mathscr{H}_i = p_i^2/2m + H_i + H_i^{\mathrm{e}}$ and where

$$\tilde{f}^{(2)}(12) \equiv \frac{N!}{(N-2)!}\int \mathrm{d}X_3 \ldots \int \mathrm{d}X_N\left(\prod_{j\geq 3} \zeta_{j,1}\right) f^{(N)}$$

$$= f^{(2)}(12) - \int_{\sigma_{1,3}} \mathrm{d}X_3 f^{(3)}(123) + \cdots \tag{1.133}$$

is the pair density conditioned by the requirement that no particles lie within the region $\sigma_{1,j}$ about molecule 1.

We now select the linear dimensions of the region σ to exceed the range of the intermolecular forces, or equivalently, that $\zeta_{j,l}\phi_{jl} = 0$ for all states of the two molecules l and j. As a consequence of this the second term on the right-hand side of equation (1.132) vanishes and we obtain

$$\partial_t \tilde{f}^{(1)}(1) + [\tilde{f}^{(1)}(1), \mathscr{H}_1] = \partial_c \tilde{f}^{(1)}(1) \tag{1.134}$$

with

$$\partial_c \tilde{f}^{(1)}(1) = \int \mathrm{d}X_2 \tilde{f}^{(2)}(12)\dot{\zeta}_{2,1} \tag{1.135}$$

and where $\dot{\zeta}_{2,1} = [\zeta_{2,1}, \mathscr{H}_1 + \mathscr{H}_2]$. The right-hand side of (1.134) is the net rate of flow across the boundaries of σ, that is, the net rate at which molecules 'break free' (at the surface of σ) into the state 1.

The space of the centre of mass variable, $\mathbf{x}_2$, can be spanned by a sequence of surfaces which are geometrically similar to the surface of $\sigma_{1,2}$ and scaled by a parameter ρ. Thus, the location of the centre of mass of molecule 2, relative to that of molecule 1, can be given in terms of ρ and the surface normal $\mathbf{k}$, with ρ specifying the convex surface on which the mass centre lies and $\mathbf{k}$ giving its position on this surface. When $\rho = 1$ the centre of mass of 2 lies on the surface of $\sigma_{1,2}$ and therefore $\zeta_{2,1} = \eta(\rho - 1)$ where η is the unit step function. It follows that $\dot{\zeta}_{2,1} = \delta(\rho - 1)\dot{\rho}$. In terms of these coordinates $\mathrm{d}^3x_2 = \rho^2 h \,\mathrm{d}\rho\, \mathrm{d}^2S$ where d^2S is the differential surface element of $\sigma_{1,2}$ and $\rho h = \mathbf{x}_{21} \cdot \mathbf{k}$. Thus,

$$\partial_c \tilde{f}^{(1)}(1) = \int \mathrm{d}2 \int \mathrm{d}^2 S \dot{\rho} h \tilde{f}^{(2)}(12) \tag{1.136}$$

with $\mathrm{d}i = \mathrm{d}^3c_i\, \mathrm{d}I_i$. Finally, $\dot{\rho}h = \mathbf{k} \cdot \mathbf{g}$ is the normal component of $\mathbf{g}$, the velocity of the centre of mass of molecule 2 relative to the point of contact on $\sigma_{1,2}$. For the moment we restrict our attention to the simplest case and select for σ a sphere whose radius is equal to the range R of the intermolecular forces. It follows that $\mathrm{d}^2S = R^2\,\mathrm{d}^2\hat{n}$, $\rho h = x_{21}$, $\dot{\rho}h = \hat{n}\,.\,\mathbf{c}_{21}$ and

$$\partial_c \tilde{f}^{(1)}(1) = R^2 \int \mathrm{d}2 \int \mathrm{d}^2\hat{n}(\hat{n}\,.\,\mathbf{c}_{21})[\tilde{f}^{(2)}(12)]_{\mathbf{x}_{21}=R\hat{n}} \tag{1.137}$$

where $\hat{\mathbf{n}} = \mathbf{x}_{21}/x_{21}$ and $\mathbf{c}_{21} = \mathbf{c}_2 - \mathbf{c}_1$.

On the hemisphere $\hat{\mathbf{n}}\,.\,\mathbf{c}_{21} < 0$ the centres of the two molecules are approaching one another, that is, the molecules are about to become engaged in a mutual collision. On the hemisphere $\hat{\mathbf{n}}\,.\,\mathbf{c}_{21} > 0$ the particles are moving away from one another on post-collisional trajectories. Thus, we can rewrite equation (1.137) in the form

$$\partial_c \tilde{f}^{(1)}(1) = R^2 \int \mathrm{d}2 \int \mathrm{d}^2\hat{n}\gamma(\hat{n}\,.\,\mathbf{c}_{21})[\tilde{f}^{(2)}(\mathbf{x}_1\mathbf{c}_1I_1\,;\mathbf{x}_1 + R\hat{\mathbf{n}}, \mathbf{c}_2I_2\,;t)_{\text{post}}$$

$$-\tilde{f}^{(2)}(\mathbf{x}_1\mathbf{c}_1I_1\,;\mathbf{x}_1 - R\hat{\mathbf{n}}, \mathbf{c}_2I_2\,;t)_{\text{pre}}] \tag{1.138}$$

where $\gamma(x) = x\eta(x)$ and where the subscripts 'pre' and 'post' indicate that the arguments of the distribution functions correspond to pre- and post-collisional state, respectively. We now select a polar axis in the direction of $\hat{c} = \mathbf{c}_{21}/c_{21}$ and denote by θ and ϕ the polar spherical coordinates of the unit vector $\hat{\mathbf{n}}$. It then follows that $\hat{\mathbf{n}}\,.\,\mathbf{c}_{21} = c\cos\theta$, $\mathrm{d}^2\hat{n} = \sin\theta\,\mathrm{d}\theta\,\mathrm{d}\phi$ and $\hat{\mathbf{n}}\,.\,\mathbf{c}_{21}R^2\,\mathrm{d}^2\hat{n} = R^2c\cos\theta\sin\theta\,\mathrm{d}\theta\,\mathrm{d}\phi = c\,\mathrm{d}^2b$ where $b = R\sin\theta$ is the magnitude and $\mathrm{d}^2b = b\,\mathrm{d}b\,\mathrm{d}\phi$ the differential element of the 'impact parameter' $\mathbf{b} = R(\boldsymbol{\delta} - \hat{c}\hat{c})\,.\,\hat{\mathbf{n}}$. Therefore, equation (1.138) becomes

$$\partial_c \tilde{f}^{(1)}(1) = \int \mathrm{d}2 \int \mathrm{d}^2bc[\tilde{f}^{(2)}(\mathbf{x}_1\mathbf{c}_1I_1\,;\mathbf{x}_1 + R\hat{\mathbf{n}}\,.\,\mathbf{c}_2I_2\,;t)_{\text{post}}$$

$$-\tilde{f}^{(2)}(\mathbf{x}_1\mathbf{c}_1I_1\,;\mathbf{x}_1 - R\hat{\mathbf{n}}\mathbf{c}_2I_2\,;t)_{\text{pre}}] \tag{1.139}$$

with $R\hat{\mathbf{n}} = \mathbf{b} + \hat{c}(R^2 - b^2)^{1/2}$.

To this point all has been rigorous, but now we introduce the first of the three approximations which will transform the identity (1.139) into the Boltzmann equation. Specifically, we assume that the 'completed' collisions to which the first term of equation (1.139) refers are isolated binary events. According to this approximation each post-collisional state $(\mathbf{x}_1\mathbf{c}_1I_1\,;\mathbf{x}_1 + R\hat{\mathbf{n}}\mathbf{c}_2I_2)$, $\hat{\mathbf{n}}\,.\,\mathbf{c}_{21} > 0$, appearing in equation (1.139) is to be identified with the termination of a binary collision which began at $t - t_c$ with the two molecules in the state $(\mathbf{x}_1{}^*\mathbf{c}_1{}^*I_1{}^*\,;\mathbf{x}_1{}^* + R\hat{\mathbf{n}}^*\mathbf{c}_2{}^*I_2{}^*)$.

This approximation, which is virtually exact for short-range repulsive interactions, permits us to replace $\tilde{f}^{(2)}(\mathbf{x}_1\mathbf{c}_1 I_1; \mathbf{x}_1 + R\hat{\mathbf{n}}\mathbf{c}_2 I_2; t)_{\text{post}}$ of equation (1.139) with $\tilde{f}^{(2)}(\mathbf{x}_1{}^\star\mathbf{c}_1{}^\star I_1{}^\star; \mathbf{x}_1{}^\star + R\hat{\mathbf{n}}^\star\mathbf{c}_2{}^\star I_2{}^\star; t-t_c)_{\text{pre}}$.

The second and most crucial approximation is that of 'molecular chaos'. In particular, we assume that for pre-collisional states the pair density factors into the product of the corresponding singlet distribution functions, that is, $\tilde{f}^{(2)}(12)_{\text{pre}} \approx \tilde{f}^{(1)}(1)\tilde{f}^{(1)}(2)$. Our final assumption has to do with 'coarse graining' in time and space and, in particular, with the variations of $f^{(1)}$ which occur during intervals of the order t_c and over distances of the order of R or $|\mathbf{x}_1{}^\star - \mathbf{x}_1| \approx c_{21}t_c$. We neglect these variations and obtain from equation (1.139)

$$\begin{aligned}\partial_c \tilde{f}^{(1)}(1) &\approx \int \mathrm{d}2 \int \mathrm{d}^2 bc[\tilde{f}^{(1)}(t-t_c, \mathbf{x}^\star 1^\star)\tilde{f}^{(1)}(t-t_c, \mathbf{x}^\star + R\hat{\mathbf{n}}^\star 2^\star) \\ &\quad - \tilde{f}^{(1)}(t, \mathbf{x}1)\tilde{f}^{(1)}(t, \mathbf{x} - R\hat{\mathbf{n}}2)] \\ &\approx \int \mathrm{d}2 \int \mathrm{d}^2 bc[\tilde{f}^{(1)}(t\mathbf{x}1^\star)\tilde{f}^{(1)}(t\mathbf{x}2^\star) - \tilde{f}^{(1)}(t\mathbf{x}1)\tilde{f}^{(1)}(t\mathbf{x}2)]\end{aligned}$$

or†

$$\begin{aligned}\partial_c f(1) &\approx \iiint \mathrm{d}1'\,\mathrm{d}2'\,\mathrm{d}2 \int \mathrm{d}^2 bc\delta(1' - 1^\star)\delta(2' - 2^\star) \\ &\quad \times [f(1')f(2') - f(1)f(2)] \\ &= \iiint \mathrm{d}1'\,\mathrm{d}2'\,\mathrm{d}2 w_{\mathrm{F}}(12R|1'2'R)[f(1')f(2') - f(1)f(2)] \qquad (1.140)\end{aligned}$$

where

$$w_{\mathrm{F}}(12R|1'2'R) = \int \mathrm{d}^2 bc\delta(1' - 1^\star)\delta(2' - 2^\star) \qquad (1.141)$$

is the specific rate (unit concentrations in the reactant states and unit extension in the space of final states) of the binary collision process $(1'2'R) \to (12R)$ and where the subscript F serves as a reminder that these events are occurring in the presence of an external field. (In the pages which follow we shall often omit this suffix unless our attention is focused

† Here and henceforth we discard the superscript (1) and the tilde from the singlet density function.

specifically upon effects associated with external fields.) Since some of the internal state variables (phases of vibrational modes and orientation of molecular axes) vary rapidly even when the molecules are in free flight, the introduction of the reference separation R is essential if the transition rate $w_F(12R|1'2'R)$ is to be well defined.

The properties of these transition rates are determined completely by the dynamics of binary encounters. From its definition as a transition rate $w(12R|1'2'R)$ must be real and non-negative. Furthermore, it is obvious that $w(12R|1'2'R) = w(21R|2'1'R)$. In addition to these there are other constraints upon the transition rates:

(i) Because of the Galilean (translational) invariance, $w(12R|1'2'R)$ can depend upon the particle velocities only in the combinations $\mathbf{c}_{21}$, $\mathbf{c}'_{21}$ and $\mathbf{G}-\mathbf{G}'$ where $\mathbf{G} = \frac{1}{2}(\mathbf{c}_1+\mathbf{c}_2)$ is the velocity of the centre of mass of the pair.

(ii) Conservation of linear momentum demands that w be diagonal with respect to the total pair momentum, i.e., that $w(12R|1'2'R)$ be proportional to the Dirac delta function $\delta_3(\mathbf{G}-\mathbf{G}')$. (This ignores the possibility of momentum exchange between the colliding molecular pairs and the external field.)

(iii) Conservation of energy implies that w is proportional to $\delta_1(E_1+E_2-E'_1-E'_2)$ where $E_i \equiv \mathscr{H}_i = \frac{1}{2}mc_i^2+H_i+H_i^e$.

(iv) Molecular systems exhibit parity and time-reversal invariance. Therefore, $w_F(12R|1'2'R) = w_{\hat{P}F}(\hat{P}1\hat{P}2R|\hat{P}1'\hat{P}2'R)$ and $w_F(12R|1'2'R) = w_{\hat{T}F}(\hat{T}1'\hat{T}2'R|\hat{T}1\hat{T}2R)$ where $\hat{P}$ and $\hat{T}$ are the parity and time-reversal operators and where $\hat{P}F$ and $\hat{T}F(\hat{T}\mathbf{E} = \mathbf{E}, \hat{T}\mathbf{H} = -\mathbf{H})$ denote the parity and time-reversed images of the external fields.

(v) As a consequence of time-reversal invariance the transition rate satisfies the condition,

$$\iint \mathrm{d}1'\,\mathrm{d}2' w(12R|1'2'R) = \iint \mathrm{d}1'\,\mathrm{d}2' w(1'2'R|12R)$$

of 'bilateral normalization'.

(vi) Because of (Galilean) rotational invariance, observers attached to coordinate frames which differ from one another by rigid rotations will record the same value for the specific rate of particular mechanical process. Thus, if we denote by $\hat{R}i$ the state which results from the application of a given rotation to all variables belonging to the set $i = (\mathbf{c}_i, I_i)$, then $w_F(12R|1'2'R) = w_{\hat{R}F}(\hat{R}1\hat{R}2R|\hat{R}1'\hat{R}2'R)$.

(vii) Finally, it can be shown[16] that the transition rate is related to the differential cross-section by the formula

$$w_F(12R|1'2'R) = \mu(c'/c)\sigma_F(\mathbf{c}'I'_1I'_2R \to \mathbf{c}I_1I_2R)$$
$$\times \delta_3(\mathbf{G}-\mathbf{G}')\delta_1(E-E') \qquad (1.142a)$$

where $\mathbf{c}' = \mathbf{c}'_2 - \mathbf{c}'_1$ and $\mathbf{c} = \mathbf{c}_2 - \mathbf{c}_1$ are the relative velocities of the incident beam and of the emergent pairs of molecules, $\mu = m/2$ is the reduced mass and $E = \frac{1}{2}\mu c^2 + (H_1 + H_2) + (H_1{}^e + H_2{}^e)$. It must be understood that in the scattering process to which the differential cross-section $\sigma(\mathbf{c}'I_1'I_2'R \to \mathbf{c}I_1I_2R)$ refers the incident molecules are in the state I'_2 at the instant when their centres of mass pierce a sphere of radius R centred on the target particles. Despite the impossibility of producing such a beam (or target), it does offer some theoretical advantages which we wish to exploit before changing our focus to cross-sections which correspond to physically accessible conditions.

As a consequence of time-reversal invariance and parity the differential cross-sections satisfy the condition

$$c'^2\sigma(\mathbf{c}'I'_1I'_2R \to \mathbf{c}I_1I_2R) = c^2\sigma(\hat{T}\{\mathbf{c}I_1I_2\}R \to \hat{T}\{\mathbf{c}'I'_1I'_2\}R)$$
$$= c^2\sigma(\mathbf{c}\hat{T}\hat{P}I_1\hat{T}\hat{P}I_2R \to \mathbf{c}'\hat{T}\hat{P}I'_1\hat{T}\hat{P}I'_2R) \quad (1.142b)$$

which commonly is referred to as microreversibility.

Although we shall not do so here, it is possible to derive the expression (1.141) for w directly from an analysis of the scattering process. From this formula and equation (1.142) it follows that

$$\sigma(c'I'_1I'_2R \to \mathbf{c}I_1I_2R) = (c/c')^2 \int d^2b\delta_2(\hat{c}' - \hat{c}^\star)$$
$$\times \delta(I'_1 - I_1{}^\star)\delta(I'_2 - I_2{}^\star) \qquad (1.143)$$

Furthermore, because of time-reversal invariance it is possible to express the transition rate and differential cross-section in the alternate forms

$$w(12R|1'2'R) = \int d^2b'c'\delta(1-1^*)\delta(2-2^*) \qquad (1.144)$$

and

$$\sigma(\mathbf{c}'I'_1I'_2R \to \mathbf{c}I_1I_2R) = \int d^2b'\delta_2(\hat{c} - \hat{c}^*)\delta(I_1 - I_1{}^*)\delta(I_2 - I_2{}^*) \qquad (1.145)$$

where the asterisks refer to the unique *final* state, on the post-collisional hemisphere, which evolves from the initial state $(\mathbf{c}'I'_1I'_2\mathbf{b}'R)$. The cross-section, as expressed by equation (1.145), can be identified with the area in the precollision $\mathbf{b}'$-plane which scatters into unit solid angle about the direction $\hat{c}$. By changing the variables of integration from $\mathbf{b}'$ to $\hat{c}^*$ (and exercising care if $\mathbf{b}'$ is not a single-valued function of $\hat{c}^*$) we obtain the relationship

$$\sigma(\mathbf{c}'I'_1I'_2R \to \mathbf{c}I_1I_2R) = \left|\frac{\partial(\mathbf{b}')}{\partial(\hat{c}^*)}\right|_{\hat{c}^*=\hat{c}} \delta(I_1 - I_1^*)\delta(I_2 - I_2^*) \quad (1.146)$$

between the differential cross-section and the Jacobian of the transformation from $\mathbf{b}'$ to $\hat{c}^*$.

Finally, as a consequence of parity and time-reversal invariance it can be shown that to each event $(\mathbf{b}'\mathbf{c}'I'_1I'_2) \to (\mathbf{b}\mathbf{c}I_1I_2)$ there corresponds the unique event $(\hat{T}\hat{P}\mathbf{b}, \hat{T}\hat{P}\mathbf{c}, \hat{T}\hat{P}I_1, \hat{T}\hat{P}I_2) \to (\hat{T}\hat{P}\mathbf{b}', \hat{T}\hat{P}\mathbf{c}', \hat{T}\hat{P}I'_1, \hat{T}\hat{P}I'_2)$ or $(-\mathbf{b}, \mathbf{c}, \hat{T}\hat{P}I_1, \hat{T}\hat{P}I_2) \to (-\mathbf{b}', \mathbf{c}', \hat{T}\hat{P}I'_1, \hat{T}\hat{P}I'_2)$. The algebraic signs of the impact parameters and of the angular momenta and axial orientations are opposite in the 'direct' and '$\hat{T}\hat{P}$-imaged' events: if the direct collision changes the relative velocity from $\mathbf{c}'$ to $\mathbf{c}$, then the $\hat{T}\hat{P}$-imaged collision causes scattering from $\mathbf{c}$ to $\mathbf{c}'$; if the direct collision changes the spins $\mathbf{L}'_1$ and $\mathbf{L}'_2$ to $\mathbf{L}_1$ and $\mathbf{L}_2$, respectively, then the $\hat{T}\hat{P}$-imaged process is accompanied by transitions from $-\mathbf{L}_1$ and $-\mathbf{L}_2$ to $-\mathbf{L}'_1$ and $-\mathbf{L}'_2$.

The transition rate and cross-section, $w(12R|1'2'R)$ and $\sigma(\mathbf{c}'I'_1I'_2R \to \mathbf{c}I_1I_2R)$, depend upon some variables whose values are not normally monitored in scattering experiments. For example, in the case of a rotor one is certainly interested in collisional alterations of the molecular spin angular momentum but one is neither concerned with nor able to measure molecular orientation. Also, when vibrations are involved it is the energy and not the phase of the motion which is of interest. The cross-sections which are experimentally accessible are averages over the unmonitored variables and, as we shall see, it is precisely these averaged cross-sections which are of importance in kinetic theory. The common characteristic of variables which can be monitored is that they are free-flight invariants, that is, variables whose values remain constant (in the absence of external fields) during the intervals between successive molecular collisions. For example, the invariants of a rigid symmetric top are the linear momentum $\mathbf{p} = m\mathbf{c}$, the rotational angular momentum $\mathbf{L}$, and the projection $K = L\cos\theta$ of $\mathbf{L}$ along the molecular symmetry axis. The variable ϕ which measures the precessional phase of the symmetry axis about the direction of $\mathbf{L}$ and ψ, the angle between the line of nodes and one of the degenerate principal axes of the inertial ellipsoid, vary linearly in time with characteristic frequencies $\omega_\phi = |\mathbf{L}|/2\pi\Gamma$ and $\omega_\psi = (K/2\pi)(\Gamma^{-1} - \Gamma'^{-1})$, respectively. Here Γ is the principal moment of inertia about the symmetry axis and Γ' the doubly-degenerate principal moment. In the case of a spherical top ($\Gamma \to \Gamma'$) the frequency ω_ψ is equal to zero and so there is an additional free-flight invariant which can be chosen equal to ψ.

One free-flight invariant can be associated with each pure vibrational mode. If this invariant is selected to be the vibrational energy, then the associated angle variable is the phase of the vibrational motion. When two vibrational frequencies are degenerate there will be an additional constant of the motion corresponding to the unvarying difference of phase between the two modes of vibration.

These few observations permit us to be rather specific concerning the types of variables which fall into the category of free-flight invariants and those which do not. Thus, we assume that the canonical variables I can be separated into a set of invariants $\mathscr{I}$, consisting of action variables and of angles with zero frequencies, and a set η, consisting of angle variables which vary linearly with time. The frequencies associated with members of the set η are typically of the order of 10^{11} sec^{-1} or greater.

Now it was essential to our derivation of the Boltzmann equation (1.140) that the distribution function $f(t\mathbf{xc}I)$ be insensitive to variations of t of the order of t_c, the duration of a collisional event, and to variations of the spatial argument $\mathbf{x}$ of the order of R or $c_{21}t_c$. Thus, the distribution functions to which the Boltzmann equation pertains is not the instantaneous density but the average over a time which is greater than the duration of an individual collision but less than the interval which elapses between successive collisions. During an interval of this magnitude a molecule will rotate and vibrate several times. Therefore, since the frequencies of the η variables exceed the collision frequency $\omega_f \approx 10^9$ sec^{-1}, we must assume that the Boltzmann distribution function is independent of these variables, i.e., that

$$f(t\mathbf{xc}I) \to \frac{1}{\Delta} f(t\mathbf{xc}\mathscr{I}) \tag{1.147}$$

where $\Delta = \int \mathrm{d}\eta$.

Then, to determine the collisional rate of change of $f(t\mathbf{x}\bar{1}) \equiv f(t\mathbf{xc}_1\mathscr{I}_1)$ we integrate equation (1.140) over the range of the variables η_1 and obtain

$$\partial_c f(t\mathbf{x}\bar{1}) = \iiint \mathrm{d}\bar{1}'\, \mathrm{d}\bar{2}'\, \mathrm{d}\bar{2} w(\bar{1}\bar{2}|\bar{1}'\bar{2}')[f(\bar{1}')f(\bar{2}') - f(\bar{1})f(\bar{2})] \tag{1.148}$$

with $\bar{i} = (\mathbf{c}_i, \mathcal{J}_i)$, $\mathrm{d}\bar{i} = \mathrm{d}^3 c_i\, \mathrm{d}\mathcal{J}_i$, and

$$w(\overline{12}\,|\,\overline{1}'\overline{2}') = \frac{1}{\Delta^2}\iiiint \mathrm{d}\eta_1\, \mathrm{d}\eta_2\, \mathrm{d}\eta'_1\, \mathrm{d}\eta'_2 w(12R\,|\,1'2'R) \qquad (1.149)$$

This transition rate and the corresponding differential cross-section,

$$\sigma(\mathbf{c}'\mathcal{J}'_1\mathcal{J}'_2 \to \mathbf{c}\mathcal{J}_1\mathcal{J}_2) = \frac{1}{\Delta^2}\iiiint \mathrm{d}\eta_1\, \mathrm{d}\eta_2\, \mathrm{d}\eta'_1\, \mathrm{d}\eta'_2 \times \sigma(\mathbf{c}'I'_1 I'_2 R \to \mathbf{c} I_1 I_2 R) \qquad (1.150)$$

are independent of the value of the parameter R and refer to scattering experiments which, at least in principle, actually can be performed. This classical description of scattering events, and especially the use of free flight invariants to specify the state of preparation of the beam and target, is completely analogous to that which is familiar from quantum mechanics.

Because of the simple relationships between the two pairs of functions it is obvious that the phase-averaged transition rate and differential cross-section exhibit the same symmetry and invariance properties as do $w(12R\,|\,1'2'R)$ and $\sigma(\mathbf{c}'I'_1 I'_2 R \to \mathbf{c} I_1 I_2 R)$.

Now that we recognize how the functional dependence of the Boltzmann distribution function is regulated by the approximations used in the derivation of the collision term, we must perform a corresponding modification upon the left-hand side of this equation. In particular, we replace $f(t\mathbf{x}1)$ with $f(t\mathbf{x}\bar{1})/\Delta$ in (1.134) and integrate over the range of η_1 to obtain

$$(\partial_t + \mathbf{c}_1 \cdot \nabla) f(t\mathbf{x}\bar{1}) + \overline{[f(t\mathbf{x}\bar{1}), H_1 + H_1{}^{\mathrm{e}}]} = \partial_c f(t\mathbf{x}\bar{1}) \qquad (1.151)$$

Here the bar above a function denotes its average over the variables η_1. Due to the definition of these variables the average of $[f(\bar{1}), H_1]$ is identically zero and so, from equations (1.148) and (1.151) we obtain the Boltzmann equation

$$\begin{aligned}
&(\partial_t + \mathbf{c}_1 \cdot \nabla) f(t\mathbf{x}\bar{1}) + \overline{[f(t\mathbf{x}\bar{1}), H_1{}^{\mathrm{e}}]} \\
&= \iiint \mathrm{d}\bar{1}'\, \mathrm{d}\bar{2}'\, \mathrm{d}\bar{2} w_{\mathrm{F}}(\overline{12}\,|\,\overline{1}'\overline{2}')[f(\bar{1}')f(\bar{2}') - f(\bar{1})f(\bar{2})] \\
&= \int \cdots \int \mathrm{d}^2 c'\, \mathrm{d}^3 c_2\, \mathrm{d}\mathcal{J}'_1\, \mathrm{d}\mathcal{J}'_2\, \mathrm{d}\mathcal{J}_2 \\
&\quad \times \left\{ \frac{c'^2}{c} \sigma_{\mathrm{F}}(\mathbf{c}'\mathcal{J}'_1\mathcal{J}'_2 \to \mathbf{c}\mathcal{J}_1\mathcal{J}_2)[f(\bar{1}')f(\bar{2}') - f(\bar{1})f(\bar{2})] \right\}_{EG}
\end{aligned} \qquad (1.152)$$

where the subscript EG indicates that the primed and unprimed states share common values of energy and momentum.

1.2.2 Equations of Change

It is easily proved that the equation of change for $n\langle\psi\rangle = \int d\bar{1} f(\bar{1})\psi(\bar{1})$, the ensemble average of the single-particle field ψ, is given by

$$n\mathrm{d}_t\langle\psi\rangle + \nabla . n\langle \mathbf{C}\psi\rangle = n\langle\overline{[\psi, H^{\mathrm{e}}]}\rangle + n\partial_c\langle\psi\rangle \tag{1.153}$$

with $\mathrm{d}_t = \partial_t + \mathbf{u} . \nabla$, $\mathbf{C} = \mathbf{c} - \mathbf{u}(t\mathbf{x})$, and where the functions

$$\begin{aligned} n\partial_c\langle\psi\rangle &= \int d\bar{1}\psi(\bar{1})\partial_c f(t\mathbf{x}\bar{1}) \\ &= \tfrac{1}{2}\iiiint d\bar{1}'\, d\bar{2}'\, d\bar{1}\, d\bar{2} w(\bar{1}'\bar{2}'|\bar{1}\bar{2}) f(\bar{1}) f(\bar{2}) \\ &\quad \times [\psi(\bar{1}') + \psi(\bar{2}') - \psi(\bar{1}) - \psi(\bar{2})] \end{aligned} \tag{1.154}$$

and $n\langle\overline{[\psi, H^{\mathrm{e}}]}\rangle$ correspond to the 'relaxation rates' χ_ψ and $\chi_\psi{}^{\mathrm{e}}$ of the previous section. To derive the second of these expressions for $n\partial_c\langle\psi\rangle$ one must invoke bilateral normalization, that is, the time-reversal invariance of collisional events, and the property $w(\bar{1}\bar{2}|\bar{1}'\bar{2}') = w(\bar{2}\bar{1}|\bar{2}'\bar{1}')$ of the transition rate.

By choosing ψ equal to $\mathbf{p} = m\mathbf{c}$ we obtain from equation (1.153) the equation of motion (1.32) with the pressure tensor $\mathsf{p} = n\langle m\mathbf{CC}\rangle$ appropriate to a dilute gas and with $\hat{\mathbf{F}}^{\mathrm{e}} = \langle\overline{[\mathbf{c}, H^{\mathrm{e}}]}\rangle = -\nabla\langle\overline{H}^{\mathrm{e}}\rangle$. Next, with $\psi = \frac{1}{2}mc^2$ (1.153) reduces to an equation of the form (1.36) which lacks the 'collisional transfer' contribution $\mathbf{Q}'_\phi$ to the energy flux and in which $\chi_K = n\partial_c\langle\frac{1}{2}mc^2\rangle$ and $\chi_K{}^{\mathrm{e}} = -\rho\langle\overline{\mathbf{C} . \nabla H^{\mathrm{e}}}\rangle$. For the density of internal energy, $e_\iota = \langle H_i\rangle$, the equation of change is (1.41) with $\chi_\iota = n\partial_c\langle H_i\rangle = -\chi_K$ and $\chi_\iota{}^{\mathrm{e}} = n\langle\overline{[H_i, H_i{}^{\mathrm{e}}]}\rangle$. Finally, the spin density $\mathbf{m} = \rho^{-1}\langle\mathbf{L}\rangle$ is found to satisfy an equation of the form (1.49) with a spin flux $\mathsf{C}_K = n\langle\mathbf{CL}\rangle$ and with a relaxation rate $\boldsymbol{\chi}_L = n\partial_c\langle\mathbf{L}\rangle$ which vanishes only if the collision dynamics are such that $w(1\bar{2}|\bar{1}'\bar{2}') \propto \delta_3(\mathbf{L}'_1 + \mathbf{L}'_2 - \mathbf{L}_1 - \mathbf{L}_2)$.

In the situations which will concern us here the interaction between an individual molecule and the external fields can be expressed as the sum

$$H_i^{\mathrm{e}} = mgz - \boldsymbol{\mu}_i . \mathbf{F} \tag{1.155}$$

of the gravitational potential and of the interaction between a molecular electric or magnetic dipole moment and an external electric or magnetic field. Therefore, $\overline{H}_i^{\mathrm{e}} = mgz - \bar{\boldsymbol{\mu}}_i \cdot \mathbf{F}$ where the averaged moment $\bar{\boldsymbol{\mu}}_i$ is necessarily parallel to the angular momentum $\mathbf{L}_i$. Consequently, it must be of the form $\bar{\boldsymbol{\mu}}_i = \bar{\gamma}(\mathscr{I}_i)\mathbf{L}_i$ where $\bar{\gamma}(\mathscr{I})$ is the gyromagnetic or gyroelectric ratio. (Later we shall consider these quantities in greater detail.) Then since $[H_i, H_i^{\mathrm{e}}] = 0$ and $[\mathbf{L}, \bar{\boldsymbol{\mu}}] = \boldsymbol{\varepsilon} \cdot \bar{\boldsymbol{\mu}}$, the appropriate equations of change are

$$\rho\, \mathrm{d}_t \mathbf{u} = -\nabla \cdot \mathsf{p} + \rho g \hat{z} + \rho \langle \bar{\boldsymbol{\mu}} \rangle \cdot (\nabla \mathbf{F})^{\dagger} \tag{1.156}$$

$$\rho\, \mathrm{d}_t e_K = -\nabla \cdot \mathbf{Q}_K - \mathsf{p} : \nabla \mathbf{u} + \chi_K - \rho \langle \mathbf{C} \bar{\boldsymbol{\mu}} \rangle : (\nabla \mathbf{F})^{\dagger} \tag{1.157}$$

$$\rho\, \mathrm{d}_t e_\iota = -\nabla \cdot \mathbf{Q}_\iota - \chi_K \tag{1.158}$$

and

$$\rho\, \mathrm{d}_t \mathbf{m} = -\nabla \cdot \mathsf{C}_K + \boldsymbol{\chi}_L + n \langle \bar{\boldsymbol{\mu}} \rangle \times \mathbf{F} \tag{1.159}$$

Furthermore, for this choice of H_i^{e} we find that

$$\overline{[f(\bar{1}), H_1^{\mathrm{e}}]} = \hat{\mathbf{F}}_1^{\mathrm{e}} \cdot \partial_{\mathbf{c}_1} f(\bar{1}) + \bar{\boldsymbol{\mu}}_1 \times \mathbf{F} \cdot \partial_{\mathbf{L}_1} f(\bar{1}) \tag{1.160}$$

with $\hat{\mathbf{F}}_1^{\mathrm{e}} = -g\hat{z} - m^{-1}\bar{\boldsymbol{\mu}}_1 \cdot (\nabla \mathbf{F})^{\dagger}$.

Let us neglect the gravitational force and assume the field $\mathbf{F}$ to be spatially uniform. Under these circumstances the force $\hat{\mathbf{F}}^{\mathrm{e}}$ is equal to zero. Then, if $f(t\mathbf{x}\bar{1})$ is assumed to be independent of t and $\mathbf{x}$, it will satisfy the equation

$$0 = \bar{\boldsymbol{\mu}}_1 \times \mathbf{F} \cdot \partial_{\mathbf{L}_1} f(\bar{1}) - \iiint \mathrm{d}\bar{1}'\, \mathrm{d}\bar{2}'\, \mathrm{d}\bar{2}\, w_{\mathbf{F}}(\bar{1}\bar{2} | \bar{1}'\bar{2}')[f(\bar{1}')f(\bar{2}') - f(\bar{1})f(\bar{2})] \tag{1.161}$$

The most general solution of this equation can be written in the form

$${}^{0}f_{\mathbf{F}}(\bar{1}) = n_0 (m/2\pi \mathbf{k} T_0)^{3/2} Z^{-1} \exp - \frac{1}{\mathbf{k} T_0} [\tfrac{1}{2} m (\mathbf{c} - \mathbf{u}_0)^2 + H_1 + H_1^{\mathrm{e}}] \tag{1.162}$$

with

$$Z(T_0) = \int \mathrm{d}\mathscr{I}_1 \exp - \frac{1}{\mathbf{k} T_0} (H_1 + H_1^{\mathrm{e}}) \tag{1.163}$$

and where T_0, n_0 and $\mathbf{u}_0$ are arbitrary constants which can, for example, be identified with reference values of the temperature, concentration and fluid velocity.

1.2.3 Linearization of the Boltzmann Equation

We now write the singlet distribution function in the form $f(\bar{1}) = {}^0f_F(\bar{1}) \times [1+\phi(\bar{1})]$ with ϕ a measure of the system's displacement from the state described by the Maxwell-Boltzmann distribution ${}^0f_F$. If this displacement is not too great we can neglect quadratic terms in ϕ and obtain for this function the equation of change

$$ {}^0f_F(\partial_t + \mathbf{c} \cdot \nabla)\phi = -{}^0f_F \bar{\boldsymbol{\mu}} \times \mathbf{F} \cdot \partial_{\mathbf{L}}\phi - n^2\hat{\Gamma}(\phi) \quad (1.164) $$

where the linear operator $\hat{\Gamma}$ is so defined that

$$ n^2\hat{\Gamma}_1(g) = \iiint d\bar{1}'\, d\bar{2}'\, d\bar{2} w_F(\bar{1}\bar{2}|\bar{1}'\bar{2}')\, {}^0f_F(\bar{1})\, {}^0f_F(\bar{2})[g(\bar{1}') + g(\bar{2}') - g(\bar{1}) - g(\bar{2})] $$

$$ = K^{(0)}(\bar{1})g(\bar{1}) + \int d\bar{2} K(\bar{1}, \bar{2})g(\bar{2}) \quad (1.165) $$

with

$$ K^{(0)}(\bar{1}) = \iiint d\bar{1}'\, d\bar{2}'\, d\bar{2} w(\bar{1}\bar{2}|\bar{1}'\bar{2}')\, {}^0f_F(\bar{1})\, {}^0f_F(\bar{2}) \quad (1.166) $$

$$ K(\bar{1}, \bar{j}) = \iiint d\bar{1}'\, d\bar{2}'\, d\bar{2} w(\bar{1}\bar{2}|\bar{1}'\bar{2}')\, {}^0f_F(\bar{1})\, {}^0f_F(\bar{2})[\delta(\bar{2}-\bar{j}) - \delta(\bar{1}'-\bar{j}) - \delta(\bar{2}'-\bar{j})] \quad (1.167) $$

and $\delta(\bar{i}-\bar{j}) = \delta_3(\mathbf{c}_i - \mathbf{c}_j)\delta(\mathscr{J}_i - \mathscr{J}_j)$.

There are several properties of the operator $\hat{\Gamma}$ and of the Fredholm kernel $K(\bar{1}, \bar{2})$ which determine the basic structure of the distortion ϕ and so also the general characteristics of relaxation and transport in polyatomic gases:

(i) The adjoint of $\hat{\Gamma}$, denoted by $\hat{\Gamma}^\dagger$ and defined by $\int d\bar{1}\psi(\bar{1})\hat{\Gamma}_1(\phi) = \int d\bar{2}\phi(\bar{1})\hat{\Gamma}_1{}^\dagger(\psi)$, differs from $\hat{\Gamma}$ (as given by equation 1.165) in that $w(\bar{1}\bar{2}|\bar{1}'\bar{2}')$ is replaced with $w(\bar{1}'\bar{2}'|\bar{1}\bar{2})$.

(ii) $\hat{\Gamma}$ commutes with the parity and rotation operators $\hat{P}$ and $\hat{R}$, that is, $\hat{P}\hat{\Gamma} = \hat{\Gamma}\hat{P}$ and $\hat{R}\hat{\Gamma} = \hat{\Gamma}\hat{R}$. Thus, the eigenfunctions of $\hat{\Gamma}$ can be chosen to have definite parities and to transform as bases for irreducible representations of the full rotation group.

(iii) $\hat{T}\hat{\Gamma} = \hat{\Gamma}^\dagger\hat{T}$. Therefore $\hat{T}$ and $\hat{\Gamma}$ commute if and only if $\hat{\Gamma}$ is self-adjoint, that is, if $\hat{\Gamma}^\dagger = \hat{\Gamma}$. Unless this is so the eigenfunctions of $\hat{\Gamma}$ cannot, in general, also be eigenfunctions of the time-reversal operator.

(iv) There is a one-to-one correspondence between the adjointness of $\hat{\Gamma}$, the symmetry of w and the symmetry of the kernel K. When $\hat{T}\hat{P}\mathscr{I}_i = \mathscr{I}_i$, $\hat{\Gamma}$ is self-adjoint and w and K are symmetric. This is the case whenever the species are structureless or whenever the molecular interactions are such that no internal variables are collisionally altered. However, we can generally expect $\hat{\Gamma}$ to be self-adjoint only if the set $\mathscr{I}$ is limited to variables with positive $\hat{T}\hat{P}$ eigenvalues, e.g. true scalars such as L^2, p^2 and $(\mathbf{L}\,.\,\mathbf{p})^2$. One may, of course, expand the 'ignorable' set η to include variables from the set $\mathscr{I}$ in which one has no interest or which one has judged to be of less importance than others. The $\hat{\Gamma}$ operator associated with the diminished set of free-flight invariants will be self-adjoint if $\hat{T}\hat{P}\mathscr{I}_i = \mathscr{I}_i$ for every variable belonging to the set.

When w is symmetric it is said that 'detailed balancing' prevails, that the specific rate of each 'elementary process' is precisely equal to that of its inverse. [From the H-theorem for the Boltzmann equation (1.152) it can be proved[16] that this detailed balancing is not essential to the maintenance of dynamic equilibrium.] Let us suppose now that one has selected a transition rate matrix w. Then, there may exist circumstances under which it is reasonable to replace this matrix with its symmetric part, i.e. to replace each element $w(\bar{1}\bar{2}|\bar{1}'\bar{2}')$ with the corresponding quantity

$$w^{(\text{sym})}(\bar{1}\bar{2}|\bar{1}'\bar{2}') \equiv \tfrac{1}{2}[w(\bar{1}\bar{2}|\bar{1}'\bar{2}')+w(\bar{1}'\bar{2}'|\bar{1}\bar{2})] = w^{(\text{sym})}(\bar{1}'\bar{2}'|\bar{1}\bar{2}) \quad (1.168)$$

To help in our search for conditions which might justify this replacement we recall that $w(\bar{1}'\bar{2}'|\bar{1}\bar{2}) = w(\hat{T}\hat{P}\bar{1}\hat{T}\hat{P}\bar{2}|\hat{T}\hat{P}\bar{1}'\hat{T}\hat{P}\bar{2}')$ wherein $\hat{T}\hat{P}\mathbf{c} = \mathbf{c}$. Thus, $w^{(\text{sym})}$ is the transition rate matrix appropriate to situations for which the states $(\mathbf{c}, \mathscr{I})$ and $(\mathbf{c}, \hat{T}\hat{P}\mathscr{I})$ can be assigned equal *a priori* probabilities. For example, under these circumstances the distribution function for a gas of symmetric top molecules must be such that $f(\bar{1}) = f(t\mathbf{x}\mathbf{c}\mathbf{L}K)$ is equal to $f(t\mathbf{x}\mathbf{c}-\mathbf{L}K)$, or equivalently that $f(\bar{1})$ must be an even function of $\mathbf{L}$. Here K is the component of the rotational angular momentum $\mathbf{L}$ in the direction of the molecular symmetry axis.

There are significant differences between this 'non-polar' ensemble and the 'isotropic' ensemble for which the appropriate transition-rate matrix is given by

$$w^{(\text{iso})}(\bar{1}\bar{2}|\bar{1}'\bar{2}') = (4\pi)^{-2}\iiiint \mathrm{d}\hat{L}_1\,\mathrm{d}\hat{L}_2\,\mathrm{d}\hat{L}'_1\,\mathrm{d}\hat{L}'_2\,w(\bar{1}\bar{2}|\bar{1}'\bar{2}') \quad (1.169)$$

Here, for simplicity, we have assumed that $\hat{\mathbf{L}} = \mathbf{L}/L$ is the only dynamical variable with a negative $\hat{T}\hat{P}$ eigenvalue. Since $w(\overline{1}\overline{2}|\overline{1}'\overline{2}') = w(\hat{T}\hat{P}\overline{1}'\hat{T}\hat{P}\overline{2}'|\hat{T}\hat{P}\overline{1}\hat{T}\hat{P}\overline{2})$ it then follows that $w^{(\mathrm{iso})}(\overline{1}\overline{2}|\overline{1}'\overline{2}') = w^{(\mathrm{iso})}(\overline{1}'\overline{2}'|\overline{1}\overline{2})$, a condition which Watanabe[17] has termed 'averaged balance'.

By adopting the non-polar ensemble one forfeits all chance of accounting for phenomena which depend upon a preferential direction of molecular rotation. The non-polar ensemble restricts the distribution function to a ray dependence upon the angular distribution of molecular spin. The isotropic ensemble admits no dependence whatsoever upon this variable. For example, the isotropic distribution function can depend upon $\mathbf{L}$ only through the rotational energy $\varepsilon(\mathscr{J})$.

These symmetry conditions have an important bearing upon the properties of the 'bracket integrals',

$$[\phi, \psi] = (\phi, \hat{\Gamma}(\psi)) \equiv \int \mathrm{d}\overline{1}\phi(\overline{1})\hat{\Gamma}_1(\psi) \tag{1.170}$$

to which the various transport and relaxation coefficients are related. In the theory of these coefficients one also encounters the related integrals, $[\phi; \psi]$, which differ from the corresponding quantities $[\phi, \psi]$ in that the integrands $\phi(\overline{1})\hat{\Gamma}_1(\psi)$ of equation (1.170) are replaced with $\phi(\overline{1}) \odot \hat{\Gamma}_1(\psi)$. Integrals of this second variety are defined only for pairs of functions ϕ and ψ both of which are tensors of the same rank. The symbol $\phi \odot \hat{\Gamma}(\chi)$ is understood to mean the appropriate scalar product, with tensor contractions formed in accordance with the 'nesting convention' of Chapman and Cowling, e.g. $\boldsymbol{\phi} \odot \hat{\Gamma}(\boldsymbol{\psi}) = \sum_i \phi_i \hat{\Gamma}(\psi_i)$, $\boldsymbol{\phi} \odot \hat{\Gamma}(\boldsymbol{\psi}) = \sum_i \sum_j \phi_{ij} \hat{\Gamma}(\psi_{ji})$, etc.

As consequences of the properties of w it can be proved that:

(i) $[\phi, \phi^\dagger] \geqslant 0$ for all functions ϕ. The equality obtains if and only if ϕ is a binary invariant. The symbol $\phi^\dagger$ here denotes the tensor with components $(\phi^\dagger)_{ij\ldots mn} = \phi_{nm\ldots ji}$ and for tensor *components* ϕ and ψ;

(ii) $[\phi, \psi] = 0$ if the parities of ϕ and ψ differ.

(iii) $[\phi, \psi] = 0$ if ϕ and ψ are components of tensors which transform as basis elements belonging to different irreducible representations of the rotation group.

(iv) $[\phi, \psi] = T_\phi T_\psi [\psi, \phi]$ where T_χ is the time-reversal eigenvalue of χ. Furthermore, if $\hat{\Gamma}$ is self-adjoint (as it is for structureless molecules) then $[\phi, \psi] = [\psi, \phi]$ and so this integral is different from zero only if $T_\phi = T_\psi$.

(v) $[\phi; \psi] = \mathscr{T}_\phi \mathscr{T}_\psi [\psi, \phi]$ where $\mathscr{T}_\phi$ and $\mathscr{T}_\psi$ are the time-reversal eigenvalues of $\hat{\Gamma}(\phi)$ and $\hat{\Gamma}(\psi)$, respectively.

(vi) $[\phi, \psi] = 0$ if either or both of ϕ and ψ is a binary invariant.

In the presence of an external, torque-producing field it is actually not $\hat{\Gamma}$ but the linear operator $\hat{\Omega}$, defined by

$$n\hat{\Omega}(\phi) = n\hat{\Theta}(\phi) + n^2\hat{\Gamma}(\phi) \tag{1.171}$$

$$n\hat{\Theta}(\phi) = {}^0f_{\mathrm{F}}\bar{\boldsymbol{\mu}} \times \mathbf{F} \,.\, \partial_{\mathbf{L}}\phi \tag{1.172}$$

which is of basic importance. For example, when the field is turned on the gas transport coefficients and characteristic relaxation times are related to the 'double bracket' integrals

$$n^2[\![\phi, \psi]\!]_{\mathrm{F}} = n(\phi, \hat{\Omega}(\psi)) \equiv \int \mathrm{d}\bar{1}\phi(\bar{1})[n\hat{\Theta}_1(\psi) + n^2\hat{\Gamma}_1(\psi)] \tag{1.173}$$

whereas in the absence of the field these quantities are related to the ordinary bracket integrals $[\phi, \psi]$. To establish the properties of $\hat{\Omega}$ we begin with the fact that $\hat{\Theta} = -\hat{\Theta}^\dagger$ is antihermitean. Next we observe that $\hat{T}\mathbf{F} = \pm\mathbf{F}$ and $\hat{T}\bar{\boldsymbol{\mu}} = \pm\bar{\boldsymbol{\mu}}$ with the plus signs applying when F is an electric field and $\bar{\boldsymbol{\mu}}$ an electric dipole moment and with the minus signs applying when F is a magnetic field and $\bar{\boldsymbol{\mu}}$ a magnetic moment. Therefore, in both cases $\hat{T}\hat{\Theta}_{\mathrm{F}} = \hat{\Theta}^\dagger_{T\mathrm{F}}\hat{T}$ and so, since $\hat{T}\hat{\Gamma}_{\mathrm{F}} = \hat{\Gamma}^\dagger_{\mathrm{F}}\hat{T}$, it follows that $\hat{T}\hat{\Omega}_{\mathrm{F}} = \hat{\Omega}^\dagger_{T\mathrm{F}}T$. From these observations and the fact that $\hat{T}[\![\mathsf{a}, \mathsf{b}]\!]_{\mathrm{F}} = [\![\mathsf{a}; \mathsf{b}]\!]_{\hat{T}\mathrm{F}}$, it follows that if $\hat{\Omega}(\mathsf{a})$ and $\hat{\Omega}(\mathsf{b})$ are eigenfunctions of $\hat{T}$ with eigenvalues τ_a and τ_b, respectively, then

$$[\![\mathsf{a}, \mathsf{b}]\!]_{\mathrm{F}} = \tau_a\tau_b[\![\mathsf{b}, \mathsf{a}]\!]_{\hat{T}\mathrm{F}}{}^\dagger \tag{1.174}$$

where the symbol † indicates bulk transposition of the two index sets associated with the tensors a and b. Later we shall see that this property leads directly to the Onsager–Casimir relations for a fluid in an external field. We note that $\hat{P}\hat{\Theta}_{\mathrm{F}} = \hat{\Theta}_{\mathrm{PF}}\hat{P}$ with $\hat{P}\,\mathbf{H} = \mathbf{H}$ and $\hat{P}\mathbf{E} = -\mathbf{E}$. $\hat{P}$ and $\hat{\Omega}$ do not commute in an electric field but do in a magnetic field.

Throughout this section we have dealt exclusively with classical mechanics but there is no difficulty whatsoever in transcribing our theory of the Boltzmann equation into the language of quantum mechanics. In place of the variables $\mathscr{J}_i$ we introduce $v_i = (\zeta_i L_i M_i$ where L_i and M_i, respectively, are the quantum numbers associated with internal angular momentum and its projection and where ζ_i is the set of whatever additional quantum numbers are needed to fully specify the state of an individual molecule. The kinetic equation for the corresponding distribution function $f(t\mathbf{x}\mathbf{c}_i\nu_i)$ follows directly from equation (1.152) provided that the integration $\int d\bar{\imath}(\ldots) = \iint d^3c_i\, d\mathscr{J}_i(\ldots)$ is replaced with the corresponding operation $\sum_{\nu_i} \int d^3c_i(\ldots)$ and the cross-section $\sigma(\mathbf{c}'\mathscr{J}'_1\mathscr{J}'_2 \to \mathbf{c}\mathscr{J}_1\mathscr{J}_2)$ with $\sigma(\mathbf{c}'\nu'_1\nu'_2 \to \mathbf{c}\nu_1\nu_2)$. Thus, we obtain the kinetic equation

$$(\partial_t + \mathbf{c}_1 \cdot \nabla + \ldots) f(\mathbf{c}_1\nu_1) = \sum_{\nu'_1}\sum_{\nu'_2}\sum_{\nu_2} \iint d^2c'\, d^3c_2 (c'^2/c)\sigma(\mathbf{c}'\nu'_1\nu'_2 \to \mathbf{c}\nu_1\nu_2) \times [f(\mathbf{c}'_1\nu'_1)f(\mathbf{c}'_2\nu'_2) - f(\mathbf{c}_1\nu_1)f(\mathbf{c}_2\nu_2)] \tag{1.175}$$

which was first proposed by Wang Chang and Uhlenbeck.[18] Their original argument was that molecules in different states of excitation could be considered as distinct chemical species and that the kinetic theory for a polyatomic gas should not, therefore, differ significantly from that for a gas mixture of structureless species which undergo binary reactions $a_i + b_j = a_{i'} + b_{j'}$. This fits perfectly into the pattern we established previously (cf. equation 1.126) provided that one identifies the distribution function $f(t\mathbf{x}\mathbf{c}\nu)$ with the diagonal element $f^{(1)}_{\nu,\nu}(t\mathbf{x}\mathbf{c})$ of the Wigner distribution function although, as we shall see, a theory of this sort does suffer from some inadequacies.

There was some criticism of the Wang Chang and Uhlenbeck equation by those who thought that detailed balancing (or equivalently, that the transition rate w be symmetric) was an essential prerequisite to the derivation of a kinetic equation of the form (1.175). However, we have seen that time-reversal invariance (bilateral normalization) alone is sufficient for this purpose. A more serious and lasting defect of the Wang Chang and Uhlenbeck equations was discovered by those who attempted to derive them directly from the Schroedinger equation[19,20,21]. The origin of this defect is the natural, spatial degeneracy of the internal eigenstates of rotating molecules. To illustrate the problem let us suppose that at one instant the microstate of the system could be described as a collection of wave packets, each representative of a molecule in some definite internal state and with a prescribed momentum. Each time two of these molecules collide scattered waves emanate from the point of their impact. These waves eventually become spatially resolved into distinct packets—an elastic component which recedes from the point of impact at the same speed as before collision and various inelastic components whose speeds generally differ from the precollision values and depend specifically upon the post-collisional states of the two molecules. The time required for this resolution to occur is dependent upon the width of the initial packet and, more particularly, upon the group velocities associated with the different components of the scattered wave. If the resolution is complete before either of the scattered particles suffers its next collision, then the situation is relatively uncomplicated. However, the group velocities associated with the several components of a degenerate post-collisional eigenstate are all the same and so, in this case, the free streaming of the particles does not result in spatial resolution of the components. The phase relationships among the components remain unchanged until the next collision. Thus, it is very reasonable to expect that the correct quantum-mechanical kinetic equations will involve the phases as well as the moduli of the scattering amplitudes. Since the (quantum) cross-sections which appear in the Wang Chang and Uhlenbeck equations depend only upon the latter we should not be surprised to find that they are of limited validity.

1.2.4 The Waldmann–Snider Equation

Our next task is to transcribe these qualitative considerations into a mathematical theory. We recall from Section 1.1 that it is the diagonal elements of the Wigner distribution functions $f^{(1)}_{\beta,\beta'}(t\mathbf{xp})$ which are the quantum-mechanical analogues of the Gibbsian distribution functions, i.e. it is the set of functions $f^{(1)}_{\beta,\beta}(t\mathbf{xp})$ for which we might hope (but perhaps fail) to derive equations of the form (1.175). As a first step toward the development of quantum kinetic equations let us recast the definition (1.125) of the Wigner function in the form

$$f^{(1)}_{\beta,\beta'}(t\mathbf{xp}) = (2\pi\hbar)^{-3}\int \mathrm{d}^3q\rho^{(1)}_{\mathrm{I}}\left(\beta, \mathbf{p}+\frac{1}{2}\mathbf{q}|\beta', \mathbf{p}-\frac{1}{2}\mathbf{q}; t\right) \times \exp\left[\frac{i}{\hbar}\mathbf{q}\,.\,(\mathbf{x}-\mathbf{p}t/m)\right]\exp\left[\frac{i}{\hbar}(\varepsilon_\beta-\varepsilon_{\beta'})t\right] \quad (1.176)$$

where

$$\rho^{(1)}_{\mathrm{I}} = \exp\left(\frac{i}{\hbar}\hat{H}^{(1)}t\right)\rho^{(1)}\exp\left(-\frac{i}{\hbar}\hat{H}^{(1)}t\right)$$

with

$$\hat{H}^{(1)} = \hat{p}^2/2m + \sum_\beta \varepsilon_\beta|\beta\rangle\langle\beta|$$

is the single-particle density matrix in the interaction picture. This function will be different from zero only in the vicinity of those points $\mathbf{x}$ at which its phase is stationary. Therefore, we conclude that $f^{(1)}_{\beta,\beta'}(t\mathbf{xp})$ is a wave packet centred at $\mathbf{x}$ and moving with the group velocity $\mathbf{p}/m$. These packets will provide a useful characterization of the state of the system only if their dispersion by collision processes proceeds sufficiently slowly, i.e. if the particle interactions are weak and/or collisions are infrequent. It is the latter situation which is of particular interest to us here. If the free-flight interval $t_{\text{free}} = \lambda_{\text{free}}/\langle c \rangle = 1/n\sigma_{\text{tot}}\langle c \rangle$ is greater than $\hbar/(\varepsilon_\beta - \varepsilon_{\beta'})$ then, on the average, only the low-frequency components of $f^{(1)}_{\beta,\beta'}$ will be different from zero, i.e. only the functions

$$f^{j}_{m,m'}(t\mathbf{xp}) \equiv \delta_1(\varepsilon_j - \varepsilon_{j'}) f^{(1)}_{jm,j'm'}(t\mathbf{xp}) = f^{(1)}_{jm,jm'}(t\mathbf{xp}) \tag{1.177}$$

will be of any real importance in accounting for the properties of the fluid. Here m denotes the projection quantum number and $j = (L\zeta)$ the set of quantum numbers upon which the internal energy depends. Since t_{free} for a gas is typically of the order of 10^{-9} seconds and since $\hbar/(\varepsilon_j - \varepsilon_{j'})$ is no greater than 10^{-12} seconds even for transitions between adjacent rotational levels, this condition invariably will be satisfied in a dilute gas.

In order that the two scattered waves associated with the processes $a_i + b_j \to a_{i'} + b_{j'}$ and $a_i + b_j \to a_{i''} + b_{j''}$ be resolved within the free-flight interval it is necessary for $t_{\text{free}}(v' - v'') \approx t_{\text{free}}(\Delta\varepsilon/m\langle c \rangle) \approx (\Delta\varepsilon/\mathbf{k}T)(3n\sigma_{\text{tot}})^{-1}$ to be greater than the range of the intermolecular forces. Here v denotes the group velocity of a packet and $\Delta\varepsilon$ the difference between the internal energies of the two final states. Since the effective range of the interactions between electrically neutral particles rarely exceeds 20 Å and since the total cross-section σ_{tot} is of the order of 10^{-15} cm^2, this condition too will be satisfied unless the states involved belong to a degenerate manifold.

Now the singlet and pair density matrices of Section 1.1 were taken to be certain contractions of the density matrix for the entire N-body system. In general, we define the density matrix for a subset of n particles by the formula

$$\rho^{(n)}(\{n\}, t) = [N!/(N-n)!] \operatorname{Tr}_{(n+1,\ldots,N)} \rho^{(N)}(\{N\}, t) \tag{1.178}$$

where $\operatorname{Tr}_{(a,b,\ldots,g)}$ indicates the trace with respect to particles a, b, . . . , g. Furthermore, we shall assume that the Hamiltonian of the system is of the form

$$\hat{H} = \sum_k \hat{H}^{(1)}_k + \sum_{k>l}\sum \hat{V}_{lk}$$

where $\hat{H}^{(1)}$ denotes the energy operator for an isolated molecule and where the pair energy $\hat{V}_{lk}$ depends upon the particle coordinates but not upon the associated conjugate momenta. By contracting the Schroedinger equation $i\hbar\partial\rho^{(N)}/\partial t = [\hat{H}, \rho^{(N)}]$ one then can verify that

$$\partial_t \rho^{(1)}(1, t) = \frac{i}{\hbar}[\hat{H}_1^{(1)}, \rho^{(1)}(1, t)] + \partial_c \rho^{(1)}(1, t) \tag{1.179}$$

where

$$\partial_c \rho^{(1)}(1, t) \equiv \frac{i}{\hbar} \mathrm{Tr}_{(2)}[\hat{V}_{12}, \rho^{(2)}(12, t)] \tag{1.180}$$

Next, from this relationship and from the definition (1.176) we obtain the set of equations

$$\left[\partial_t + \frac{\mathbf{p}_1}{m} . \nabla + \frac{i}{\hbar}(\varepsilon_{\beta_1} - \varepsilon_{\beta'_1})\right] f^{(1)}_{\beta_1, \beta'_1}(t\mathbf{x}\mathbf{p}_1) = \partial_c f^{(1)}_{\beta_1, \beta'_1}(t\mathbf{x}\mathbf{p}_1) \tag{1.181}$$

wherein

$$\begin{aligned}
\partial_c f^{(1)}_{\beta_1, \beta'_1}(t\mathbf{x}\mathbf{p}_1) & \\
&= (2\pi\hbar)^{-3} \int \mathrm{d}^3 q \exp\left[\frac{i}{\hbar}\mathbf{q} . \mathbf{x}\right] \partial_c \rho^{(1)}(\beta_1, \mathbf{p}_1 + \tfrac{1}{2}\mathbf{q} | \beta'_1, \mathbf{p}_1 - \tfrac{1}{2}\mathbf{q}; t) \\
&\equiv \frac{i}{\hbar}(2\pi\hbar)^{-3} \int \mathrm{d}^3 \mathbf{q} \exp\left[\frac{i}{\hbar}\mathbf{q} . \mathbf{x}\right] \langle \beta_1, \mathbf{p}_1 \\
&\quad + \tfrac{1}{2}\mathbf{q} | \mathrm{Tr}_{(2)}[\hat{V}_{12}, \rho^{(2)}(12, t)] | \beta'_1; \mathbf{p}_1 - \tfrac{1}{2}\mathbf{q}\rangle
\end{aligned} \tag{1.182}$$

Our objective is to transform this exact relationship between the singlet and pair density matrices into an approximate 'kinetic equation' for the former. Several of the approximations upon which this derivation depends are summarized by the following:

$$\begin{aligned}
\rho^{(2)}(12, t) &\approx \exp[-i\hat{H}_{12}(t - t_0)/\hbar]\rho^{(2)}(12, t_0) \exp[i\hat{H}_{12}(t - t_0)/\hbar] \\
&= \hat{U}(0, t - t_0) \exp[-i\hat{K}_{12}(t - t_0)/\hbar]\rho^{(2)}(12, t_0) \exp[i\hat{K}_{12}(t - t_0)/\hbar] \\
&\quad \times \hat{U}(0, t - t_0)^\dagger \\
&\approx \hat{\Omega} \exp[-i\hat{K}_{12}(t - t_0)/\hbar]\rho^{(1)}(1, t_0) \otimes \rho^{(1)}(2, t_0) \\
&\quad \times \exp[i\hat{K}_{12}(t - t_0)/\hbar]\hat{\Omega}^\dagger \\
&\approx \hat{\Omega}\rho^{(1)}(1, t) \otimes \rho^{(1)}(2, t)\hat{\Omega}^\dagger
\end{aligned} \tag{1.183}$$

with $\hat{H}_{12} = \hat{K}_{12} + \hat{V}_{12}$ and $\hat{K}_{12} = \hat{H}_1^{(1)} + \hat{H}_2^{(1)}$. Here $\hat{U}(t, t_0) = \exp(i\hat{K}t/\hbar) \times \exp[-i\hat{H}(t - t_0)/\hbar] \exp(-i\hat{K}t_0/\hbar)$ is the evolution operator which transforms wave functions in the interaction picture from one time t_0 to another t, i.e., $\psi_I(12, t) = \hat{U}(t, t_0)\psi_I(12, t_0)$ and $\rho_I(12, t) = \hat{U}(t, t_0)\rho_I(12, t_0) \times \hat{U}(t, t_0)^\dagger$ where $\psi_I(t) = \exp(i\hat{K}t/\hbar)\psi(t)$ and $\rho_I(t) = \exp(i\hat{K}t/\hbar)\rho(t) \times \exp(-i\hat{K}t/\hbar)$. In order for the commutator $[\hat{V}_{12}, \rho^{(2)}(12, t)]$, which appears in equation (1.182), to be different from zero the amplitude of the system wave function must be large at time t in the region $|\mathbf{x}_{21}| < a$ where a denotes the range of the pair potential, that is, the molecules must be intimately involved in a collisional encounter at time t. Therefore, at an earlier instant t_0 (where $t - t_0$ usually need be no greater than 10^{-12} sec) the coordinate picture of the wave function consists of two freely translating molecules or wave packets proceeding toward one another on a collision course but as yet beyond range of their mutual interaction. Then, because a variation of state can occur in the interaction picture only as the result of a collision, the system wave function is constant for all times prior to t_0. It is this line of reasoning which accounts for our replacement, in the third line of equation (1.183), of the operator $\hat{U}(0, t - t_0)$ with its asymptotic value

$$\hat{\Omega} = \lim_{\tau \to -\infty} \hat{U}(0, \tau) = 1 - \mathrm{i} \lim_{\varepsilon \to 0^+} \int_{-\infty}^{0} \mathrm{d}t' \, \mathrm{e}^{\varepsilon t'} \, \mathrm{e}^{i\hat{H}t'/\hbar} \hat{V} \mathrm{e}^{-\mathrm{i}\hat{K}t'/\hbar} \tag{1.184}$$

The assumption of molecular chaos embodied in our factorization of $\rho(12, t_0)$ into the direct product $\rho(1, t_0) \otimes \rho(2, t_0)$ involves a neglect of many-body correlations. This approximation is a common ingredient of all dilute-gas kinetic theories. Finally, the replacement of $\rho(k_{t-t_0}, t) \equiv \exp[-\mathrm{i}\hat{H}_k^{(1)}(t - t_0)/\hbar]\rho(k, t_0) \exp[\mathrm{i}\hat{H}_k^{(1)}(t - t_0)/\hbar]$ with $\rho(k, t)$ implies that the singlet density function is only weakly dependent upon the location of the molecular centre of mass.

From equation (1.184) it follows that

$$\hat{\Omega}|\alpha\rangle = |\alpha\rangle + (E_\alpha - \hat{H} + \mathrm{i}\varepsilon)^{-1}\hat{V}|\alpha\rangle \equiv |\alpha^+\rangle \tag{1.185}$$

where $|\alpha\rangle$ denotes an eigenket of $\hat{K}$ and $|\alpha^+\rangle$ the associated solution of the Lippman–Schwinger equation. Thus, we are able to identify $\hat{\Omega}$ with $1 + G_0\hat{t}$ where $G_0(E) = (E - \hat{K} + \mathrm{i}\varepsilon)^{-1}$ is the free-particle propagator and

$\hat{t} = \hat{V}\hat{\Omega}$ the familiar scattering operator. Then since $G_0{}^+ - G_0 = 2\pi i\delta_1(E-\hat{K})$ and $\hat{t}^\dagger - \hat{t} = 2\pi i\hat{t}^\dagger\delta_1(E-\hat{K})\hat{t}$ we find that

$$\begin{aligned}[\hat{V}_{12}, \rho(12, t)] &\approx \hat{t}\rho(1)\rho(2)\hat{\Omega}^\dagger - \hat{\Omega}\rho(1)\rho(2)\hat{t}^\dagger \\ &= \hat{t}\rho(1)\rho(2) - \rho(1)\rho(2)\hat{t}^\dagger + \hat{t}\rho(1)\rho(2)\hat{t}^\dagger G_0{}^\dagger - G_0\hat{t}\rho(1)\rho(2)\hat{t}^\dagger \\ &= \hat{t}\rho(1)\rho(2) - \rho(1)\rho(2)\hat{t}^\dagger + 2\pi i\hat{t}\rho(1)\rho(2)\hat{t}^\dagger\delta_1(E-K) \\ &\quad + [\hat{t}\rho(1)\rho(2)\hat{t}^\dagger, G_0] \end{aligned} \tag{1.186}$$

The only non-zero elements of $\mathrm{Tr}_{(2)}[\hat{t}\rho(1)\rho(2)\hat{t}^\dagger, G_0]$ are those connecting states of different internal energy and so the last term of equation (1.186) contributes nothing to $\partial_c f^{j_1}_{m_1m'_1}(t\mathbf{x}\mathbf{p}_1)$.

To complete the derivation we adopt the approximation implicit within all derivations of the classical Boltzmann equation and neglect the influence of spatial variations of the density matrices upon the collision term. Thus, we replace $\rho^{(1)}(\beta\mathbf{p}\,|\,\beta'\mathbf{p}', t)$ with $(2\pi h)^3 f^{(1)}_{\beta,\beta'}(t\bar{\mathbf{x}}\mathbf{p})\delta_3(\mathbf{p}-\mathbf{p}')$ where $\bar{\mathbf{x}} \approx \mathbf{x}$ denotes a point selected arbitrarily from within the region where the collision occurs. Then, since the $\hat{t}$-matrix is diagonal in the momentum representation we adopt Snider's notation

$$\langle \beta_1\mathbf{p}_1\beta_2\mathbf{p}_2|\hat{t}|\beta'_1\mathbf{p}'_1\beta'_2\mathbf{p}'_2\rangle \equiv \left\langle \begin{matrix}\beta_1\beta_2\\(\mathbf{p}_1,\mathbf{p}_2)\end{matrix}\right|\hat{t}\left|\begin{matrix}\beta'_1\beta'_2\\(\mathbf{p}'_1,\mathbf{p}'_2)\end{matrix}\right\rangle\delta_3(\mathbf{p}_1+\mathbf{p}_2-\mathbf{p}'_1-\mathbf{p}'_2)$$

and so obtain the equation

$$\begin{aligned}\partial_c f^j_{mm'}(t\mathbf{x}\mathbf{p}) \approx{}& \frac{2\pi h^3}{\hbar}\sum_{\substack{j_1m_1JJ_1\\nn_1n'n'_1}}\iint \mathrm{d}^3p_1\,\mathrm{d}^3p' \\ &\times\left\langle\begin{matrix}jmj_1m_1\\(\mathbf{p},\mathbf{p}_1)\end{matrix}\right|\hat{t}\left|\begin{matrix}JnJ_1n_1\\(\mathbf{p}',\mathbf{p}+\mathbf{p}_1-\mathbf{p}'_1)\end{matrix}\right\rangle f^J_{nn'}(t\bar{\mathbf{x}}\mathbf{p}) \\ &\times f^{J_1}_{n_1n'_1}(t\bar{\mathbf{x}}\mathbf{p}+\mathbf{p}_1-\mathbf{p}')\left\langle\begin{matrix}Jn'J_1n'_1\\(\mathbf{p}',\mathbf{p}+\mathbf{p}_1-\mathbf{p}')\end{matrix}\right|\hat{t}^\dagger\delta_1(E-\hat{K})\left|\begin{matrix}jm'j_1m_1\\(\mathbf{p},\mathbf{p}_1)\end{matrix}\right\rangle \\ &+\frac{h^3}{i\hbar}\sum_{\substack{j_1m_1\\nn_1}}\int\mathrm{d}^3p_1\left[\left\langle\begin{matrix}jmj_1m_1\\(\mathbf{p},\mathbf{p}_1)\end{matrix}\right|\hat{t}\left|\begin{matrix}jnj_1m_1\\\mathbf{p},\mathbf{p}_1)\end{matrix}\right\rangle f^j_{nm'}(t\bar{\mathbf{x}}\mathbf{p})f^{j_1}_{n_1m_1}(t\bar{\mathbf{x}}\mathbf{p}_1)\right. \\ &\left.-f^j_{mn}(t\bar{\mathbf{x}}\mathbf{p})f^{j_1}_{m_1n_1}(t\bar{\mathbf{x}}\mathbf{p}_1)\left\langle\begin{matrix}jnj_1n_1\\(\mathbf{p},\mathbf{p}_1)\end{matrix}\right|\hat{t}^\dagger\left|\begin{matrix}jm'j_1m_1\\(\mathbf{p},\mathbf{p}_1)\end{matrix}\right\rangle\right]\end{aligned} \tag{1.187}$$

of Waldman and Snider. Here, to emphasize the energy diagonality of the Wigner functions we have replaced the single symbol β with the pair (jm). The index j denotes the quantum number (or set thereof) upon which the energy $\varepsilon_\beta = \varepsilon_j$ depends and m is the projection quantum number associated with the molecular angular momentum. Finally, $f^j_{mm'} \equiv f^{(1)}_{jm,jm'}$.

Now since the density matrix is hermitean there does exist a unitary matrix $\hat{U}(t\mathbf{xp})$ by which $f^j(t\mathbf{xp})$ can be brought into diagonal form. However, this is of less practical value than one might expect for in order to construct $\hat{U}(t\mathbf{xp})$ it first is necessary to solve the W-S equation. We can, of course, simply discard the off-diagonal elements and so ignore the quantum-phase conditions associated with scattering into a degenerate manifold. According to this approximation $f^j_{mm'} = \delta_{mm'} f^j_m$ and equation (1.187) reduces to the Wang Chang and Uhlenbeck equation (1.175) with σ given by the usual quantum-mechanical formula

$$\sigma^{j_1m_1j_2m_2}_{j'_1m'_1j'_2m'_2}(\mathbf{c}'_{21} \to \mathbf{c}_{21}) = (2\pi)^4 \mu^2 \hbar^2 (c_{21}/c'_{21}) \times \left| \left\langle \begin{matrix} j_1m_1j_2m_2 \\ (\mathbf{p}_1, \mathbf{p}_2) \end{matrix} \right| \hat{t} \left| \begin{matrix} j'_1m'_1j'_2m'_2 \\ (\mathbf{p}'_1, \mathbf{p}'_2) \end{matrix} \right\rangle \right|^2 \tag{1.188}$$

for the differential cross-section. This kinetic equation also can be written in the generic form of equation (1.152) with

$$f(\bar{\imath}) = f^{j_i}_{m_i}(t\mathbf{xp}_i), \qquad \mathrm{d}\bar{\imath} = \sum_{j_im_i} \int \mathrm{d}^3 p_i$$

and

$$w(12|1'2') = \mu^{-2}\delta_1(E_1+E_2-E'_1-E'_2)\delta_3(\mathbf{p}_1+\mathbf{p}_2-\mathbf{p}'_1-\mathbf{p}'_2) \times (c'_{21}/c_{21})\sigma^{j_1m_1j_2m_2}_{j'_1m'_1j'_2m'_2}(\mathbf{c}'_{21} \to \mathbf{c}_{21}) \tag{1.189}$$

For this case it can easily be proved that $\hat{T}\hat{P}$-invariance of collisional events, as implied by the relationship

$$\hat{T}\hat{P}w(12|1'2') \equiv w(\hat{T}\hat{P}1\hat{T}\hat{P}2|\hat{T}\hat{P}1'\hat{T}\hat{P}2') = w(1'2'|12)$$

assumes the more familiar form,

$$c'^2_{21}\sigma^{j_1m_1j_2m_2}_{j'_1m'_1j'_2m'_2}(\mathbf{c}'_{21} \to \mathbf{c}_{21}) = c^2_{21}\sigma^{j'_1-m'_1j'_2-m'_2}_{j_1-m_1j_2-m_2}(\mathbf{c}_{21} \to \mathbf{c}'_{21}) \tag{1.190}$$

which is, of course, the quantal analogue of the classical relationship (1.142).

The presence of an external field alters the Waldmann–Snider equation in two ways: (i) the collision operator is modified by the level shifts of the molecular eigenstates (a study of this has been conducted by Hess[22]); (ii) the left-hand side of the equation is augmented by terms descriptive of the response to the field of individual molecules. The second of these alterations is connected with the additional term, $(i/\hbar)[\hat{H}_1{}^e, \hat{\rho}^{(1)}(1,t)]_-$ which the field-interaction operator $\hat{H}^e$ contributes to the right-hand side of equation (1.179). To the left side of equation (1.181) one must then add the corresponding contribution

$$\begin{aligned}
\hat{\theta}^e f^{(1)}_{\beta,\beta'}(tx p) &= \frac{i}{\hbar}(\pi\hbar)^{-3}\sum_{\beta''}\iint d^3q\, d^3y[\langle\beta|\hat{H}^e(\mathbf{y})|\beta''\rangle f^{(1)}_{\beta''\beta'}(t\mathbf{xp}-\mathbf{q}) \\
&\quad - f^{(1)}_{\beta,\beta''}(t\mathbf{xp}+\mathbf{q})\langle\beta''|\hat{H}^e(\mathbf{y})|\beta'\rangle] \exp\frac{2i}{\hbar}\mathbf{q}\,.\,(\mathbf{x}-\mathbf{y}) \\
&= \frac{i}{\hbar}\sum_{\beta''}\left[\left(\exp-\frac{\hbar}{2i}\nabla_{\mathbf{x}}\,.\,\nabla_{\mathbf{p}}\right)\langle\beta|\hat{H}^e(\mathbf{x})|\beta''\rangle f^{(1)}_{\beta'',\beta'}(t\mathbf{xp})\right. \\
&\quad \left. -\left(\exp\frac{\hbar}{2i}\nabla_{\mathbf{x}}\,.\,\nabla_{\mathbf{p}}\right) f^{(1)}_{\beta,\beta''}(t\mathbf{xp})\langle\beta''|\hat{H}^e(x)|\beta'\rangle\right] \\
&= \frac{i}{\hbar}\cos\left(\frac{\hbar}{2}\nabla_{\mathbf{x}}\,.\,\nabla_{\mathbf{p}}\right)\langle\beta|[\hat{H}^e, f^{(1)}]_-|\beta'\rangle \\
&\quad -\frac{1}{\hbar}\sin\left(\frac{\hbar}{2}\nabla_{\mathbf{x}}\,.\,\nabla_{\mathbf{p}}\right)\langle\beta|[\hat{H}^e, f^{(1)}]_+|\beta'\rangle
\end{aligned} \tag{1.191}$$

where $[a,b]_\pm = ab \pm ba$. Finally, since

$$\hat{\theta}^e f^{(1)} = (i/\hbar)[\hat{H}^e, f^{(1)}]_- - (1/2)[\nabla_{\mathbf{x}}\hat{H}^e, \nabla_{\mathbf{p}} f^{(1)}]_+ : \boldsymbol{\delta} + 0(\hbar) \tag{1.192}$$

it is obvious that $\hat{\theta}^e f^{(1)}$ reduces to the single commutator $(i/\hbar)[\hat{H}^e, f^{(1)}]_-$ whenever the external field is spatially uniform.

Waldmann and Snider and their coworkers have shown how one can learn to live with their more complicated kinetic equation. In particular, Snider has[23] established modified variation principles for this equation and Snider and McCourt[24] have succeeded in deriving from it formal expressions for the transport coefficients of a polyatomic gas. The second of these accomplishments was performed independently by Waldmann and Hess[25] who also linearlized the equation and expanded its solutions in irreducible tensors. This leads to a set of moment equations which are quite analogous to those obtained from more conventional kinetic equations. Thus, we see that tools with which to handle this equation have been identified and applied to a few problems of interest. There are, however, many phenomena which do not depend sensitively upon the subtleties which distinguish the Waldmann–Snider equation from that of Wang Chang and Uhlenbeck. Since these are, in fact, the subjects of greatest interest to us here we shall usually employ the Waldmann–Snider equation only in those situations where the Boltzmann equation of Wang Chang and Uhlenbeck is inapplicable. The interested reader is encouraged to acquaint himself with the very excellent and elegant papers of Waldmann, Hess and Snider.

The main purpose of the last few sections has been to introduce three kinetic equations—the classical master or Boltzmann equation and the quantum-mechanical equations of Wang Chang and Uhlenbeck and of Waldmann and Snider—which are particularly appropriate to the description of transport and relaxation processes in dilute polyatomic gases. These are not the only equations which have been devised for this purpose but they do appear to be the most general and best founded of the lot and so we shall concern ourselves with no others.

Once one of these equations has been selected, the next problem we face is that of constructing a solution for it which will be descriptive of the non-equilibrium state of the gas. Possession of this solution is the final and essential prerequisite to computing the various fluxes and relaxation rates which characterize the fluid's macroscopic behaviour. This final step, that of solving the kinetic equation, can be carried out in considerable generality—without explicit knowledge of the collision cross-sections and, in fact, even without a prior commitment as to particle kinematics*. However, we are less interested here in such broad generalities than in the particulars of a gas composed of rotating molecules. Therefore, we restrict our considerations to physical conditions which are unfavourable to the production of vibrationally excited and/or chemically reactive molecules. Under these circumstances the molecules can be treated as rigid, non-deformable bodies whose intrinsic properties include mass, moments of inertia and distributions of electrical charge, current and spin. We assume that the values of these parameters depend only upon the electronic state of the molecule and so neglect centrifugal

* In the case of the isotropic ensemble this first was done by Wang Chang and Uhlenbeck. Modifications and extensions of their results have been presented by Waldmann[26], by Mason and Monchick[27], by Monchick, Yun and Mason[28], and by Monchick, Pereira and Mason[29]. A similar analysis of the non-polar ensemble has been conducted by Snider and McCourt[24].

stretching of bonds and other second-order kinematic effects. Furthermore, we shall ignore changes of these properties which might occur during the progress of a collision or as the result of interaction with an external field—unless, of course, these perturbations cause electronic transitions to occur.

1.2.5 Intermolecular Forces and Quantum-Mechanical Collision Cross-Sections

The internal (rotational) state of a molecule can be altered either by absorption or emission of field (electromagnetic) quanta or as the result of a collisional event. The first of these possibilities will be dealt with more fully in a subsequent section. The collisional processes occupy our attention here. The qualitative and quantitative aspects of these events may depend very sensitively upon the precise nature of the particle interactions. (For example, if the interactions are assumed to be independent of molecular orientation and/or spin then the rotational states will not be altered at all. Collisions of this sort are said to be elastic.) Ideally, we should be able to select a molecular system for study and then consult an expert who would provide us with a suitable interaction and/or a complete tabulation of the scattering cross-sections appropriate to the species under consideration. And the day is fast approaching when this ideal will be realized. But despite the great efforts which have been expended and the very considerable body of knowledge amassed, quantum theorists are only now reaching the stage where they can perform reliable *ab initio* computations of the interactions between small polyatomic molecules. This need not distress us too much, for the lack of a complete quantum-mechanical theory of intermolecular forces constitutes more of an inconvenience than a truly significant obstacle to our understanding of transport and relaxation phenomena. If the need for an interaction potential arises then we can surely invent one which will be sufficient for our purposes. For there are among us chemists and physicists who possess highly developed albeit intuitive understandings of intermolecular forces. Only the quantification of their impressions is lacking. This suggests that one adopt a semiempirical approach which places strong reliance upon the intuition of the chemist and uses for its building blocks the atom–atom and electromagnetic-multipole interactions about which so much empirical and theoretical information is available. For example, it often should be (and is) possible to represent the intermolecular force by a sum of terms each pertaining to the interaction between a specific pair of constituent atoms. In other cases the total interaction is represented better by a sum of 'group contributions' the fundamental elements of which are to be identified either as chemical (e.g. methyl or hydroxyl groups) or electromagnetic (e.g. the static multipole(s) associated with some particular electronic state of a molecule or the transition moments of the London theory). If handled with sufficient care an approach of this sort could produce representations of the (non-central) forces between diatomic and polyatomic molecules which would be of the same, relatively high, quality as are the representations of central forces provided by familiar potential functions such as those of Lennard-Jones or of Buckingham.

It is not our presumption that these few, superficial remarks provide a proper introduction to the subject of intermolecular forces nor that they permit one to see the various issues in proper perspective. Our point is simply that a good theory of intermolecular forces is on the way and that in the meantime, until it arrives, we are unlikely to face demands which will exceed the capabilities of present-day semiempiricism. In fact, at their current levels of sophistication the theories and experiments pertaining to transport and relaxation processes do not make very exacting demands upon the intermolecular forces. In most cases it is enough to be able to identify the dominant contribution to the non-central part of the force. And indeed, it is not obvious that much greater subtlety will be needed for the qualitative elucidation and possibly even for the quantitative prediction of the various phenomenological coefficients. This comes as no surprise for the macroscopic parameters of interest (such as viscosity, thermal conductivity and spin relaxation times) are functionals of the differential cross-sections and as such can be expected to depend much less sensitively upon the details of the particle interactions than will the cross-sections themselves. From the standpoint of one who is interested in the prediction of gas transport properties this insensitivity is all to the good. But if one's search is for information about intermolecular forces then one must examine those coefficients which depend most sensitively upon the choice of particle interactions. In the sections which follow we shall encounter examples of both extremes.

The basic data we require are not the intermolecular forces themselves but the related set of differential cross-sections. (The latter can, in principle, be taken directly from the output of sophisticated beam experiments but the practical obstacles to this are immense.) With central forces, the problem of computing the scattering cross-section can be reduced to one of quadrature. However, in all but a very few special cases the solution to the problem of scattering by a non-central force cannot be rendered in closed form. Here then is the real problem which we face, that of computing the differential cross-sections for inelastic scattering. Now under gas kinetic conditions the mean values of the rotational angular velocity and of the relative velocity of impact are given by $\bar{\omega} \equiv |\bar{\mathbf{L}}|/\Gamma \approx \sqrt{3\mathbf{k}T/\Gamma}$ and by $\bar{c} \approx \sqrt{3\mathbf{k}T/\mu}$, respectively. This means that the period of molecular rotation $\bar{\omega}^{-1}$ will be very nearly equal to the duration of a collision, $R/\bar{v} \approx 10^{-7}$ cm/$\bar{c}$. Consequently, the rotational transitions with which we are concerned can be categorized neither as adiabatic nor as impulsive events. This is bad news for these are the limiting cases with which quantum-scattering theory is able to cope most easily.

One of the most significant efforts to adapt quantum-scattering theory to these specific and trying demands of kinetic theory have been those of Curtiss and his coworkers[30]. The emphasis of their work has been on rotational degrees of freedom and on scattering by potentials which are only slightly non-spherical. The product of this long and difficult programme of studies is a formal theory of considerable generality, one which has been described by its architects as an extended distorted wave treatment. This theory separates the problem of computing the scattering amplitudes into the performance of two well-defined but difficult tasks. First one must develop means for dealing with the enormous algebraic complexities which stem from the collisional alterations of the molecular angular momenta. Then, as with all multichannel scattering processes one must solve sets of coupled and linear, ordinary differential (or integral) equations in order to determine the phase shifts appropriate to the various open channels. Modern high-speed computing machines are well adapted to handling the second of these tasks. However, even with the aid of these fantastic tools the sheer magnitude of the problem remains impressive. At moderate temperatures of a few hundred degrees Kelvin, a large number ($5 < \bar{L} < 15$) of rotational states contribute directly to the properties of the gas. Furthermore, there is some evidence that the collisional alterations of angular momentum can be rather large in real gases and so, associated with each initial state there may be a considerable number of possible final states. Very little is known about the degree of coupling (between initial and final state channels) which must be retained in order to ensure accurate estimates of the cross-sections.

Progress toward the solution of the first problem, that of mastering the algebraic complexities of non-central scattering events, has been more impressive. Thus, Mueller and Curtiss[31] recently have shown how the scattering theory is to be applied to the actual computation of transport coefficients for a dilute gas of rigid loaded spheres. The generalization to soft, nearly central interactions has been given by Curtiss and Hardisson[32]. In both cases the authors succeeded in constructing explicit formulas for the degeneracy-averaged cross-sections which appear in the master equation appropriate to the isotropic ensemble. As yet no computations based upon these formulas of Curtiss and his coworkers have been reported. However, it has been demonstrated that in the classical limit the predictions of the Mueller–Curtiss quantum-kinetic theory for rigid loaded spheres coincide with those of the corresponding (isotropic) classical theory.

At the time of writing it seems beyond doubt that numerical techniques for the efficient handling of many channel scattering processes are already available or soon will be. These problems are being attacked vigorously by many very capable scientists—John Light[33], Roy Gordon[34], Lester and Bernstein[35], Johnson and Secrest[36], to mention but a few—and so the whole situation with regard to the future availability of theoretical cross-sections is one of rapid evolution. Furthermore, considerable effort is being directed toward the further development of the semi-classical JWKB method[37], which is based upon approximations which are excellent for the conditions encountered in most gas kinetic collisions.

1.2.6 The Theory of Mason and Monchick

One can judge from these remarks that the time is very nearly here when we can look to quantum mechanics for a complete and rigorous treatment of the inelastic scattering amplitudes which are needed for kinetic theory calculations. However, instead of waiting for such results to become available, Mason and Monchick[27] began in 1962 to investigate alternative means for estimating the contributions of inelastic events to gas transport coefficients. Their aim was to invent a scheme which would reduce the computational problems to manageable proportions. Of course, this had to be done in such a way that the final results would faithfully reflect the qualitative and quantitative features associated with molecular internal degrees of freedom. Without some prior numerical experimentation it is difficult to predict precisely which set of approximations will ensure this outcome. However, they were guided by the empirical fact that reasonably accurate values for the viscosity and diffusion coefficients of polyatomic fluids can be computed from a theory which neglects completely all internal degrees of freedom. Furthermore, they knew that even the thermal conductivity can be obtained from such a theory provided that one includes a correction (Eucken) factor to account for the internal energy which is transferred by molecular diffusion. This suggested that they adopt a perturbation approach which in first approximation totally ignored the possibility of inelastic collisions. To this approximation the actual cross-sections would be replaced with those characteristic of an average 'effective' central force. The corrections due to inelastic scattering are assumed to be small and so can be evaluated with rather less care than is required for the larger elastic contributions.

As their starting point Mason and Monchick selected the formulas which Wang Chang and Uhlenbeck had derived (in the isotropic approximation) for the gas transport coefficients. For example, the coefficients of shear and bulk viscosity (η and η_b) and of thermal conductivity ($\lambda = \lambda_{tr} + \lambda_{int}$) were assumed to be given by

$$\eta^{-1} = \tfrac{8}{5}\beta W; \qquad \eta_b = \tfrac{2}{3}\tau\left(\frac{n\mathbf{k}}{\beta}\frac{c_{int}}{c_V^2}\right) \tag{1.193}$$

and

$$\lambda_{tr}(1 - Y^2/XZ) = \frac{75}{8}\frac{\mathbf{k}}{m\beta}X^{-1} + \frac{15}{4}\frac{c_{int}}{m\beta}\frac{Y}{XZ} \tag{1.194}$$

$$\lambda_{int}(1 - Y^2/XZ) = \frac{3}{2}\frac{c_{int}^2}{m\beta\mathbf{k}}Z^{-1} + \frac{15}{4}\frac{c_{int}}{m\beta}\frac{Y}{XZ} \tag{1.195}$$

with the integrals W, X, Y and Z defined according to

$$W = \left(\sum\int\right) d\tau[\gamma^4 \sin^2\chi + \tfrac{1}{2}(\beta\Delta\varepsilon)^2(\tfrac{2}{3} - \sin^2\chi)] \tag{1.196}$$

$$X = 4\left(\sum\int\right) d\tau[\gamma^4 \sin^2\chi + \tfrac{1}{2}(\beta\Delta\varepsilon)^2(\tfrac{11}{4} - \sin^2\chi)]$$

$$Y = \frac{5}{2}\left(\sum\int\right) d\tau[(\beta\Delta\varepsilon)^2] \equiv \frac{5c_{int}}{4n\mathbf{k}\tau}$$

$$Z = 4\left(\sum\int\right) d\tau[\beta^2(\varepsilon_i - \bar{\varepsilon})\{-\tfrac{3}{2}\Delta\varepsilon + \gamma^2(\varepsilon_i - \varepsilon_j) - \gamma(\gamma^2 - \beta\Delta\varepsilon)^{1/2} \times (\varepsilon_k - \varepsilon_l)\cos\chi\}$$

and where

$$\left(\sum \int\right) \mathrm{d}\tau[g] \equiv (\pi\beta m)^{-1/2} q^{-2} \sum_{ijkl} \mathrm{e}^{-\beta(\varepsilon_i+\varepsilon_j)} \int \mathrm{d}\gamma\gamma^3 \,\mathrm{e}^{-\gamma^2} \int \mathrm{d}^2\Omega \bar{\sigma}_{ij}^{kl}[g] \quad (1.197)$$

In these formulas $\beta = 1/\mathbf{k}T$, $\gamma^2 = \frac{1}{2}\beta\mu c_{21}^2$, $\Delta\varepsilon = \varepsilon_k + \varepsilon_l - \varepsilon_i - \varepsilon_j$, $\chi = \sphericalangle(\mathbf{c}'_{21}, \mathbf{c}_{21})$ is the scattering angle, and $q = \sum_i \exp(-\beta\varepsilon_i)$ the partition function associated with the molecular internal degrees of freedom. Finally, $\bar{\varepsilon} = q^{-1} \sum_i \varepsilon_i \exp(-\beta\varepsilon_i)$ is the thermal mean internal energy, $c_{\text{int}} = \mathbf{k}\beta^2[\overline{\varepsilon^2} - \bar{\varepsilon}^2]$ the corresponding heat capacity, and $c_{\text{V}} = c_{\text{V tr}} + c_{\text{int}} \equiv \frac{3}{2}\mathbf{k} + c_{\text{int}}$.

In first approximation the collisions are assumed to be elastic. Therefore, we must set $\Delta\varepsilon$ equal to zero and replace $\bar{\sigma}_{ij}^{kl}$ with σ_{el}. To this approximation $X = \frac{5}{2}(\mathbf{k}T/\eta_{\text{el}})$, $Y = 0$ and $Z = \frac{3}{2}(c_{\text{int}}T/\rho D)$. Here

$$D^{-1} = \tfrac{8}{3}\rho(\beta/\pi m)^{1/2} \int_0^\infty \mathrm{d}\gamma\gamma^5 \,\mathrm{e}^{-\gamma^2} \int_{(4\pi)} \mathrm{d}^2\Omega\sigma_{\text{el}}(1 - \cos\chi) \quad (1.198)$$

and

$$\eta_{\text{el}}^{-1} = \tfrac{8}{5}(\beta/\pi m)^{1/2} \int_0^\infty \mathrm{d}\gamma\gamma^7 \,\mathrm{e}^{-\gamma^2} \int_{(4\pi)} \mathrm{d}^2\Omega\sigma_{\text{el}}(1 - \cos^2\chi) \quad (1.199)$$

is the familiar formula (of classical kinetic theory) for the viscosity of a gas of structureless molecules. In this 'elastic limit' the thermal conductivity can be written

$$\lambda = (\eta_{\text{el}}/m)(f_{\text{tr}}c_{\text{V tr}} + f_{\text{int}}c_{\text{int}}) \quad (1.200)$$

with $f_{\text{tr}} = \frac{5}{2}$ and $f_{\text{int}} = \rho D/\eta_{\text{el}}$. Eucken's original expression for the thermal conductivity of a polyatomic gas was of this same form but with f_{int} taken equal to unity. For most central force laws the value of $\rho D/\eta_{\text{el}}$ is approximately equal to 1·3. Neither value results in truly satisfactory estimates of the thermal conductivity. Agreement with theory is particularly poor at low temperatures. Finally, with $\Delta\varepsilon$ set equal to zero the relaxation time τ (and so also the coefficient of bulk viscosity) is infinite, i.e. in the elastic limit there is no mechanism for the collisional exchange of internal and translational energy.

To correct these obvious inadequacies of the first-order theory one must include the effects of inelastic collisions. However, it was Mason and Monchick's avowed intention to avoid entanglement in the dynamics of molecular encounters. To escape this dilemma they devised approximations which permitted all of the inelastic contributions to be expressed in terms of a single function of the cross-section, namely, the (rotational) relaxation time τ.. [When contributions from other internal degrees of freedom are included one must introduce the corresponding relaxation times.] This relaxation time can be treated as an adjustable parameter or it can be extracted from experiment or independent theoretical studies.

One approximation used by Mason and Monchick was the replacement of the integrand term $(\Delta\varepsilon)^2 \sin^2 \chi$ with its unweighted average $\frac{2}{3}(\Delta\varepsilon)^2$. From this it follows that $W \approx (\sum \int) \,d\tau\,[\gamma^4 \sin^2 \chi]$ and $X \approx (\sum \int) \,d\tau[4\gamma^4 \sin^2 \chi + (\frac{25}{6})(\Delta\varepsilon)^2]$. Then if the differences between $\bar{\sigma}$ and σ_{el} are ignored one finds that the viscosity is unaffected by inelastic collisions and that, in second approximation, $X = \frac{5}{2}(\mathbf{k}T/\eta_{\text{el}}) + (\frac{25}{12})(c_{\text{int}}/n\mathbf{k}\tau)$ and $Y = \frac{5}{4}(c_{\text{int}}/n\mathbf{k}\tau)$. The other simplifying assumption introduced by Mason and Monchick consisted of replacing the factor $(\gamma^2 - \beta\Delta\varepsilon)^{1/2}$ which appears in the integrand of Z with the first two terms of the expansion $\gamma - \frac{1}{2}\beta\Delta\varepsilon + \cdots$. The estimate of Z which corresponds to this is

$$Z = \tfrac{3}{2}(c_{\text{int}}T/\rho D) + \tfrac{3}{4}(c_{\text{int}}/n\mathbf{k}\tau) \tag{1.201}$$

From these results one finds that in second approximation $\eta = \eta_{\text{el}}$ and that (to first order in τ^{-1})

$$f_{\text{tr}} = \frac{5}{2}\left[1 - \frac{5}{6}\frac{c_{\text{int}}\eta}{p\mathbf{k}\tau}\left(1 - \frac{2}{5}\frac{\rho D}{\eta}\right)\right] \tag{1.202}$$

$$f_{\text{int}} = (\rho D/\eta)\left[1 + \frac{5}{4}\frac{\eta}{p\tau}\left(1 - \frac{2}{5}\frac{\rho D}{\eta}\right)\right] \tag{1.203}$$

with $p = n\mathbf{k}T$. Prerequisites for the use of these formulas are theoretical or experimental estimates of the characteristic relaxation times. Mason and Monchick have shown that when such data are available the thermal conductivities computed from equations (1.202) and (1.203) invariably agree better with experiment than those computed from the Eucken–Hirschfelder formulas which we have already cited. The qualitative and quantitative successes of this simple theory are truly remarkable. Furthermore, the same approach has been applied to mixtures of polyatomic gases[28,29] and satisfactory agreement between theory and experiment obtained in almost all cases. For details concerning the comparison of theory with experiment and for various modifications and embellishments on the theory itself the reader is referred to the original papers[27,28,29,38].

Now on the one hand we have described the progress which has been made toward the development of a rigorous quantum-scattering theory for rotating molecules. The other approach we have mentioned is the semiempirical procedure of Mason and Monchick which owes its simplicity to several not unreasonable but largely untested approximations and which is deserving of our attention primarily because of its considerable numerical success. This scheme was advanced on the faith that inelastic collisions would provide only minor corrections to the coefficients which characterize the macroscopic phenomenology of a polyatomic gas. We also have mentioned that the formal scattering theory and associated quantum-kinetic theory are best understood and most highly developed for cases where the interactions are nearly central. Therefore, both approaches seem best suited to situations where the probability of occurrence of grossly inelastic collisional events is small. This observation is not intended so much as an indictment of theoretical limitations but as a reminder that in most cases of interest these conditions of 'almost elastic' collisions are, in fact, satisfied. But if this is so— if the inelastic contributions to the cross-sections are truly small in comparison with elastic effects— then we should be searching for theories which capitalize on the fact rather than accommodate to it. For example, one might adopt a sort of 'weak-coupling' method whereby the inelastic part of $\partial_c f$ is expanded in powers of $\sigma_{inel}/\sigma_{el}$ (or some other more suitable 'small' quantity) and only a few terms of the series retained. Provided that the state of the system is not displaced too far from some well-defined reference state (such as that of local equilibrium) it is then possible to transform the inelastic part of $\partial_c f$ into a linear operator for which the classical limit is of the Lenard–Balesque or Fokker–Planck (differential) variety. In operators of this kind the rôle of the non-central interactions is to fix the numerical values of the friction coefficients or relaxation times associated with the internal (rotational) degrees of freedom. With regard to these coefficients the available options are precisely the same as those of the Mason–Monchick approach. Some very preliminary studies of this sort already have been conducted[39] and others are in progress. In a similar but somewhat more empirical vein are the investigations of Morse and his coworkers[40] who have constructed Krook equations (of the form $\partial_c f = -\sum_k (f - F_k)/\tau_k$) which include a characteristic time parameter and reference distribution function for each mode of relaxation. We mention these developments and emphasize the desirability of kinetic equations which are adapted to the specific conditions of almost-elastic collisional encounters. However, so little attention has been given to studies of this type that we shall not consider them at further length.

1.2.7 Classical Scattering Cross-sections

In the previous sections we have surveyed the present status of the rigorous and semiempirical approaches to the quantum-kinetic theory of rotating molecules. We have also drawn attention to weak-coupling methods for dealing with slightly inelastic collisions. The levels of activity and directions of current research assure us of swift and almost certain achievements in each of these directions. In particular, we can look forward with confidence to the development of a significantly improved and more tractible quantum theory of inelastic collisions. However, there is reason to expect that real progress toward an understanding of collisions between rotating molecules might come first from classical rather than from quantum mechanics. The reason for this is simple. Since a solution in closed form does not exist for the general problem of inelastic scattering, numerical methods must be used to solve the appropriate dynamical equations. The classical equations of motion lend themselves to this approach much more readily than do those of quantum mechanics. The difference in complexity is roughly that which distinguishes ray from wave optics.

The uncertainty principle does place limits on the validity of the classical analysis of collision problems. Violations of this principle occur for very small momentum transfers and/or small impact parameters, i.e. for cases where $|\mathbf{b}|\,|\Delta\mathbf{p}| \ll \frac{1}{2}\hbar$. In particular, the classical approach may lead to unrealistic predictions concerning small angle scattering. It would not appear to be too difficult to correct for this inadequacy since the conditions which favour forward scattering are precisely those to which quantum perturbation techniques are best adapted.

Up to now only three attempts have been made to solve inelastic scattering problems by integrating the appropriate classical equations of motion. In all three cases the problem studied was the collision between an atom and a rigid rotor. In most of their calculations Cross and Herschbach[41] used a potential of the form

$$\phi(x, \gamma) = 4\varepsilon[(\sigma/x)^{12} - (\sigma/x)^{6}][1 + \mathrm{a}\,P_2(\cos\gamma)] \qquad (1.204)$$

where $x = |\mathbf{x}|$ is the distance between the atom and the centre of mass of the rotor and where γ denotes the angle made by $\mathbf{x}$ and the axis of the rotor. The scattering angle $\chi = \sphericalangle(\mathbf{c}'_{21}, \mathbf{c}_{21})$ and the collisional alteration of molecular spin (tumbling angular momentum) were computed for a variety of impact parameters, relative velocities and initial spins.

Of the many conclusions which can be drawn from these studies we mention but a few : (i) the scattering angle exhibits a very weak dependence upon the non-sphericity of the interaction except for conditions specific to rainbow scattering and to orbiting of the colliding atom and the rotor ; (ii) the collisional transfer of spin is of the order of 0–10 $\hbar$ for a wide range of initial spins (0 to 100 $\hbar$); (iii) it is incorrect to treat the molecular orientation as fixed throughout the collision interval. That this comes as somewhat of a surprise and/or disappointment can be judged from the following. The major contributions to the value of χ are associated with the relatively small portion of the collision trajectory which lies near the distance of closest approach. During the short but decisive interval when the molecules are close to this separation they will scarcely be able to rotate at all. Thus, for the purpose of computing the scattering angle one might have expected that he would be able to associate a single fixed orientation with—and so assign an effective central potential to—each collisional event. By proceeding on this assumption Mason and Monchick[42] were able to fashion a very simple theory for the viscosity of polar gases. And it works! Apparently the out-of-plane scattering which is neglected by the approximation of fixed orientation has no great effect upon the cross-section for momentum transfer. (iv) Over a broad range of conditions the computed values of scattering angle and spin transfer can be reproduced with fair accuracy by a perturbation procedure which Cross and Herschbach dubbed the 'flywheel model'. According to this scheme one assumes that in first approximation the rotational motion of the molecule is collisionally unaltered. The next step is the computation of the atomic trajectory as it moves in the fluctuating field generated by the tumbling rotor. Then, to determine the change of spin one computes the change of relative orbital angular momentum from one's previously acquired knowledge of the atomic trajectory and invokes the conservation law for angular momentum.

This model is reminiscent in many ways of a procedure which was developed and tested by J. G. Parker[43]. To simplify the analysis of a collision between two rigid rotors Parker assumed that the non-central portion of the interaction could be treated as a perturbation and that the solutions of the classical equations of motion could be expanded in terms of this perturbation parameter. The equations of the first-order theory are those for a central interaction. The associated scattering events are elastic. The equations appropriate to the second-order theory are far more complex. However, Parker discovered conditions which greatly reduced this complexity and which permitted him to construct (almost exact) solutions without undue labour. These conditions included the assumption that the planes of rotation of the two rotors be coincident and that the potential of interaction be given by

$$\phi(\mathbf{x}, \gamma_1, \gamma_2) = \mathrm{A\,e}^{-\alpha x}\{1+\mathrm{a}P_2(\cos\gamma_1)\}\{1+\mathrm{a}\,P_2(\cos\gamma_2)\} - \mathrm{B\,e}^{-1/2\alpha x} \quad (1.205)$$

where γ_i is the angle between the axis of molecule i and the vector $\mathbf{x}$ which joins the molecular centres of mass. The rotational relaxation times which Parker got from his theory have since been found to agree tolerably well with experiment. He also applied this same method to the evaluation of vibrational relaxation times and met with a similar success. It should be possible to use this method for solving other problems of kinetic theory. We had hoped to be able to report on progress in this direction but at the time of writing our studies simply were not completed.

Very recently van der Ree[44] has performed calculations similar to those of Cross and Herschbach but for an off-centre potential of the form

$$\phi(x, \gamma) = 4\varepsilon[(\sigma/R)^{12} - (\sigma/R)^6] \quad (1.206)$$

with $\mathbf{R} = \mathbf{x} + \boldsymbol{\xi}$ and $\gamma = \measuredangle(\mathbf{x}, \boldsymbol{\xi})$. This function is intended to represent the interaction between an atom (^{4}He) and a diatomic molecule (HT) which is very nearly spherical but which has its centre of mass displaced a distance ξ from its centre of force, i.e. from its geometrical centre, the centre of its electron cloud. A very large number of collisions were examined. From these van der Ree was able to compute the contributions of non-sphericity to the cross-sections for diffusion and momentum transfer. Furthermore, he could analyse in fair detail the dependence of these cross-sections upon energy and spin but the inherent limitations of his numerical procedure did introduce considerable uncertainties.

These studies of Cross and Herschbach and of van der Ree have already done much to dispel the pall of ignorance surrounding scattering by non-central potentials. They will surely stimulate further research of a similar kind for, provided that one is willing and able (has a sufficiently large computing budget) to construct enough orbits, this method is capable of yielding results of great accuracy and detail. Furthermore, it is not limited in its applicability to interactions of any specific sort. However, it is a costly business and one where the computational effort is focused upon individual orbits rather than upon the observables of real interest, namely the cross-sections.

This encourages one to search for techniques which would permit the direct, but almost necessarily approximate evaluation of the cross-sections themselves. In particular, it would seem worthwhile to see if one can develop perturbation techniques, perhaps inspired by and patterned after those already known and tested in quantum-scattering theory. Some studies of this sort already have been reported[45] and others will surely appear. Furthermore, there are special models for which it is possible to reduce the task of calculating cross-sections to one of quadrature. The only one of these which has been studied so far is that of rigid loaded spheres, a model similar to that considered by van der Ree but with the off-centre 12–6 potential of equation (1.206) replaced with

$$\phi(x, \gamma) = \begin{Bmatrix} 0; R > \sigma \\ +\infty; \mathbf{R} \leqslant \sigma \end{Bmatrix} \tag{1.207}$$

Miles and Dahler[46] have reported numerical values of loaded-sphere cross-sections of $\sigma(\mathbf{c}|\mathbf{L}'| \rightarrow \mathbf{C}|\mathbf{L}|$ and $\sigma_{\text{tot}}(|\mathbf{L}'| \rightarrow |\mathbf{L}|)$ for excitation by Helium atoms of ground state ($\mathbf{L}' = 0$) HD and HT. They also derived an integral expression for the helicity ($\zeta = \mathbf{L}\,.\,\mathbf{c}/Lc$) dependent cross-section but have not yet published numerical results based upon this formula. Their methods can and will be applied to a few other models such as the 'loaded spherocylinder' which is discussed below.

1.2.8 The Rough Sphere, Loaded Sphere and Spherocylinder Models

Although the future of the kinetic theory for rotating molecules probably lies with one or more of the procedures already described in this section, few of these theories have been tested numerically or even developed to the point where such tests are practical. Indeed, among the theories mentioned only those of Mason and his coworkers have yielded numerical results of any real consequence. Despite their many successes one may feel the need for exercising caution when dealing with these theories for they do rest upon approximations which cannot be tested fully until accurate quantum-mechanical cross-sections become available. The situation is rather better if one accepts classical in place of quantum dynamics. The classical counterpart of the Mason–Monchick theory differs scarcely at all from the quantal version. But in view of the recent studies of Cross and Herschbach and of van der Ree the possibility of testing the adequacy of this simple theory is probably not so remote as it is in the quantum case. Furthermore, there do exist a few model systems —rough spheres, loaded spheres and spherocylinders—for which one can perform simple and exact (or almost exact) analyses of inelastic collisional events. With these solvable models one escapes the great complexities of classical scattering theory and is able to study the effects of inelastic collisions in great detail. The price one pays for this must be reckoned in terms of the limited realism of the models with which one deals. Thus, studies based upon these models should be viewed as diagnostic rather than representative of real fluids.

The derivation of the classical Boltzmann equation which we presented earlier relied upon approximations which were exclusively statistical, that is, the theory incorporated a rigorous treatment of the binary scattering process. In the cases of interest to us here the inelastic scattering events will be treated as if the colliding molecules were hard, non-deformable convex objects. The approximations involved in the kinetic theory of these particles are rather special. Thus, a collision between two rigid non-spherical objects may be 'simple' in the sense that the entire event consists of a single impulsive encounter, or it may be a complex 'chattering' event consisting of a sequence of highly correlated impulses. Since collisions of this second variety are so very difficult to analyse, it is the natural approximation simply to ignore their existence. The probability of occurrence of a chattering collision is dependent upon the degree of eccentricity of the non-spherical interactions and upon the values of various kinetmatic parameters such as masses, moments of inertia and characteristic frequencies of vibrational modes. The neglect of chatter will, of course, be a less severe approximation for some values of these variables than for others.

Once we agree to neglect chatter it becomes convenient to define the beginning and end of a collision as the times just prior to and immediately following the instant of contact between the surfaces of the two rigid bodies. We have described elsewhere[16] the method of proceeding from this point to the derivation of the Boltzmann equation for rigid non-spherical particles. Here we shall refer to these results rather than reproduce the detailed derivations.

Although the assumption of impulsive molecular interactions is certainly somewhat less than realistic, there is little reason to suspect that it will lead to grossly inaccurate conclusions about fluids composed of relatively small, weakly polar molecules. This is particularly true at temperatures for which the value of $\mathbf{k}T$ greatly exceeds the strength of the actual interactions.

1.2.9 Rough Spheres

We attend first to the Bryan–Pidduck model of perfectly rough spheres. Although it is the least realistic of the three we shall consider, this model has two points in its favour. Firstly, the relations between the pre- and post-collisional states are significantly less complicated for this than for the other models. This reduces to a minimum the difficulties attendant to the evaluation of the many-dimensional collision integrals which are involved in the various transport and relaxation coefficients. Furthermore, even the very slightest of grazing collisions between rough spheres is grossly inelastic. This is the second advantage of the model for it so drastically overemphasizes the importance of non-central forces that it permits us to put upper limits upon the contributions which stem from inelastic collisional events.

The dynamical state of a rough sphere is fully specified by the location of its centre of mass, $\mathbf{x}_i$, by its linear momentum, $\mathbf{p}_i = m\mathbf{c}_i$, and by its internal angular momentum, $\mathbf{L}_i = I\boldsymbol{\omega}_i$. Here m is the mass of a sphere, I its moment of inertia and $\boldsymbol{\omega}_i$ its rotational velocity relative to a space-fixed frame of reference. The internal (and spherically symmetrical) mass distribution characterized by I is conveniently represented by the dimensionless moment of inertia $\kappa = 4I/m\sigma^2$ where σ denotes the diameter of a sphere. The numerical value of κ ranges from zero, when the mass of the molecule is localized at its centre, to 2/3 when the mass is uniformly distributed over the surface of the sphere.

Let us now consider the changes of velocities which accompany a collisional encounter between two rough spheres. Such an event occurs when the distance between their centres is just equal to σ. Thus, if the unit vector $\mathbf{k}$ indicates the direction of the line of centres from molecule 1 to molecule 2, their precollision configuration can be represented as $(\mathbf{x}_1, \mathbf{c}_1, \boldsymbol{\omega}_1; \mathbf{x}_1 + \sigma\mathbf{k}, \mathbf{c}_2, \boldsymbol{\omega}_2)$ and the condition $\mathbf{k}\,.\,\mathbf{c}_{21} < 0$ will ensure that they are proceeding *toward* a collision. Just prior to this collision the relative velocity of the points of contact on the two spheres is $\mathbf{g}_{21} = \mathbf{c}_{21} + \frac{1}{2}\sigma\mathbf{k} \times (\boldsymbol{\omega}_1 + \boldsymbol{\omega}_2)$.

The collisional alterations of state must conform to the equations of impact

$$\Delta\mathbf{p}_1 \equiv \mathbf{p}_1{}^* - \mathbf{p}_1 = -(\mathbf{p}_2{}^* - \mathbf{p}_2) = \mathbf{K} \tag{1.208}$$

$$\Delta\mathbf{L}_1 \equiv \mathbf{L}_1{}^* - \mathbf{L}_1 = \mathbf{L}_2{}^* - \mathbf{L}_2 = \tfrac{1}{2}\sigma\mathbf{k} \times \mathbf{K} \tag{1.209}$$

and

$$\Delta E = -\mathbf{K} \cdot (\mathbf{g}_{21}{}^* + \mathbf{g}_{21}) = 0 \tag{1.210}$$

which are, in order of their presentation, statements of the conservation laws for the linear momentum, angular momentum and energy of the two colliding molecules. **K** is the collisional impulse imparted to molecule 2 as a result of its encounter with 1. The rough-sphere model is defined by the requirement that $\mathbf{g}_{21}{}^*$, the postcollisional value of the relative velocity, be the negative of $\mathbf{g}_{21}$. From this and from the equations of impact it follows that

$$\mathbf{K} = [m\kappa/(\kappa+1)]\left[\mathbf{g}_{21} + \frac{1}{\kappa}\mathbf{kk} \cdot \mathbf{c}_{21}\right] \tag{1.211}$$

and

$$\begin{aligned} \mathbf{c}_1{}^* &= \mathbf{c}_1 + \frac{\kappa}{\kappa+1}\left(\mathbf{g}_{21} + \frac{1}{\kappa}\mathbf{kk} \cdot \mathbf{c}_{21}\right); \qquad \boldsymbol{\omega}_1{}^* = \boldsymbol{\omega}_1 + \frac{2/\sigma}{\kappa+1}\mathbf{k} \times \mathbf{g}_{21} \\ \mathbf{c}_2{}^* &= \mathbf{c}_2 - \frac{\kappa}{\kappa+1}\left(\mathbf{g}_{21} + \frac{1}{\kappa}\mathbf{kk} \cdot \mathbf{c}_{21}\right); \qquad \boldsymbol{\omega}_2{}^* = \boldsymbol{\omega}_2 + \frac{2/\sigma}{\kappa+1}\mathbf{k} \times \mathbf{g}_{21} \end{aligned} \tag{1.212}$$

The rough sphere model is somewhat peculiar in that it is the internal mass distribution κ and not the degree of collisional inelasticity which is the adjustable parameter. There is no value of κ for which the collisions are truly inelastic. Thus, even when $\kappa \to 0$ and no transfer between the rotational and translational degrees of freedom can occur, the mechanism for collisional exchange of angular momentum still remains.

The collision terms of the Boltzmann equation for rough spheres can be written in either of the forms

$$\begin{aligned} \partial_c f(\bar{1}) &= \iiint d^2k\, d^3c_2\, d^3L_2 \sigma^2 \gamma(\mathbf{k} \cdot \mathbf{c}_{21})[f(\bar{1}')f(\bar{2}') - f(\bar{1})f(\bar{2})] \\ &= \iiint d1'\, d2'\, d2\, w(\overline{12}|\overline{1'2'})[f(\bar{1}')f(\bar{2}) - f(\bar{1})f(\bar{2})] \end{aligned} \tag{1.213}$$

with $d\bar{i} = d^3c_i\, d^3L_i$ and where

$$\begin{aligned} w(\overline{12}|\overline{1'2'}) = \sigma^2 \int d^2k \gamma(\mathbf{k} \cdot \mathbf{c}_{21}) \delta_3(\mathbf{c}_1{}^* - \mathbf{c}'_1) \delta_3(L_1{}^* - L'_1) \delta_3(\mathbf{c}_2{}^* - \mathbf{c}'_2) \\ \times \delta_3(L_2{}^* - L'_2) \end{aligned}$$

Finally, $\gamma(x)$ is the product of x with the unit step function. It is readily verified that $w(\overline{12}|\overline{1'2'}) = w(\hat{T}\hat{P}\bar{1}'\ \hat{T}\hat{P}\bar{2}'|\hat{T}\hat{P}\bar{1}\ \hat{T}\hat{P}\bar{2})$ where $\hat{T}\hat{P}\bar{i} = (\mathbf{c}_i, -L_i)$.

Furthermore, since $\mathbf{L}_1 - \mathbf{L}_2$ is collisionally invariant (cf. equation 1.212), it follows that the transition rate matrix and cross-section must be proportional to $\delta_3[(\mathbf{L}_1 - \mathbf{L}_2) - (\mathbf{L}'_1 - \mathbf{L}'_2)]$.

1.2.10 Loaded Spheres

Loaded spheres were first introduced into kinetic theory by Jeans[47] in an effort to estimate the rate of equilibration of rotational and translational degrees of freedom. This appears to have been the first serious attempt to develop a model which could account for some of the characteristic features of non-equilibrium processes in polyatomic fluids. The reason for the recent resurgence of interest in this model can be traced directly to the stimulating experiments performed by Becker and his coworkers[48], by de Vries, Haring and Slots[49], and by Schirdewahn, Klemm and Waldmann[50]. These experiments proved that thermal diffusive enrichment occurs in gas mixtures such as $^{16}O^{13}C^{16}O:{}^{17}O^{12}C^{16}O$, $^{18}O^{12}C:{}^{16}O^{14}C$ and $D_2:HT$. Now it is well known that the classical kinetic theory of Chapman and Enskog is capable of accounting for thermal diffusion in mixtures such as $^{19}Ne:{}^{20}Ne$ or $^{14}N^{16}O:{}^{18}O^{12}C$ by representing the molecules to be point masses which interact with one another through central force fields. In the first of these two cases the atomic species are chemically identical but differ in mass; in the second the masses are essentially equivalent but the forces between like and unlike molecules are quite different. It is clear that neither of these 'mechanisms' can be responsible for thermal diffusion in the CO_2, CO or hydrogen mixtures mentioned above. The species differ only very slightly in mass and the interactions among the different molecules are almost completely lacking in specificity; the mechanism underlying the separation must be closely connected with the rotational degrees of freedom of the molecules and with the single most dramatic distinction between the species involved —their different internal distributions of mass.

The loaded-sphere model (or the off-centre potential model as it is called by van der Ree) incorporates this distinction and exploits the fact that many of the molecules which concern us are themselves very nearly spherical. Thus, the force between two loaded spheres acts along a line $\boldsymbol{\rho}_{ij} = \boldsymbol{\rho}_i - \boldsymbol{\rho}_j$ which is not coincident with $\mathbf{x}_{ij} = \mathbf{x}_i - \mathbf{x}_j$, the vector which connects the centres of mass of the two interacting molecules. Here $\boldsymbol{\rho}_i$ is the radius vector of a point 'interior to molecule i' but not necessarily coincident with $\mathbf{x}_i$, the location of its centre of mass. Therefore, $\boldsymbol{\rho}_i = \mathbf{x}_i + \boldsymbol{\xi}_i$ where the length of $\boldsymbol{\xi} = \xi\mathbf{e}$, the vector which extends from the mass centre to the centre of force, is of the order of molecular dimensions. For example, let us suppose this model were applied to a mixture of the two equi-mass molecular species D_2 and HT. To an extremely high degree of accuracy the electronic wavefunctions for these isotopes of molecular hydrogen are identical. In the case of D_2 the centre of mass is coincident with the centre of symmetry of the electronic charge cloud but in the case of HT the centres of mass and of symmetry are displaced from one another by approximately one-fourth of the bond length, i.e. $\xi = 0$ for D_2 and $\xi = \frac{1}{4}$ (bond length) for HT. In a less pathological system such as a mixture of $^{16}O^{13}C^{16}O$ and $^{16}O^{12}C^{17}O$ the value of ξ for the unsymmetrical isotopic species is approximately equal to one-forty-fifth of the CO bond length.

Interactions of the type we have chosen do not give rise to torques directed along the molecular body axes. Therefore, $\mathbf{L}\,.\,\mathbf{e} = L_3 = \Gamma'\dot{\eta}_3$ is a collisional invariant. From this it might appear that L_3 would be totally devoid of physical significance and therefore could be set equal to zero. This is usually true. However, one might wish to associate a magnetic moment with L_3 in which case there would be great interest in the collisional realignment of this invariant component of spin. For simplicity we assume the internal distributions of mass to be symmetric about the molecular body axes $\mathbf{e} = \boldsymbol{\xi}/\xi$. The inertial tensors then will be of the form $\mathsf{I}(\mathbf{e}) = \Gamma(\boldsymbol{\delta} - \mathbf{ee}) + \Gamma'\mathbf{ee}$ and, regardless of the value of L_3, the orientation of the molecule will be fully specified by the direction of $\mathbf{e}$. This means that we can, if we so desire, dismiss from our considerations the 'third Euler angle' ψ which measures rotation about the body axis.

In general it is no easier to analyse the collisions of loaded spheres than those of molecular species which interact with some other non-central force. However, there is a special case, that of *rigid* loaded spheres, which is quite tractible. This model is defined by the choice $\varphi(\rho) = 0$, $\rho > \sigma$; $\varphi(\rho) = +\infty$, $\rho \leqslant \sigma$ for the potential which acts between the molecular force centres. In this special limit the collisions become impulsive events and the instantaneous alterations of linear and angular momentum are trivially easy to compute. Thus, from the conservation laws for energy and for linear and angular momentum it follows that

$$\begin{aligned}
\Delta\mathbf{p}_1 &= -\Delta\mathbf{p}_2 = \mathbf{k}K \\
\Delta\mathbf{L}_1 &\equiv \mathsf{I}_1\,.\,(\boldsymbol{\omega}_1{}^* - \boldsymbol{\omega}_1) = \boldsymbol{\xi}_1 \times \mathbf{k}K \\
\Delta\mathbf{L}_2 &= \mathsf{I}_2\,.\,(\boldsymbol{\omega}_2{}^* - \boldsymbol{\omega}_2) = -\boldsymbol{\xi}_2 \times \mathbf{k}K
\end{aligned} \tag{1.215}$$

where $K = m\mathbf{k}\,.\,\mathbf{g}_{21}[1+(M\xi^2/2\Gamma)\{(\mathbf{e}_1\times\mathbf{k})^2+(\mathbf{e}_2\times\mathbf{k})^2\}]^{-1}$ is the magnitude of the collisional impulse. Here $\mathbf{k}$ is the unit vector directed from the symmetry centre of molecule 1 to that of 2 at the moment of impact; $\mathbf{g}_{21} \equiv \mathbf{c}_{21}+\boldsymbol{\omega}_2\times\boldsymbol{\sigma}_2-\boldsymbol{\omega}_1\times\boldsymbol{\sigma}$, with $\boldsymbol{\sigma}_1 = \boldsymbol{\xi}_1+\frac{1}{2}\sigma\mathbf{k}$ and $\boldsymbol{\sigma}_2 = \boldsymbol{\xi}_2-\frac{1}{2}\sigma\mathbf{k}$, is the relative velocity of the points of contact on the two spheres. The component of this velocity which lies parallel to $\mathbf{k}$ is reversed upon collision.

From these equations it is readily verified that to each dynamical event

$$(\mathbf{x}, \tau_1{}^*; \mathbf{x}+\boldsymbol{\sigma}_1-\boldsymbol{\sigma}_2, \tau_2{}^*) \xrightarrow{\mathbf{k}\,.\,\mathbf{g}_{21}^* < 0} (\mathbf{x}, \tau_1; \mathbf{x}+\boldsymbol{\sigma}_1-\boldsymbol{\sigma}_2, \tau_2)$$

there corresponds a unique 'inverse collision',

$$(\mathbf{x}, \tau_1; \mathbf{x}+\tilde{\boldsymbol{\sigma}}_1-\tilde{\boldsymbol{\sigma}}_2, \tau_2) \xrightarrow{\mathbf{k}\,.\,\mathbf{g}_{21} > 0} (\mathbf{x}, \tau_1{}^*; \mathbf{x}+\tilde{\boldsymbol{\sigma}}_1-\tilde{\boldsymbol{\sigma}}_2, \tau_2{}^*)$$

Here $\tilde{\boldsymbol{\sigma}}_i = \boldsymbol{\sigma}_i(-\mathbf{k})$ and the symbol τ_i refers to the collection of dynamical variables $(\mathbf{c}_i, \mathbf{e}_i, \mathbf{L}_i)$.

It is usual when dealing with molecules of this sort (linear arrays of mass points) to select for coordinates the two polar spherical angles ϕ and θ which describe the orientation of the molecular axis $\mathbf{e}$. The remaining dynamical variables are then taken to be the conjugate momenta p_ϕ and p_θ. However, we prefer to use in place of these four variables the orientation vector $\mathbf{e}$ and the angular momentum $\mathbf{L}$. Of course, these are not independent variables, for since the rotational angular momentum is perpendicular to the axis of the rotor, $\mathbf{e}$ and $\mathbf{L}$ must satisfy the constraint $\mathbf{e}\,.\,\mathbf{L} = 0$. It is convenient to select as independent variables the three components of $\mathbf{L}$ in the laboratory frame and the phase angle φ, which fixes the location of the molecular axis in the plane perpendicular to $\mathbf{L}$.

Because $\mathbf{e}$ and $\mathbf{L}$ are not independent variables some care must be exercised with regard to the definitions of the transition rates and differential cross-sections: for details the reader is referred to Reference 46. The collisional rate of change of the phase-averaged distribution function $f(\bar{1}) = f(t\mathbf{x}\mathbf{c}_1\mathbf{L}_1)$ is given by

$$\begin{aligned}\partial_c f(\bar{1}) &= \iiint \frac{\mathrm{d}^2 e_1}{2\pi}\frac{\mathrm{d}^2 e_2}{2\pi}\,\mathrm{d}^2 S \int \mathrm{d}\bar{2}\,\delta(\mathbf{e}_1\,.\,\hat{\mathbf{L}}_1)\delta(\mathbf{e}_1\,.\,\hat{\mathbf{L}}_2) \\ &\quad \times[\gamma(\mathbf{k}\,.\,\mathbf{g})f(\bar{1}^*)f(\bar{2}^*)-\gamma(-\mathbf{k}\,.\,\mathbf{g})f(\bar{1})f(\bar{2})] \\ &= \int \frac{\mathrm{d}\varphi_1}{2\pi}\int \mathrm{d}^2 S \int \mathrm{d}\bar{2}[\gamma(\mathbf{k}\,.\,\mathbf{g})f(\bar{1}^*)f(\bar{2}^*)-\gamma(-\mathbf{k}\,.\,\mathbf{g})f(\bar{1})f(\bar{2})]\end{aligned} \tag{1.216}$$

with $d\bar{i} = d^3c_i\, d^3L_i/L_i$ and $d^2S = \sigma^2\, d^2k$ and where it is to be understood that in the second line of (1.216) $\mathbf{e}_1$ and $\mathbf{e}_2$ are perpendicular to $\mathbf{L}_1$ and $\mathbf{L}_2$, respectively.

1.2.11 Loaded Spherocylinders

It was Curtiss[51] who first demonstrated the practicality of a kinetic theory involving collisions between rigid and smooth convex bodies. In a recent paper by Hoffman[52] the computational problems associated with this model are systematized to the point where it is now feasible to treat a considerable variety of convex shapes. However, the only shape for which calculations are presently available is the spherocylinder, a smooth right-circular cylinder of length L with hemispherical caps of diameter σ. The mass of the molecule is assumed to be distributed along the cylinder axis and the centre of mass is displaced by a distance ξ from the centre of the cylinder. This model is kinematically identical with that of loaded spheres.

The collisional contacts between two spherocylinders fall into three categories; cap–cap, cap–cylinder and cylinder–cylinder encounters. These three situations are characterized uniquely by the associated values for the pair of vectors $\boldsymbol{\zeta}_1$ and $\boldsymbol{\zeta}_2$ which extend from the mass centres of the two molecules to their point of mutual contact. The Boltzmann equation for this model is given by (1.216) provided that d^2S is identified with the differential element of surface swept by the centre of molecule 2 as it slides over molecule 1 in such a way that the orientations of both remain fixed. The collisional alterations of velocity and angular momentum are given by the formulas

$$\begin{aligned} \mathbf{c}_i^* - \mathbf{c}_i &= \mathbf{k}_i(K/m) \\ \mathbf{L}_i^* - \mathbf{L}_i &= \boldsymbol{\zeta}_i \times \mathbf{k}_i K \end{aligned} \tag{1.217}$$

with $K = m(\mathbf{k} \cdot \mathbf{g})[1 + (m/2\Gamma)\{(\boldsymbol{\zeta}_1 \times \mathbf{k})^2 + (\boldsymbol{\zeta}_2 \times \mathbf{k})^2\}]^{-1}$ and where $\mathbf{k}_i$ is the outward directed unit normal to the surface of molecule i at the point of its common tangency with its collision partner. Finally, $\mathbf{k} = \mathbf{k}_1$ and $\mathbf{g} = \mathbf{c}_{21} + [\boldsymbol{\zeta}_1 \times \boldsymbol{\omega}_1 - \boldsymbol{\zeta}_2 \times \boldsymbol{\omega}_2]$.

In discussing the rough-sphere, loaded-sphere and spherocylinder models we have confined our attention to a single-component fluid. It is a trivial matter to generalize our results to particles with different sizes, shapes, masses and moments of inertia. Furthermore, it is tedious but not very difficult to extend the theory to rough spheres with non-isotropic distributions of mass and to loaded spheres and spherocylinders with mass distributions which are not axially symmetric. Of course, it would be extremely interesting to compare the cross-sections for our three 'basic models' with the results of Cross and Herschbach and of van der Ree. We had hoped to present such comparisons here but at the time of writing our results were incomplete and included only the two examples of Miles and Dahler cited earlier.

1.2.12 Electric and Magnetic Dipole Moments

The effects of interaction between individual molecules and external electromagnetic fields may be categorized either as level shifts (of the molecular states) or as the absorption and emission of electromagnetic radiation. Our attention will be confined to phenomena which involve electric and magnetic dipole moments of electrically neutral molecules. The methods we shall use in dealing with these phenomena will be heuristic instead of rigorous and the topics with which we shall concern ourselves will be representative rather than exhaustive.

The electric and magnetic multipoles associated with molecular distributions of charge and current are classical constructs. In a quantum-mechanical theory they are represented by operator functions of the rotational angular momentum. On the other hand, the magnetic moments which owe their existence to electron and nuclear spins can be fully understood only in a quantum-mechanical context. However, there are times when these spin moments can be treated as functions of classical dynamical variables. The molecular dipole moment will be denoted by the symbol $\boldsymbol{\mu}$ and the corresponding quantal operator by the symbol $\hat{\boldsymbol{\mu}}$. The strength of the external (electric or magnetic) field with which this dipole interacts is denoted by $\mathbf{F} = \mathbf{F}(t\mathbf{x})$. Therefore, the interaction energy is given by $H^{e} = -\boldsymbol{\mu} \cdot \mathbf{F}$ and the corresponding quantal operator is $\hat{H}^{e} = -\hat{\boldsymbol{\mu}} \cdot \mathbf{F}$. We shall restrict our studies to fields which are limited by the conditions $\beta|\mathbf{F}| = |\mathbf{F}|/\mathbf{k}T \ll 1$. This often is called the 'high temperature limit'. Frequently the fields are very nearly homogeneous so that one can exclude from consideration the very interesting dielectrophoretic force $\mathbf{F}^{e} = \boldsymbol{\mu} \cdot \nabla \mathbf{E}$ and its magnetic analogue.

The 'level shifts' to which we have alluded can be indicated in the manner $\varepsilon \to \varepsilon + H^{e}$ or $\hat{H}^{(1)} \to \hat{H}^{(1)} + \hat{H}^{e}$. If the molecular collision processes are *not* accompanied by absorption or emission then the only consequences of the field's presence will be these level shifts. Because of them the factor

$\delta(E_1+E_2-E'_1-E'_2)$ which occurs in the field-free transition rate $w(12|1'2')$ must be replaced with $\delta(E_1{}^{\mathrm{F}}+E_2{}^{\mathrm{F}}-E_{1'}{}^{\mathrm{F}}-E_{2'}{}^{\mathrm{F}})$ wherein $E_i^{\mathrm{F}} = E_i-\boldsymbol{\mu}_i\,.\,\mathbf{F}$. As we shall see this modification alone is sufficient for the theory of non-resonant absorption and line shapes. Indeed, it is only for very rare circumstances that one must consider absorption and/or emission by the 'collision complex': despite the intrinsic interest of these rare events we shall consider them no further.

By way of illustration let us examine the effect of level shifts upon the collision dynamics of rough spheres. The collisional alterations of linear and angular momentum are given by the equations of impact, (1.208) and (1.209), respectively. According to these equations the molecules do not exchange linear or angular momentum with the field. Furthermore, we assume there to be no exchange of energy with the field and so require that $\Delta(E_1{}^{\mathrm{F}}+E_2{}^{\mathrm{F}}) = E_1{}^{\mathrm{F}}+E_2{}^{\mathrm{F}}-E_{1*}{}^{\mathrm{F}}-E_{2*}{}^{\mathrm{F}} = 0$. To simplify the analysis we assume that $\boldsymbol{\mu} = \bar{\gamma}\mathbf{L}(\mathbf{L} = I\boldsymbol{\omega})$ where $\bar{\gamma}$ is a numerical constant. It then follows that $\mathbf{K}\,.\,(\mathbf{g}'_{21}+\mathbf{g}_{21}+2\sigma\bar{\gamma}\mathbf{F}\times\mathbf{k}) = 0$ where $\mathbf{g}_{21}$ is the relative velocity of the points of contact on the surfaces of the two colliding spheres. From the many solutions of this equation we select

$$\mathbf{K} = m\frac{\kappa}{1+\kappa}\left(\mathbf{g}_{21}+\frac{1}{\kappa}\mathbf{kk}\,.\,\mathbf{c}_{21}+\sigma\bar{\gamma}\mathbf{F}\times\mathbf{k}\right) \tag{1.218}$$

as the simplest non-trivial extension of the usual rough-sphere model. When either or both of $\mathbf{F}$ and $\bar{\gamma}$ tends to zero this formula for $\mathbf{K}$ reduces to that for ordinary rough spheres and the connection $\mathbf{g}_{21}{}^*+\mathbf{g}_{21}+2\sigma\bar{\gamma}\mathbf{F}\times\mathbf{k} = 0$ reduces to the expression $\mathbf{g}_{21}{}^* = -\mathbf{g}_{21}$ which is the accepted definition for the rough-sphere model.

In order that a molecule experience a torque it must exchange angular momentum with the field, that is, it must absorb or emit field quanta. If the frequency of the field is not resonant with any of the frequencies associated with transitions between molecular states then there will be no torque. Of course, in the classical limit the spectrum of internal (rotational) states becomes continuous and so the field always gives rise to a torque, $\mathbf{N} = \boldsymbol{\mu}\times\mathbf{F}$. The operator analogue of this torque is $\hat{\mathbf{N}} = \hat{\boldsymbol{\mu}}\times\mathbf{F}$. Now according to the arguments presented in the preceding section it is $\bar{\boldsymbol{\mu}}$, the phase-average of the dipole moment, which should appear in the kinetic equation (1.152) (see equation 1.160 as well) for the phase-averaged function $f(t\mathbf{xcL})$. The remainder of this section is devoted to the description of how one forms rough estimates of these average moments.

We focus first upon an electric dipole moment of magnitude μ_0. This moment can be written $\boldsymbol{\mu} = \mu_0 \mathbf{e}$ where $\mathbf{e} = \mathbf{x} \sin\theta \cos\phi + \mathbf{y} \sin\theta \sin\phi + \mathbf{z} \cos\theta$ is a unit vector fixed to the molecular frame of reference. In the case of a diatomic molecule the direction of $\mathbf{e}$ coincides with that of the internuclear axis. If this diatomic molecule is in a $^1\Sigma$ state, then between collisions—and in the absence of an external field— its orientation will change in accordance with the relationship

$$\mathbf{e}(t) = \mathbf{e}(0) \sin \omega_r t + \mathbf{l} \times \mathbf{e}(0) \cos \omega_r t \tag{1.219}$$

Here $\omega_r = |\mathbf{L}|/\Gamma$ and $\mathbf{l} = \mathbf{L}/L$ is the direction of the (constant) rotational angular momentum. The time average of this dipole moment will obviously vanish provided that it is permissible to assume that the dipole rotates freely between collisions. In fact, the field does modify the motion of the dipole and generates an average moment which, in first approximation, is proportional to the field strength. This 'second-order Stark effect' gives rise to a torque which depends quadratically upon $\mathbf{E}$. A classical treatment of this effect and of its bearing upon some aspects of gas kinetic theory has been presented by Borman, Nikolaev and Nikolaev[53]. Under most circumstances one can ignore these second-order effects and so assume that $^1\Sigma$ states do not respond to the imposition of electric fields. However, the Stark effect is the key to understanding certain optical phenomena such as flow birefringence and Raleigh scattering, topics which we consider in the following section. To prepare for these applications let us express the induced dipole moment of a linear non-polar molecule in the form $\boldsymbol{\mu} = \mathsf{a} \,.\, \mathbf{E}$ where the polarizability tensor a is related to $\mathbf{e}$ $(=\mathbf{e}(0) \sin \varphi + \hat{l} \times \mathbf{e}(0) \cos \varphi)$ of equation (1.219) by

$$\mathsf{a} = \alpha_{\|} \mathbf{ee} + \alpha_{\perp} (\boldsymbol{\delta} - \mathbf{ee})$$

and where $\alpha_{\|}$ and $\alpha_{\perp}$ are the component polarizabilities parallel and perpendicular to the axis of symmetry. The phase average of a is

$$\overline{\mathsf{a}} = \frac{1}{2\pi} \int_0^{2\pi} \mathrm{d}\varphi \mathsf{a} = \tfrac{1}{2} \alpha_{\|} (\boldsymbol{\delta} - \hat{l}\hat{l}) + {}^{10}_{\ 2} \alpha_{\perp} (\boldsymbol{\delta} + \hat{l}\hat{l}) \tag{1.220}$$

and so the electric displacement vector can be written in the form

$$\mathbf{D} = \mathbf{E} + 4\pi \mathbf{P} = \mathbf{E} + 4\pi n \langle \overline{\mathsf{a}} \rangle \,.\, \mathbf{E} = \boldsymbol{\varepsilon} \,.\, \mathbf{E} \tag{1.221}$$

with $\boldsymbol{\varepsilon} = \varepsilon \boldsymbol{\delta} + \overset{\circ}{\overline{\overline{\mathsf{e}}}}$ and where $\varepsilon = 1 + \tfrac{4}{3} \pi n (2\alpha_{\perp} + \alpha_{\|})$ and

$$\overset{\circ}{\overline{\overline{\mathsf{e}}}} = -2\pi n (\alpha_{\|} - \alpha_{\perp}) \, \langle \overset{\circ}{\overline{\overline{\hat{l}\hat{l}}}} \rangle \tag{1.222}$$

respectively, are the dielectric constant and the anisotropic portion of the electric permeability tensor. We shall have use for these results in the next section.

If the molecule in question is a symmetric (or spherical) top and if the direction of **e** coincides with a principal axis of the internal mass distribution, then

$$\mathbf{e}(t) = [\mathbf{ll}+(\boldsymbol{\delta}-\mathbf{ll})\sin\omega_r t+\boldsymbol{\delta}\times\mathbf{l}\cos\omega_r t]\,.\,\mathbf{e}(0) \tag{1.223}$$

(The 'free motion' of **e** is only slightly more complicated for a general rigid body.) The average dipole-moment for the symmetric top is then

$$\bar{\boldsymbol{\mu}} = \text{avg}\,(\mu_0\mathbf{e}) = \mu_0\mathbf{e}(0)\,.\,\mathbf{ll} = \mu_0 K\mathbf{L}/L^2 \tag{1.224}$$

where $K = \mathbf{L}\,.\,\mathbf{e}$ is the component of rotational angular momentum along the principal axis of the molecular mass distribution. The quantum analogue of this moment is the operator $\hat{\bar{\boldsymbol{\mu}}} = \mu_0\hbar^{-1}K\hat{\mathbf{L}}/L(L+1)$ where $\hbar K$ and $\hbar^2 L(L+1)$ are eigenvalues of $\hat{\mathbf{L}}\,.\,\mathbf{e}$ and of $\hat{L}^2$, respectively. In the correspondency limit of large quantum numbers (e.g., $L \gg 1$) the operator $\hat{\mathbf{L}}$ is to be replaced with the classical variable $\mathbf{L}$ and $\hat{\bar{\boldsymbol{\mu}}}$ with $\bar{\gamma}(K, L)\mathbf{L}$ where by $\bar{\gamma}(K, L) = \mu_0 K/L^2$ we denote the classical gyroelectric ratio. Whether our appeal is to classical or quantum mechanics it is important to remember that the introduction of an averaged dipole moment $\bar{\boldsymbol{\mu}}$ implies a frequency of tumbling which greatly exceeds the frequency of inelastic collisional events. No attempt will be made to analyse cases for which this approximation is invalid.

The dipole moment of a symmetric top has a component directed along the axis of rotation. To the neglect of the quadratic Stark effect the value of this component will be invariant unless the particle interactions depend upon the 'third Euler angles' ψ associated with rotations about the symmetry axes of the two colliding molecules. For example, the values of $K = \mathbf{L}\,.\,\mathbf{e}$ are constant for the spherocylinder and loaded-sphere models which we have discussed previously. The situation is very much the same for a heteronuclear diatomic molecule provided that its state is one (such as a $^1\Pi$ or $^1\Delta$) for which there is a component of electronic orbital angular momentum directed along the internuclear axis. The average moment operator then is given (at least approximately) by the formula $\hat{\bar{\boldsymbol{\mu}}} = \mu_0\hbar^{-1}\Lambda\hat{\mathbf{J}}/J(J+1)$ wherein $\hat{\mathbf{J}} = \hat{\mathbf{L}}+\hat{\Lambda}\mathbf{e}$ denotes the total angular momentum of the molecule and $\Lambda\hbar$ the component of electronic orbital angular momentum along the internuclear axis. Even at rather moderate temperatures the most probable value for the rotational quantum number L will generally exceed that of Λ by quite a margin.

Therefore, it is often permissible to replace the operator $\hat{\bar{\boldsymbol{\mu}}}$ with its classical counterpart $\bar{\boldsymbol{\mu}} = \bar{\gamma}\mathbf{L}$ where $\bar{\gamma}(\Lambda, L) \equiv \mu_0\hbar\Lambda/L^2$.

The electric dipole moment is a measurable property of the charge distribution associated with some particular electronic state of a molecule. We have ignored the dependence of this charge distribution upon vibrational and rotational motions although the latter obviously has a profound effect upon the *average* of $\boldsymbol{\mu}$, that is, upon the value of $\bar{\boldsymbol{\mu}}$. Our approach has been to accept these moments as intrinsic molecular properties. Likewise, we recognize that the sources of molecular magnetism are circulating charges and/or electronic and nuclear spins but assume that the values of the various gyromagnetic ratios are among our input data. The total magnetic moment of a molecule is the sum of the moments due to the electronic orbital angular momentum, $\hat{\boldsymbol{\Lambda}}$, the electron spin, $\hat{\mathbf{S}}$, the nuclear spin $\hat{\mathbf{I}}$, and $\hat{\mathbf{L}}$, the rotational angular momentum due to the tumbling of the molecule. Thus $\hat{\boldsymbol{\mu}} = \gamma_{er}\hat{\boldsymbol{\Lambda}} + \gamma_{es}\hat{\mathbf{S}} + \gamma_{ns}\hat{\mathbf{I}} + \gamma_{nr}\hat{\mathbf{L}}$ where the individual coefficients of proportionality, the gyromagnetic ratios, are taken as known. Now when an electrical current I flows about a closed circular loop of area A, it generates a magnetic moment whose magnitude is given by IA/c (c = velocity of light) and whose orientation is normal to the plane of the loop. If this current is caused by the motion of a single particle of mass m and charge e, then the magnitude of the associated moment is given by $\mu = I(\pi a^2/c) = e(2\pi a/v)^{-1}(\pi a^2/c) = (e/2mc)ap = (e/2mc)L$. Here a is the radius of the circular orbit and $p = mv$ is the particle momentum. This simple and familiar argument establishes that the gyromagnetic ratio $\gamma = e/2mc$ for circulating charges should be inversely proportional to the mass of the charge carriers. Furthermore, it leads us to expect that $\gamma_{er} = (e\hbar/2m_e c)\hbar^{-1}g_{er}$ where m_e and e are the mass and charge of an electron and where the g-factor is very nearly equal to unity. The intrinsic spin angular momentum of the electron has associated with it a magnetic moment of this same order of magnitude. In particular, $\gamma_{es} = \mu_B g_{es}/\hbar$ where $\mu_B = e\hbar/2m_e c$ is the Bohr magneton and where $g_{es} \approx -2$ for an individual electron.

This line of reasoning suggests that $\hbar\gamma_{nr}$ should be of the order of the nuclear magneton $\mu_N = e\hbar/2m_p c$ where m_p is the mass of a proton. The magnetic moment associated with rotation of the positively charged molecular nuclei is offset by the oppositely directed moment generated by the concomitant rotation of the electronic charge cloud. It is quite difficult to construct accurate estimates of the rotational g-factors $g_{nr} = \hbar\gamma_{nr}/\mu_N$ since the balance between these two contributions to $\gamma_{nr}\mathbf{L}$

is delicately controlled by the coupling between the electronic and nuclear motions.

In analogy with the case of an electron the gyromagnetic ratio for the intrinsic nuclear spin is of the same order of magnitude as that for nuclear rotation. Hence, $\gamma_{ns} = \mu_N g_{ns}/\hbar$ where the factor g_{ns} is of the order of unity.

Strictly speaking not one of the separate angular momenta $\hat{\mathbf{\Lambda}}$, $\hat{\mathbf{S}}$, $\hat{\mathbf{I}}$ and $\hat{\mathbf{L}}$ is an eigenoperator of the isolated molecule. Only the total angular momentum $\hat{\mathbf{J}} = \hat{\mathbf{\Lambda}} + \hat{\mathbf{S}} + \hat{\mathbf{I}} + \hat{\mathbf{L}}$ is an integral of the motion. Therefore, we should like to determine the relationship between the magnetic moment and this total angular momentum. Since this connection must be linear, the problem amounts to that of determining the effective gyromagnetic ratio which occurs in the proportionality $\hat{\boldsymbol{\mu}} = \bar{\gamma}\hat{\mathbf{J}}$. In the event that all but one of the contributing moments is zero this can be accomplished trivially. In all other cases one must attempt to calculate $\bar{\gamma}$. It is usual to select some 'coupling scheme', invoke the Wigner–Eckert theorem, and compute the corresponding gyromagnetic ratio. The success of this programme depends upon how well one is able to guess the relative strengths of the coupling among the various angular momenta and upon whether all of these couplings are weak enough to justify the use of first-order perturbation theory. Here it is assumed that both these criteria are satisfied.

By way of example let us suppose that the nuclei of the molecule with which we are dealing have no spin and that the electronic state of the molecule is one for which the electronic orbital angular momentum is zero. Therefore, $\hat{\mathbf{J}} = \hat{\mathbf{L}} + \hat{\mathbf{S}}$ and $\hat{\boldsymbol{\mu}} = \gamma_{nr}\hat{\mathbf{L}} + \gamma_{es}\hat{\mathbf{S}}$. Classically, one describes the motion of the free molecule as involving the nutation of $\mathbf{L}$ and $\mathbf{S}$ about the fixed direction of $\mathbf{J}$. Thus, we write $\mathbf{S} = \mathbf{S}\,.\,\mathbf{JJ}/J^2 + \mathbf{S}_\perp$ and $\mathbf{L} = \mathbf{L}\,.\,\mathbf{JJ}/J^2 + \mathbf{L}_\perp$ where $\mathbf{S}_\perp$ and $\mathbf{L}_\perp$ are the components of $\mathbf{S}$ and $\mathbf{L}$ which are perpendicular to $\mathbf{J}$ and which swing with a nutational frequency of the order of $\omega_r = |\mathbf{L}|/\Gamma \approx 10^{12}$ or 10^{13} cps. The averages of $\mathbf{S}$ and $\mathbf{L}$ over a time which is large in comparison with ω_r^{-1} are given by $\bar{\mathbf{S}} = \overline{\mathbf{S}.\mathbf{J}}\mathbf{J}/J^2 = \mathbf{J}(J^2 + S^2 - L^2)/2J^2$ and $\bar{\mathbf{L}} = \overline{\mathbf{L}.\mathbf{J}}\mathbf{J}/J^2 = J(J^2 - S^2 + L^2)/2J^2$. If it is now assumed that the operators $\hat{\mathbf{L}}^2$ and $\hat{\mathbf{S}}^2$ are at least approximately constants of the motion, then to first order we obtain the familiar Landé result

$$\begin{aligned}\hat{\boldsymbol{\mu}} &= \hat{\mathbf{J}}\frac{1}{2J(J+1)}[\gamma_{nr}\{J(J+1) + L(L+1) - S(S+1)\} \\ &\quad + \gamma_{es}\{J(J+1) + S(S+1) - L(L+1)\}] \\ &\equiv \gamma(JLS)\hat{\mathbf{J}}\end{aligned}$$

This same formula can also be obtained by application of the Wigner–Eckert theorem. Since $\gamma_{nr}/\gamma_{es} \approx m_e/m_p$ is of the order of 10^{-3} we can neglect the first of the two contributions to $\gamma(JLS)$ in comparison with the second.

To be specific let us now consider the $^3\Sigma$ ground state of molecular oxygen. The quantum number L must assume one of the three values $J-1$, J or $J+1$ and so for each choice of J there are the three values of the magnetic moment,

$$\hat{\boldsymbol{\mu}} = 2\mu_B \hbar^{-1} \left\{ \begin{array}{ll} -\hat{\mathbf{J}}/J; & m_s = +1 \\ -\hat{\mathbf{J}}/J(J+1); & m_s = 0 \\ +\hat{\mathbf{J}}/(J+1); & m_s = -1 \end{array} \right\} \tag{1.225}$$

corresponding to the three allowable values for the projection of the molecular electron spin in the direction of the total angular momentum. If one adopts a quantal description of the gas no further simplification of equation (1.225) is required. However, the conditions may permit one to adopt a (semi) classical approach. At an absolute temperature T the mean value of the quantum number L is given by $L_{avg} \approx \hbar^{-1}(2\mathbf{k}T\Gamma)^{1/2}$. For oxygen at 0°C this number is approximately equal to twelve. Therefore, the replacement of $\hat{\mathbf{J}}$ with $\hat{\mathbf{L}}$ only introduces an error of about 8%. If we are willing to accept this error—in addition to that which is implicit in our reliance upon a first-order perturbative treatment of the couplings among the various angular moments—we can then replace equation (1.225) with $\boldsymbol{\mu} = \gamma(L, m_s)\mathbf{L}$ where $\mathbf{L}$ is the classical rotational angular momentum and where

$$\gamma(L, m_s) = -2\mu_B m_s/L - (2\mu_B \hbar/L^2)\delta_{m_s 0}; \qquad m_s = 0, \pm 1 \tag{1.226}$$

This last result shows how it is possible to represent a magnetic moment which originates from the intrinsic electron spin in terms of the classical dynamical variable associated with the tumbling of the molecule. Of course, in this somewhat special case of a molecular triplet one must append to each classical dynamical state ($\mathbf{pL}$) an additional lable m_s. For particle interactions which do not alter the values of this spin projection, it is rather as though the fluid were composed of three immutable species.

We have examined this one case in considerable detail. Others can be handled with no more difficulty. For example, the magnetic moments associated with the low-lying Π states of the NO molecule can be written as

$$\hat{\boldsymbol{\mu}} = \gamma_{\rm nr}\hat{\mathbf{L}} + \gamma_{\rm es}\hat{\mathbf{S}} + \gamma_{\rm er}\hat{\boldsymbol{\Lambda}} = \gamma_{\rm nr}(\hat{\mathbf{J}} - \hat{\boldsymbol{\Omega}}) + \gamma_{\rm es}\hat{\mathbf{S}} + \gamma_{\rm er}\hat{\boldsymbol{\Lambda}} \tag{1.227}$$

where $\hat{\boldsymbol{\Omega}} = \hat{\boldsymbol{\Lambda}} + \hat{\mathbf{S}}$ is the total electronic angular momentum. Because of the axially directed magnetic field generated by the orbiting of the electrons, the spin is strongly coupled to the internuclear axis. If we assume (Hund's case *a*) that this coupling is much stronger than that between either of $\hat{\mathbf{S}}$ or $\hat{\boldsymbol{\Lambda}}$ and the rotational momentum $\hat{\mathbf{L}}$, then in place of equation (1.227) we obtain

$$\hat{\boldsymbol{\mu}} = \hat{\mathbf{J}}\frac{1}{J(J+1)}[\gamma_{\rm nr}\{J(J+1) - \Omega^2\} + \Omega(\gamma_{\rm er}\Lambda + \gamma_{\rm es}\Sigma)] \tag{1.228}$$

Here, Ω, Λ and S are the quantum numbers associated with the projections of $\hat{\boldsymbol{\Omega}}$, $\hat{\boldsymbol{\Lambda}}$, $\hat{\mathbf{S}}$ along the internuclear axis.

The value of Ω is zero for the $^2\Pi_{1/2}$ ground state of NO; it is equal to unity for the first excited, $^2\Pi_{3/2}$ state. In the classical limit ($\hat{\mathbf{J}} \to \mathbf{L}$) the gyromagnetic ratios of these two states are $\mu_{\rm N}\hbar^{-1}g_{\rm nr}$ and $3\mu_{\rm B}\hbar/L^2$, respectively. At room temperature the mean value of the latter is approximately $45/g_{\rm nr}$ as great as that of the former. The magnetic moment $\boldsymbol{\mu} = (\mu_{\rm N}\hbar^{-1}g_{\rm nr})\mathbf{L}$ of the $^2\Pi_{1/2}$ ground state of NO is the same as that for other diamagnetic species such as N_2.

It is important to recognize the implications* of using one of these formulas for the effective magnetic moment. The arguments used in establishing a relationship between the magnetic moment and the total momentum are valid only if the effective moment $\boldsymbol{\mu} = \bar{\gamma}\hat{\mathbf{J}}$ is interpreted as the average over very many periods of the associated (classical) nutational motion. If collisions are so prevalent that this nutation is frequently interrupted, a more elaborate theory of the magnetic moments must be sought. Although we may wish to apply our kinetic theory to conditions which exceed these limits, we nevertheless assume that the effective gyromagnetic ratios are independent of collision frequency.

1.3 APPLICATIONS OF THE THEORY

In the previous sections we have tried to assemble some of the most important elements of the statistical mechanics of irreversible processes in polyatomic fluids. Also, we have included in those sections brief

* See appendix 1.

accounts concerning items such as scattering cross-sections, intermolecular forces and molecular dipole moments, all of which must be available if the statistical theory is to be truly useful. In this the final portion of the chapter we shall put the theory to work. Of course, there are far more applications of the theory than we can hope to consider here and even those topics which we do examine must, of necessity, be dealt with briefly and in little depth.

1.3.1 Chapman–Enskog Method and Transport Coefficients

Among the most important applications is that of relating the phenomenological coefficients of macroscopic fluid mechanics to the kinematic parameters and scattering cross-sections descriptive of the molecular constituents of the fluid. In many cases molecular internal degrees of freedom do not have profound effects upon the values of these transport coefficients or upon their temperature dependence. For this reason interest in the theoretical prediction of gas transport properties was flagging until some six or seven years ago when a vast assortment of new transport experiments was begun. The common feature of all these experiments has been that the measurements are performed in the presence of external electric or magnetic fields. In 1930 Senftleben had observed that the thermal conductivity of a paramagnetic gas (oxygen) was diminished by the imposition of a field and in the early nineteen sixties Beenakker and his colleagues[54] commenced a reinvestigation of this phenomenon as well as others. The Leiden group soon established that the field effects observed by Senftleben were not restricted to paramagnetic species but could be measured in diamagnetic gases as well. In the past few years their studies and those of groups in the Soviet Union, the Netherlands, the United States and elsewhere have been extended to the measurement and theory of the large number of phenomenological coefficients associated with transport and relaxation in gases rendered anisotropic by the imposition of external electric or magnetic fields.

Although one should not discount the intrinsic interest[55] of these exotic phenomena, it is of even greater importance that they invariably depend quite sensitively upon kinematic properties of the molecules associated exclusively with internal degrees of freedom (moments of inertia, magnetic or electric moments, rotational *g*-factors, polarizabilities) and with the non-central parts of the intermolecular forces. Thus, there are close connexions between these external field effects and the inelastic contributions to scattering cross-sections. Not only does this mean that the experiments can be used to help judge the adequacy of various collision models but that they also provide means for testing the kinetic theory of rotating molecules with much greater sensitivity and subtlety than is possible with the more standard transport and relaxation experiments.

With these considerations in mind we turn now to the application of the theory of Section 1.2 to the evaluation of gas transport coefficients. Our task is that of constructing a solution of the Boltzmann equation which corresponds to the conditions under which transport experiments are performed. Transport coefficients such as the viscosity and thermal conductivity are extracted from fluid mechanical experiments with characteristic scales of time which are very much greater than those connected with molecular events. Furthermore, except in the vicinity of a shock front or near a fluid interface the conditions of these experiments are such that properties such as fluid density, velocity and temperature are well-defined local variables which vary only slightly from point to point. Thus, in fluid mechanics the length $L_{\text{hy}} = P/|\nabla P|$, which characterizes the rate of variation of a local field $P = P(t, \mathbf{x})$, is invariably much greater than R the range of the intermolecular forces, of $n^{-1/3}$ the mean spacing between molecules, and/or of the free path length $\lambda_{\text{fp}} \approx 1/nR^2$. The units of time characteristic of macroscopic and free-path phenomena are, accordingly, $t_{\text{hy}} = L_{\text{hy}}/\bar{c}$ and $t_{\text{fp}} = \lambda_{\text{fp}}/\bar{c}$ where $\bar{c} \sim (\mathbf{k}T/m)^{1/2}$. Therefore, one expects the magnitudes of the terms $(\partial_t + \mathbf{c}\,.\,\nabla)f$ and $-\overline{[f, H^{\text{e}}]} + \partial_{\text{c}} f$ to be in the ratio of t_{fp} to t_{hy} or, equivalently, of λ_{fp} to L_{hy} : a more fully developed argument to this point can be found in the appendix of Reference 16. This suggests that we adopt a perturbation procedure and replace the Boltzmann equation (1.152) with

$$\varepsilon(\partial_t + \mathbf{c}\,.\,\nabla)f = -\overline{[f, H^{\text{e}}]} + \partial_{\text{c}} f \tag{1.229}$$

Here the purpose of the dimensionless marker ε (which we later set equal to unity) is to indicate that the left-hand side of the equation is smaller by a factor $\varepsilon(\approx \lambda_{\text{fp}}/L_{\text{hy}} = \lambda_{\text{fp}}|\nabla P|/P)$ than the terms on the right. The next step is to assume that the solution of equation (1.229) can be written as a series (in gradients of the macroscopic field variables)

$$f = f^{(0)} + \varepsilon f^{(1)} + \cdots = \sum \varepsilon^k f^{(k)} \tag{1.230}$$

where each of the expansion coefficients $f^{(k)} = f^{(0)}\phi^{(k)}$ is taken to be a 'slowly varying' function of time and position. Specifically, it is assumed that the time and position dependence of $f^{(k)}$ is governed exclusively by the time and position dependences of the fluid mechanical variables ρ which consist of the density, temperature and mass average velocity. Furthermore, it is assumed that the equations of motion of each of these variables is expandable in a series $\partial_t \mathsf{p} = \partial_0 \mathsf{p} + \varepsilon \partial_1 \mathsf{p} + \cdots$, where the coefficients $\partial_k \mathsf{p}$ are determined by conditions which we soon shall describe.

It is now a simple matter to show that the function $f^{(0)}$ must satisfy the non-linear equation

$$0 = -\overline{[f^{(0)}(\bar{1}), H_1{}^{e}]} + \iiint d\bar{1}' \, d\bar{2}' \, d\bar{2} w(\bar{1}\bar{2} | \bar{1}'\bar{2}')$$
$$\times [f^{(0)}(\bar{1}') f^{(0)}(\bar{2}') - f^{(0)}(\bar{1}) f^{(0)}(\bar{2})]$$

which differs inconsequentially from equation (1.161). Therefore, we can identify the first term of the series (1.230) with the Maxwell–Boltzmann function ${}^0 f_F(\bar{1})$ given by equations (1.162) and (1.163). The associated contributions to the fluxes of energy and linear momentum are $\mathbf{Q}^{(0)} = 0$ and $\mathsf{p}^{(0)} = \mathbf{\delta} p_0$ with $p_0 = n_0 \mathbf{k} T_0$.

To assure the uniqueness of the functions $f^{(k)}$ we set the constants n_0, T_0 and $\mathbf{u}_0$, which appear in $f^{(0)}$ (see equation 1.162) equal to the local and instantaneous values of the concentration, temperature and fluid velocity. Therefore, each of the functions $f^{(k)}$, $k > 0$, must satisfy the 'subsidiary conditions',

$$\int d\bar{1} \chi(\bar{1}) f^{(k)}(\bar{1}) = 0 \tag{1.231}$$

with $\chi(\bar{1})$ taken equal to m, $\mathbf{c}_1$ and $\frac{1}{2} m c_1{}^2 + H_{1'}$.

The equations satisfied by the second and higher order contributions to the series (1.230) can be written in the forms

$$(\partial_0 + \mathbf{c}_1 \cdot \nabla) f^{(0)}(\bar{1}) = -n\hat{\Omega}_1(\phi^{(1)}) \tag{1.232}$$

$$(\partial_0 + \mathbf{c}_1 \cdot \nabla) f^{(1)}(\bar{1}) + \partial_1 f^{(0)}(\bar{1}) - I_1[f^{(1)}, f^{(1)}] = -n\hat{\Omega}_1(\phi^{(2)}) \tag{1.233}$$

$$\cdots$$

with

$$I_1(g, h) = \iiint d\bar{1}' \, d\bar{2}' \, d\bar{2} w(\bar{1}\bar{2} | \bar{1}'\bar{2}') [g(\bar{1}')h(\bar{2}') - g(\bar{1})h(\bar{2})] \tag{1.234}$$

and where $\hat{\Omega}$ is the linear operator defined by equation (1.171). In these equations it is to be understood that the symbol $\partial_k f^{(m)}$ refers to the chain of derivatives

$$\sum_n \left[(\partial_k \rho_n)(\partial \rho_n f^{(m)}) + \sum_p \{(\partial_k \nabla_p \rho_n)(\partial_{\nabla_p \rho_n} f^{(m)})\} + \cdots \right]$$

If we assume that the operator $\hat{\Omega}$ is completely continuous (this is known to be the case only for a few special choices of the particle interactions and, even then, only for structureless particles[56]), the equations (1.232), (1.233), ..., then will be soluble if and only if their left-hand members are orthogonal to every solution of the homogeneous adjoint equation $\hat{\Omega}^\dagger(\chi) = 0$. It is not difficult to show that the functions χ are limited to the binary invariants m, $\mathbf{c}_1$ and $\frac{1}{2}mc_1{}^2 + H_1 + H_1{}^e$. The solubility conditions (orthogonality of χ's to inhomogeneous terms) are then just sufficient to establish values for the expansion coefficients $\partial_k \rho$. Consequently, the relationships (1.232), (1.233), ..., become a sequence of soluble linear integrodifferential equations for the functions $\phi^{(k)}$.

This scheme was developed originally, and independently, by Sidney Chapman and David Enskog: see Reference (56) for a historical survey. A theorem due to Hilbert[57] established that the many assumptions which we listed above are internally consistent. However, it was not until 1963 that Harold Grad[58] rigorously established the asymptotic nature of this 'normal' solution of the Boltzmann equation. This proof was only one of the many important results included in a paper which surely ranks as one of the most significant contributions to kinetic theory in many years.

In the high temperature limit with which we are concerned the level shifts are so small that it is reasonable to neglect their effects upon ordinary gas transport properties. To this approximation the binary invariants are m, $\mathbf{c}$ and $\frac{1}{2}mc^2 + H$ and the solution of equation (1.161) is the field-free Maxwell–Boltzmann function

$$f_i^{(0)} = n(m/2\pi\mathbf{k}T)^{3/2}(2\pi\Gamma\mathbf{k}T)^{-r/2} c_r \exp\left\{-\frac{1}{\mathbf{k}T}\left(\tfrac{1}{2}mC_i{}^2 + \frac{L_i{}^2}{2\Gamma}\right)\right\} \tag{1.235}$$

with the integer r denoting the number of degrees of rotational freedom and $c_2 = \frac{1}{2}$, $c_3 = 1$. Here, and henceforth we neglect vibrations and assume that the molecules either are spherical or symmetric tops: in the former instance $\mathrm{d}\bar{i} = \mathrm{d}^3c_i\,\mathrm{d}^3L_i$ and in the latter $\mathrm{d}\bar{i} = \mathrm{d}^3c_i\,\mathrm{d}^3L_i/L_i$.

The solubility conditions for equation (1.232) can now be written in the forms

$$\begin{aligned} \partial_0 n + \nabla \cdot n\mathbf{u} &= 0 \\ \rho(\partial_0 u + \mathbf{u} \cdot \nabla\mathbf{u}) &= -\nabla p + \rho\hat{\mathbf{F}}^e \end{aligned} \tag{1.236}$$

and

$$nc_v(\partial_0 T + \mathbf{u} \cdot \nabla T) = -p\nabla \cdot \mathbf{u}$$

with $c_v = \frac{1}{2}(3+r)\mathbf{k}$ and $p = n\mathbf{k}T$. When these are substituted into equation (1.232) the result is

$$f_1^{(0)}\left\{2\overline{\overline{\mathbf{W}_1\mathbf{W}_1}}^{\circ} : \overline{\overline{\nabla\mathbf{u}}}^{\circ} + \frac{2}{3+r}\left(\frac{r}{3}\mathbf{W}_1{}^2 - \Omega_1{}^2\right)\nabla \cdot \mathbf{u} + \sqrt{\frac{2\mathbf{k}T}{m}}\mathbf{W}_1\left[\left(\mathbf{W}_1{}^2 - \frac{5}{2}\right)\right.\right.$$

$$\left.\left. + \left(\Omega_1{}^2 - \frac{r}{2}\right)\right] \cdot \frac{\nabla T}{T}\right\} = -n\hat{\Omega}_1(\phi^{(1)}) \tag{1.237}$$

where $\mathbf{W} = (m/2\mathbf{k}T)^{1/2}\mathbf{C}$ and $\Omega = (2\Gamma\mathbf{k}T)^{-1/2}\mathbf{L}$.

The solution of equation (1.237) is of the form

$$\phi^{(1)} = -n^{-1}(2\mathbf{k}T/m)^{1/2}\mathbf{A} \cdot \frac{\nabla T}{T} - n^{-1}\mathsf{B} : \overline{\overline{\nabla\mathbf{u}}}^{\circ} - n^{-1}D\nabla \cdot \mathbf{u} \tag{1.238}$$

where **A**, B and D are determined by the soluble equations

$$f_1^{(0)}\left[\mathbf{W}_1\left(W_1{}^2 + \Omega_1{}^2 - \frac{5+r}{2}\right)\right] = \hat{\Omega}_1(\mathbf{A}) \tag{1.239}$$

$$f_1^{(0)}[2\overline{\overline{\mathbf{W}_1\mathbf{W}_1}}^{\circ}] = \hat{\Omega}_1(\mathsf{B}) \tag{1.240}$$

and

$$f_1^{(0)}\left[\frac{2}{3+r}\left(\frac{r}{2}W_1{}^2 - \Omega_1{}^2\right)\right] = \hat{\Omega}_1(D) \tag{1.241}$$

The solutions of these equations are related to the first-order contributions to the fluxes of energy and momentum by the formulas

$$\mathbf{Q}^{(1)} \equiv \int \mathrm{d}\bar{\Gamma} f_1^{(1)}\mathbf{C}_1\left(\frac{1}{2}mC_1{}^2 + \frac{L_1{}^2}{2\Gamma}\right) = -\boldsymbol{\lambda} \cdot \nabla T - \boldsymbol{\gamma} \cdot \overline{\overline{\nabla\mathbf{u}}}^{\circ} - \boldsymbol{\nu}\nabla \cdot \mathbf{u} \tag{1.242}$$

$$\mathsf{p}^{(1)} \equiv \int \mathrm{d}\bar{\Gamma} f_1^{(1)} m\mathbf{C}_1\mathbf{C}_1 = -2\boldsymbol{\eta} : \overline{\overline{\nabla\mathbf{u}}}^{\circ} - \boldsymbol{\eta}_b\nabla \cdot \mathbf{u} - \boldsymbol{\gamma}' \cdot \frac{\nabla T}{2T} - \boldsymbol{\delta\nu}' \cdot \frac{\nabla T}{2T} \tag{1.243}$$

$$\lambda(\mathbf{F}) = \frac{2\mathbf{k}^2 T}{\rho} \int \mathrm{d}\bar{1} f_1^{(0)} \mathbf{W}_1 \mathbf{A}_1 (W_1{}^2 + \Omega_1{}^2) = \frac{2\mathbf{k}^2 T}{m} [\![\mathbf{A}, \mathbf{A}]\!]_{\mathrm{F}}{}^{\dagger} \tag{1.244}$$

$$\eta(\mathbf{F}) = \frac{\mathbf{k}T}{n} \int \mathrm{d}\bar{1} f_1^{(0)} \mathbf{W}_1 \mathbf{W}_1 \mathsf{B}_1 = \tfrac{1}{2}\mathbf{k}T [\![\mathsf{B}, \mathsf{B} + \delta D]\!]_{\mathrm{F}}{}^{\dagger} \tag{1.245}$$

$$\eta_b(\mathbf{F}) = \frac{2\mathbf{k}T}{n} \int \mathrm{d}\bar{1} f_1^{(0)} \mathbf{W}_1 \mathbf{W}_1 D_1 = \mathbf{k}T [\![D, \mathsf{B} + \delta D]\!]_{\mathrm{F}}{}^{\dagger} \tag{1.246}$$

where the double bracket integrals, $[\![a, b]\!]_{\mathrm{F}}$, are defined by equation (1.173). From the properties of these integrals which were established previously we now obtain the Onsager–Casimir relations $\lambda(\hat{T}\mathbf{F}) = \lambda^{\dagger}(\mathbf{F})$, $\eta(\hat{T}\mathbf{F}) = \eta^{\dagger}(\mathbf{F})$, and $\eta_b(\hat{T}\mathbf{F}) = \eta_b{}^{\dagger}(\mathbf{F})$.

$\gamma(\mathbf{F}) = \mathbf{k}T(2\mathbf{k}T/m)^{1/2}[\![\mathsf{B}, \mathbf{A}]\!]^{\dagger}{}_{\mathrm{F}}$, $\mathbf{v}(\mathbf{F}) = \mathbf{k}T(2\mathbf{k}T/m)^{1/2}[\![D, \mathbf{A}]\!]^{\dagger}{}_{\mathrm{F}}$, $\gamma'(F) = \gamma^{\dagger}(\hat{T}F)$ and $\mathbf{v}'(F) = \mathbf{v}^{\dagger}(\hat{T}F)$ are not zero only for the case of an electric field and owe their existence to the 'parity mixing' which arises from the anti-commutation of $\hat{P}$ and $\hat{\Theta}_E$.[33]

Since our interests lie more with thermal conduction than with viscosity we shall only consider $\lambda(\mathbf{F})$ in detail. In the absence of the external field this tensor reduces to the isotropic form $\lambda(\mathbf{0}) = \lambda_0 \boldsymbol{\delta}$ with $\lambda_0 = (2\mathbf{k}^2 T/3m)[\mathbf{A}; \mathbf{A}]$. However, when the field is on, this tensor must transform according to the totally symmetric representation of the group of rotations about the direction of the external field. The space of second rank tensors which transform in this manner is spanned by the three basis elements $\hat{h}\hat{h}$, $\boldsymbol{\delta} - \hat{h}\hat{h}$ and $\boldsymbol{\varepsilon} . \hat{h} = \hat{i}\hat{j} - \hat{j}\hat{i}$ where $\hat{i}$, $\hat{j}$ and $\hat{h} \equiv \mathbf{F}/F$ form a right-handed orthonormal set. Consequently,

$$\begin{aligned} \mathbf{Q}^{(1)} &= -\lambda(\mathbf{F}) . \nabla T \\ &= -\lambda_{\parallel}(\mathbf{F})\hat{h}\hat{h} . \nabla T - \lambda_{\perp}(\mathbf{F})(\nabla T - \hat{h}\hat{h} . \nabla T) - \lambda_{\mathrm{tr}}(F) \nabla T \times \hat{h} \end{aligned} \tag{1.247}$$

$$\lambda_{\parallel}(\mathbf{F}) = (2\mathbf{k}^2 T/m)[\![\hat{h} . \mathbf{A}, \hat{h} . \mathbf{A}]\!]_{\mathrm{F}} = \lambda_{\parallel}(\hat{T}\mathbf{F}) \tag{1.248}$$

$$\lambda_{\perp}(\mathbf{F}) = \tfrac{1}{2}\{(2\mathbf{k}^2 T/m)[\![\mathbf{A}; \mathbf{A}]\!]_{\mathrm{F}} - \lambda_{\parallel}(\mathbf{F})\} = \lambda_{\perp}(\hat{T}\mathbf{F}) \tag{1.249}$$

$$\lambda_{\mathrm{tr}}(\mathbf{F}) = (\mathbf{k}^2 T/m)\{[\![\hat{j} . \mathbf{A}, \hat{i} . \mathbf{A}]\!]_{\mathrm{F}} - [\![\hat{i} . \mathbf{A}, \hat{j} . \mathbf{A}]\!]_{\mathrm{F}}\} \tag{1.250}$$

is equal to $-\lambda_{\mathrm{tr}}(\hat{T}F)$ if F is a magnetic field and to zero if F is an electric field. As the field strength shrinks to zero $\lambda_{\parallel}$ and $\lambda_{\perp}$ both tend to the common value of λ_0; $\lambda_{\mathrm{tr}}(\mathbf{0}) = 0$.

From equation (1.247) it follows that the heat flux in the direction of the temperature gradient is given by $\lambda(\alpha)|\nabla T|$ where α denotes the angle between $\mathbf{F}$ and ∇T and where $\lambda(\alpha) = \lambda_{\parallel} \cos^2 \alpha + \lambda_{\perp} \sin^2 \alpha$. Therefore, $\lambda(\alpha)$ is equal to $\lambda_{\parallel}$ when $\mathbf{F}$ and ∇T are parallel and to $\lambda_{\perp}$ when $\mathbf{F}$ and ∇T are mutually perpendicular. According to Gorelik and Sinitsin[59] the experiments of the Leiden Group (Korving[60], Beenakker, *et al.*) are performed in such a way that they yield $\lambda_{Av} = \frac{1}{2}(\lambda_{\parallel} + \lambda_{\perp})$, the unweighted average of $\lambda(\alpha)$ over all possible values of α. On the other hand, in the experiments of Senftleben and Pietzner[61] and in those of Gorelik and his coworkers[62] $\mathbf{H}$ is perpendicular to ∇T so that the measured conductivity must be identified with $\lambda_{\perp}$. Actually, in both cases the measured quantities are the changes of conductivity caused by the field. We denote these 'field shifts' by the symbols $\Delta\lambda_{\parallel} = \lambda_0 - \lambda_{\parallel}$, $\Delta\lambda_{\perp} = \lambda_0 - \lambda_{\perp}$ and $\Delta\lambda = \lambda_0 - \lambda_{Av} = \frac{1}{2}(\Delta\lambda_{\parallel} + \Delta\lambda_{\perp})$.

Formulas similar to those derived here also can be obtained from the Waldmann–Snider equation. The reason we have selected the classical theory and that we shall comment so specifically upon the rough sphere and spherocylinder models is that only for these have calculations actually been performed. However, much has been learned[54] (by the Leiden Group, by Waldmann and Hess, by Tip, and by Snider and his coworkers) from the formal theory about the general characteristics of the transport coefficients and, in particular, it was known long before model calculations were performed that the theory was capable of explaining the field dependence of $\lambda_{\parallel}$, $\lambda_{\perp}$ and λ_{tr}. Furthermore, it has been found that the same integrals (of the type $[a, b]$) of the scattering cross-section arise in the theories of distinct experimental phenomena, e.g. Senftleben effects and n.m.r. measurements, so that even when it is impossible or impractical to evaluate these integrals one still may be able to derive (approximate) theoretical relations among the various observations. Later we shall briefly touch upon a few examples of this sort but for a full account the reader should consult the original papers cited here as references. The point of view we shall promote (not because it is better but simply because it is that which recently has guided the direction of our own research) is that whenever possible one should adapt the theory to those models for which numerical calculations can be conducted. Of course, as our stock of information concerning inelastic cross-sections increases, so also will the number of 'solvable models'.

We can associate with each molecular species a 'skeletal length', d. The skeletal length of a diatomic molecule is its bond length: in the case of a linear polyatomic molecule it is the separation of the two outermost nuclei. The dimensions of the thermal conductivity coefficients $\lambda_i(i = \parallel, \perp, \mathrm{tr})$ are the same as those of $\mathrm{d}^{-2}(\mathbf{k}^3 T/m)^{1/2}$. Therefore, λ_i may be written as the product of this factor with a function Υ_i which depends only upon the set of dimensionless variables which can be constructed from the parameters which characterize the local thermodynamic state and the kinematics and collision dynamics of the molecules. In the case of rough spheres these parameters are σ, I, m, $\bar{\gamma}H$, the temperature T and the gas pressure p. For the corresponding set of dimensionless variables one can select $\kappa = 4I/m\sigma^2$ and $\mathscr{B} \propto H/p$. For diamagnetic species $\mathscr{B} \to \mathscr{B}_\mathrm{D} = (m\mathbf{k}T/\pi)^{1/2}(\mu_\mathrm{N} g/d^2\hbar)(H/p)$; for a paramagnetic molecule such as the (Hund's case a approximation) $^2\Pi_{3/2}$ state of NO, $\mathscr{B} \to \mathscr{B}_\mathrm{P} = (m/\pi\mathbf{k}T)^{1/2}(\mu_\mathrm{B}\hbar/\mathrm{d}^2 I)(H/p)$. With loaded spherocylinders the single rough-sphere parameter σ is replaced with σ, L and ξ: in place of κ a natural set of dimensionless variables consists of L/d, σ/d, $g_1(= 2\xi/d)$ and $g_2(= [md^2/2I]^{1/2})$.

In the case of a diatomic molecule the values of the 'load parameters' $g_1 = |m_1 - m_2|/(m_1 + m_2)$ and $g_2 = (m_1 + m_2)/(2m_1 m_2)^{1/2}$ are determined solely by the masses of the two nuclei. However, for a linear polyatomic molecule g_1 and g_2 depend upon the masses of all the nuclei and upon the bond lengths d_{ij} as well. For example, in the case of a linear triatomic molecule $g_1 = (M_1 - M_3) - M_2\delta$ and $g_2 = [\frac{1}{2}\{(M_1 + M_3) + (M_2 - g_1{}^2)\delta^2\}]^{-1/2}$ where $M_i = m_i/\Sigma_1{}^3 m_j$ and $\delta = |d_{12} - d_{13}|/(d_{12} - d_{13})$. The quantities g_1, g_2 and $\mathscr{B}$ are determined by the properties of the individual molecules and by the nature of their interactions with the magnetic field. The quantities L/d and σ/d are characteristic of the interactions between the molecules.

In the absence of an external α field the vector $\mathbf{A}$ can be written in the form

$$\mathbf{A} = A_1\mathbf{W} + A_2\mathbf{W}\,.\,\boldsymbol{\Omega}\boldsymbol{\Omega} + A_3\boldsymbol{\Omega} \times \mathbf{W} \tag{1.251}$$

where $A_i = A_i[W^2, \Omega^2, (\mathbf{W}\,.\,\boldsymbol{\Omega})^2]$. Although the subsidiary conditions (1.231) impose no restrictions upon A_3 one does find that the functions A_1 and A_2 must satisfy the constraint

$$\int \mathrm{d}\bar{1} f^{(0)}(\bar{1})[A_1\mathbf{W}_1\mathbf{W}_1 + A_2\mathbf{W}_1\mathbf{W}_1\,.\,\boldsymbol{\Omega}_1\boldsymbol{\Omega}_1] = 0 \tag{1.252}$$

In our calculations we have neglected the dependence of A_i upon the helicity $\mathbf{W}\,.\,\boldsymbol{\Omega}$ and expanded each of these functions in double series of

Sonine polynomials,

$$A_i = \sum_j \sum_k A_i^{jk} S_{3/2}^{(j)}(W^2) S_{n_i}^{(k)}(\Omega^2) \tag{1.253}$$

with $n_1 = \frac{1}{2}(r-2)$ and $n_2 = n_3 = \frac{1}{2}r$. The subsidiary condition (1.252) then reduces to $6A_1^{00} + rA_2^{00} = 0$ and the thermal conductivity can be written in the form

$$\begin{aligned}\lambda_0 = (\mathbf{k}^2 T/12m)[6r(A_1^{00} - A_1^{01}) + r(r+2)(A_2^{00} - A_2^{01}) \\ - 5(6A_1^{10} + rA_2^{10})]\end{aligned} \tag{1.254}$$

To determine the coefficients A_i^{jk} we use a moment method which amounts to approximating the kernel of the integral operator $\hat{\Gamma}$ by a degenerate kernel. The particular degenerate forms we employ are finite sums of dyads, each element of which corresponds to a product $S_{3/2}^{(j)}(W^2) S_{n_i}^{(k)}(\Omega^2)$ of orthogonal polynomials. In the case of loaded spheres there is an inverse to each collisional event. Therefore, $\hat{\Gamma}$ is self-adjoint and so commutes with the time reversal operator. From this we conclude that for loaded spheres $\hat{T}\mathbf{A} = -\mathbf{A}$ and $A_3 = 0$. For a Fredholm equation with a symmetric kernel (self-adjoint integral operator) each increase in the number of terms of the polynomial expansion is accompanied with a corresponding rise in the value of the bracket integral, $[\mathbf{A}; \mathbf{A}]$. Furthermore, the sequence of integrals so obtained converges monotonically to the value associated with the exact solution of the equation. The situation will be different from this for rough spheres and spherocylinders for in these cases the operator $\hat{\Gamma}$ is not self-adjoint. Thus, while the associated unsymmetric kernels can also be uniformly approximated by a sequence of degenerate kernels, we are no longer assured that the convergence of the sequence of related bracket integrals will be monotonic. However, it is important to note that if one were to exclude from equation (1.251) the terms $A_3 \mathbf{\Omega} \times \mathbf{W}$ which are odd functions of the spin, $\mathbf{\Omega}$, one would observe monotone convergence of the corresponding sequence of bracket integrals. Indeed, this exclusion amounts to a neglect of the antisymmetric part of the kernel, or equivalently, to replacement of the actual ensembles for these model systems with non-polar ensembles.

The moment equations for the scalar coefficients A_i^{jk} are obtained by inserting the expansion (1.251) into the right-hand side of (1.239), terminating the series after some finite number of terms, and then forming the inner product of the resulting expression with each of the 'basis functions', $\mathcal{G}_i^{jk} = \mathcal{W}_i S_{3/2}^{(j)}(W^2) S_{n_i}^{(k)}(\Omega^2)$ [where $\mathcal{W}_1 = \mathbf{W}$, $\mathcal{W}_2 = \mathbf{W} \cdot \mathbf{\Omega\Omega}$ and $\mathcal{W}_3 = \mathbf{\Omega} \times \mathbf{W}$], which was retained in the expansion. This leads to the set of linear algebraic equations

$$n^{-1} \int \mathrm{d}\bar{\Gamma} f_1^{(0)} \left\{ \mathcal{G}_{i'}^{j'k'} \cdot \mathbf{W}_1 \left(W_1{}^2 + \Omega_1{}^2 - \frac{5+r}{2} \right) \right\}$$
$$= \sum_i \sum_j \sum_k [\mathcal{G}_{i'}^{j'k'}; \mathcal{G}_i^{jk}] A_i^{jk} \tag{1.255}$$

The task of computing the coefficients A_i^{jk} and the related transport coefficient λ_0 then reduces to that of evaluating the set of collision integrals $[\mathcal{G}_{i'}^{j'k'}; \mathcal{G}_i^{jk}]$.

The most extensive calculations have been conducted for rough spheres[63] where as many as ten terms in the series (1.251) were retained. It was found that:

(i) values of λ_0 computed with the four term trial function ($A_1^{00}, A_1^{10}, A_1^{01}$; A_2^{00}) and the ten term function (A_1^{00}, A_1^{01}, A_1^{20}, A_1^{11}, A_1^{02}; A_2^{00}, A_2^{10}, A_2^{01}; A_3^{00}) differ by no more than 2% over the entire range (0, 2/3) of the variable κ.

(ii) the term $A_3^{00}\mathbf{\Omega} \times \mathbf{W}$ has very little effect upon the numerical value of λ_0.

(iii) the term $A_2^{00}\mathbf{W} \cdot \mathbf{\Omega\Omega}$ makes a significant (approximately 10%) contribution to the thermal conductivity.

(iv) the value of λ_0 increases with that of κ and exhibits a full-range variation of about 10%.

From all this we conclude that the lack of symmetry of the kernel, as measured by the importance of the terms $A_3\mathbf{\Omega} \times \mathbf{W}$, is of minor consequence to the description of energy transport in an isotropic gas. Furthermore, it is clear that for this model the approximation of a polar ensemble is significantly better than that of an isotropic (in spin space) ensemble.

The most ambitious calculations for loaded spheres were based upon the trial function (A_1^{00}, A_1^{10}, A_1^{01}; A_2^{00})[64]. For this model the dimensionless parameter $a = m\xi^2/2 \equiv (g_1 g_2/2)^2$ provides a convenient measure of the mass eccentricity. The value of a is equal to 0.00034 for $^{12}C^{16}O^{17}O$, 0·0022 for $^{14}C^{16}O$, 0·0208 for $^{12}C^{18}O$, 0·0625 for $^{1}H^{2}H$ and 0·167 for $^{1}H^{3}H$. Over this entire range of a values the term $A_2^{00}\mathbf{W} \cdot \mathbf{\Omega\Omega}$ provides a contribution to λ_0 of less than 2·3%. From this and other evidence it is clear that anisotropy of the spin distribution is of little consequence to the reckoning of the thermal conductivity of loaded spheres. On the other hand distortions of the form $\mathbf{W}\,.\,\mathbf{\Omega\Omega}$ have a considerable effect ($\approx$ 10–20%) upon the predicted values of the thermal diffusion ratio[65]. [Using no adjustable parameters whatsoever, this model leads to thermal diffusion ratios for D_2:HT mixtures with the correct algebraic sign and with values which differ from experiment by less than a factor of two. When the load parameter a is 'adjusted' so that the experimental and theoretical values of the rotational relaxation time are equal, the theoretical and experimental values of the thermal diffusion ratio are found to coincide.]

The thermal conductivity of (unloaded) spherocylinders was first calculated[66] with a trial function of the same form as that described previously for loaded spheres. Recent studies by Klein and others[67] have shown that the terms $A_3\mathbf{\Omega}\times\mathbf{W}$, omitted from these earlier calculations, are of minor consequence, at least for N_2, CO and NO. Sandler and Dahler reported their results for spherocylinders in terms of the dimensionless variables $\beta = L/2\sigma \equiv \frac{1}{2}(L/d)(\sigma/d)^{-1}$ and $\Delta = mL^2/8\Gamma \equiv (g_2L/2d)^2$. They found that to a very good approximation λ_0 was inversely proportional to $A_p = \pi\sigma^2 + \frac{1}{2}\pi\sigma L = \pi\sigma^2(1+\beta)$, the mean projected area or geometric cross-section of a spherocylinder. Also, the differences between calculations based upon the isotropic and polar approximations were never greater than 5%.

The imposition of an external (magnetic) field destroys the spatial isotropy of a polyatomic gas and so, in place of equation (1.251), one introduces an expansion

$$\mathbf{A} = \sum_k \psi^k\,.\,\mathsf{T}^k \tag{1.256}$$

where ψ^k is a tensor function of $\mathbf{W}$ and $\mathbf{\Omega}$ and T^k a tensor (with a rank greater by one than that of ψ^k) descriptive of the local thermodynamic state of the (physically anisotropic) gas. In principle the series (1.256) should be extended over a complete set of tensors ψ^k but, in practice, we expect that a very few terms will do. In the three calculations which have been reported[67,68,69] the trial functions were limited to sums over the six functions

$$\left\{\begin{array}{ll} \psi^1 = \mathbf{W}(W^2 - 5/2); & \psi^4 = \mathbf{W}\times\mathbf{\Omega} \\ \psi^2 = \mathbf{W}(\Omega^2 - r/2); & \psi^5 = \mathbf{W}\,.\,\mathbf{\Omega} \\ \psi^3 = \mathbf{W}\overset{\circ}{\overline{\overline{\mathbf{\Omega\Omega}}}}; & \psi^6 = \overset{\circ}{\overline{\overline{\mathbf{W\Omega}}}} \end{array}\right\} \tag{1.257}$$

To this approximation the thermal conductivity tensor is given by

$\lambda = (\mathbf{k}^2T/2m)(5\mathsf{T}^1 + r\mathsf{T}^2)^\dagger$ and the moment equations, analogous to equation (1.255), by

$$n^{-1}\int \mathrm{d}\bar{1} f_1^{(0)} \psi_1{}^i \mathbf{W}_1\left(W_1{}^2+\Omega_1{}^2-\frac{5+r}{2}\right) = \sum_j [\![\psi^i, \psi^j]\!] \cdot \mathsf{T}^j. \quad (1.258)$$

Each T^k can be separated into the sum of a 'field-off' part X^k, which is an isotropic tensor, and a part Y^k which vanishes in the absence of the field. Each of the Y^k's can be expanded in the complete set of irreducible tensors (with the same rank and parity) which transform according to the totally symmetric representation of the group (C_∞) of rotations about the direction of the field. All of this leads finally to a set of linear, inhomogeneous equations for the scalar expansion coefficients of the Y^k's. Not only is this an extremely messy business but, once again, to get numbers it is first necessary to evaluate a rather large number of complicated collision integrals.

Originally[68] the equations (1.258) were solved for rough spheres with the functions ψ^5 and ψ^6 omitted from the basis set; a more recent study[69] has dealt with all six of the basis functions $\psi^1, \ldots, \psi^6$. Solutions for the full set of six have also been constructed for the loaded spherocylinder model[67].

The data to be explained are the following:

(i) $\Delta\lambda_\parallel$, $\Delta\lambda_\perp$, $\Delta\lambda$ and λ_{tr} depend upon the magnetic field strength and pressure in the combination H/p. Dimensional arguments alone lead to this conclusion.

(ii) $\Delta\lambda_\parallel$, $\Delta\lambda_\perp$ and $\Delta\lambda$ are 'S-shaped' functions of H/p, each with a saturation value, $(\Delta\lambda_i)_{sat}$, and a characteristic field strength, $\mathscr{B}_{1/2}$, associated with a value of $\Delta\lambda_i$ equal to $\frac{1}{2}(\Delta\lambda_i)_{sat}$.

(iii) the value of λ_{tr} rises from zero at $\mathscr{B} = 0$ to a maximum at $\mathscr{B}_{max}$ and then falls again to zero as $\mathscr{B}$ becomes larger.

During the early days of research in this area it was found that even a relatively crude theory could account quite well for the general shapes (functional forms) of $\Delta\lambda_i(\mathscr{B})$ and $\lambda_{tr}(\mathscr{B})$. This aspect of the theory has been highly refined. To make a more quantitative comparison between experiment and theory one must have numerical values of the collision integrals.

Because the rough sphere description of inelastic collisions is so grossly unrealistic, calculations based upon this model can only provide us with a very crude and imperfect understanding of the detailed relationship between inelastic collisions and experimental observations of heat conduction in the presence of a field. Indeed, our real purpose in studying this model was simply to gain experience about the computational procedure without getting over-involved in the evaluation of collision integrals. For values of κ corresponding to real molecules, e.g. 0·091 for CO, 0·0950 for N_2, 0·0457 for CH_4, the shapes of the theoretical and experimental curves for $\Delta\lambda(\mathscr{B})$ and $\lambda_{tr}(\mathscr{B})$ were found to be remarkably similar. However, the theoretical estimates for $(\Delta\lambda)_{sat}$ generally ran from two to three times greater than experiment and the predicted values for $\mathscr{B}_{1/2}$ were sometimes greater and sometimes less, by factors of as much as two, than those observed experimentally. The situation with regard to λ_{tr} and $\mathscr{B}_{max}$ was much the same. Although none of this could be viewed as surprising, it also did little to bolster our enthusiasm for the rough sphere model. Among the other items of interest which resulted from this investigation were the following:

(i) the contributions to $\Delta\lambda_i$ and λ_{tr} from the separate terms (classified according to their symmetries) $\psi^1 . \mathsf{T}^1 + \psi^2 . \mathsf{T}^2$, $\psi^3 . \mathsf{T}^3$ and $\psi^4 . \mathsf{T}^4$ were very nearly additive and, in the case of $\Delta\lambda$, the algebraic sign of the contribution from $\psi^4 . \mathsf{T}^4$ was negative whereas those due to the other two were positive.

(ii) for large values of κ ($\kappa \gtrsim 0{\cdot}3$) the curves of $\Delta\lambda$ versus $\mathscr{B}$ exhibited small overshoots or humps near $\mathscr{B} \approx 2\mathscr{B}_{1/2}$. These humps did not appear when the term $\psi^4 . \mathsf{T}^4$ was excluded from the trial function. However, it has subsequently been shown by McCourt, Knaap and Moraal[69] that this odd behaviour is an artifact which disappears when the basis set is expanded to include ψ^5 and ψ^6, that is, all three of the irreducible tensors associated with the dyad $\mathbf{W\Omega}$.

Our recent calculations for the loaded spher-cylinder model[67] permit one to focus much more sharply upon the connexions between individual collisional events and experimental observations. Although this model completely disregards attractive contributions to the intermolecular forces, it does mirror rather faithfully the gross characteristics of those portions of the interactions which dominate at high temperature. We expect it to provide a reasonable description of the collision events which occur at moderate and high temperatures in gases composed of non-polar or only slightly polar linear and diatomic species such as N_2, O_2, CO, NO and CO_2.

Since it is only for N_2, CO and NO that there are currently sufficient data to permit an extensive comparison with theory, we limited our initial considerations to those species. The case of O_2 was not studied because it would have required the introduction of a three-component distribution function $f_{m_s}(txcL)$ with $m_s = 0, \pm 1$ and a significantly greater computational effort.

The shapes of the experimental and theoretical curves are, as always, very similar. Therefore, we can concern ourselves exclusively with the seven numbers λ_0, $(\Delta\lambda_\perp)_{sat}$, $(\Delta\lambda)_{sat}$, $(\lambda_{tr})_{max}$, $\mathscr{B}_{1/2}$, $\mathscr{B}^{\perp}_{1/2}$ and $\mathscr{B}_{max}$ which characterize the thermal conductivity tensor of a specific gas. Unfortunately the data on λ_{tr} ($\mathscr{B}$) invariably have been obtained at much lower temperatures than those for the other properties. So far as we are aware no experimental studies of the temperature dependence of any of the Senftleben phenomena have been reported.

Our calculations reveal some very pleasing facts:

(i) the values of λ_0, $\mathscr{B}_{1/2}$, $\mathscr{B}^{\perp}_{1/2}$ and $\mathscr{B}_{max}$ are strongly dependent upon the value of the cylinder radius σ but only weakly upon the sheath length L.

(ii) $(\Delta\lambda_\perp)_{sat}$, $(\Delta\lambda)_{sat}$ and $(\lambda_{tr})_{max}$ are weak functions of σ but depend sensitively upon the value of L.

(iii) for reasonable values of L the terms $\psi^4 . \mathsf{T}^4$, $\psi^5 . \mathsf{T}^5$ and $\psi^6 . \mathsf{T}^6$, which are odd functions of the spin, account for less than 2% of the experimental values of $(\Delta\lambda_\perp)_{sat}$, $(\Delta\lambda)_{sat}$ and $(\lambda_{tr})_{max}$. This fits very nicely into the pattern observed by Sandler and Dahler[66] for the field-free transport coefficients.

(iv) for the species which we examined the importance of mass asymmetry was very much less than that of molecular non-sphericity. Thus, in the case of CO the mass asymmetry contribution to $(\Delta\lambda)_{sat}$ was only 4% of the whole.

(v) Knapp and Beenakker's prediction (on the basis of a very much simplified theory) that $(\Delta\lambda_\perp)_{sat}/(\Delta\lambda_\parallel)_{sat} = 3/2$ is excellently borne out by our calculations: for N_2, CO and NO we obtain values of 1·511, 1·509 and 1·508, respectively.

The practical consequences of (i) and (ii) are that we can use experimental values of λ_0 and $\mathscr{B}_{1/2}$ to fix the value of σ and experimental values of $(\Delta\lambda)_{sat}$ to assign values of L. This is what we have done. Then, with these values of σ and L we have constructed theoretical estimates of $(\Delta\lambda_\perp)_{sat}$, $\mathscr{B}^{\perp}_{1/2}$, $(\Delta\lambda_\parallel)_{sat}$ and $\mathscr{B}^{\parallel}_{1/2}$. In all cases the agreement between theory and experiment was satisfactory enough to convince us that nothing was amiss and that, in particular, the spherocylinder model was capable of providing an almost quantitative theory of the thermal conductivity Senftleben phenomena (see appendix 2).

As we mentioned previously the experimental measurements of λ_{tr} have been conducted at much lower temperatures than have those of $\Delta\lambda$. At low temperatures molecular collisions are less energetic and penetrating than at high temperatures. Consequently, it is to be expected that the collision cross-section, which is determined primarily by the value of σ, will diminish with increasing temperature. This expectation is confirmed by the fact that smaller values of σ are needed to fit the data for $(\Delta\lambda)_{sat}$ than for λ_{tr}.

Our present plans call for further comparisons between the spherocylinder model and available experimental data and for an extension of the calculations to other rigid molecular models, for example, to 'smoothed' tetrahedra representing spherical top molecules such as CH_4 and CF_4. We shall also conduct a detailed study into the sensitivity of the theoretical predictions to the model chosen for the magnetic moment.

In addition to the consideration of energy transfer presented here there are, of course, other linear transport processes to which the Chapman–Enskog theory can be applied (see equations 1.245 and 1.246 for the shear and bulk viscosity tensors). For these the phenomenology is more complex but the general principles, problems and theoretical techniques differ from those presented above. An excellent survey of the entire area of external-field effects is contained in the previously cited paper by Beenakker[54].

With regard to model calculations the situation is as follows:

(i) The coefficients of shear and bulk viscosity have been calculated for rough spheres[63], for loaded spheres[64] and for spherocylinders[66]. Calculations of the viscous Senftleben effects for the spherocylinder model are currently nearing completion[70].

(ii) Binary and thermal diffusion coefficients have been calculated for rough spheres[63,71] and for loaded spheres[65]. An extensive study of diffusion and thermal diffusion for spherocylinders is currently in progress[72].

(iii) There have been several attempts to extend the kinetic theory of rotating molecules to higher densities. Thus, to deal with transport in a dense gas of rough spheres McCoy, Sandler and Dahler[1] applied the Chapman–Enskog method to a kinetic equation which was a generalization of that which had been used much earlier by Enskog[56] in his theory of transport for smooth spheres. Although the original Enskog theory and its rough sphere variant both suffer from several conceptual deficiencies, the remarkable empirical success of the theory for smooth spheres makes it likely that the corresponding theory for rough spheres will not be altogether unreliable. A similar approach was used by Sandler and Dahler[1] to handle the problems of transport in a dense gas of loaded spheres. For detailed accounts of these theories and of their predictions the reader is referred to the original papers.

(iv) At high densities the flux of each dynamical quantity is the sum of a 'diffusion' and of a 'collisional transfer' contribution (see the first section of this chapter). At very high densities the second of these contributions becomes dominant, and so it is not an unreasonable approximation to totally neglect the former. By adopting this optimistic point of view and by ignoring distortions of the singlet distribution functions Sather and Dahler[73] were able to obtain theoretical estimates for the high density limits of the thermal conductivity and the coefficient of shear, bulk and vortex viscosity (the quantity L_{66} of equation 1.67) for rigid molecular species (rough spheres and spherocylinders) with non-central square-well interactions.

More recently McCoy and Dahler[74] have adapted the dense gas theory of Rice and Allnatt[75] to a fluid composed of rough spheres which have in addition long-range central forces of interaction. The Rice–Allnatt approach should be applicable whenever the molecular interactions can be separated into the sum of a short-range repulsion and a long-range slowly varying interaction. The argument is then that a time trace of the force (and/or torque) experienced by a molecule will consist of a fluctuating background of small random impulses which occur because each molecule is immersed in the overlapping long-range force field of its many neighbours. Superimposed upon this noise will be a few sharp spikes representing the occasional large transfers of momentum (or angular momentum) which accompany collisions between two molecular 'cores'. To implement these ideas one introduces for the distribution functions kinetic equations with collision operators which are sums of Fokker–Planck operators to account for the weak coupling associated with the slowly varying long-range forces and Enskog dense gas terms to account for the core events.

A more interesting variation on this theme is the 'polar liquid' model originally suggested by Condiff and Dahler[39] and currently under development by Luks, Davis and Dahler[76]. In this theory one identifies the short-range interactions with those of smooth spheres and for the long-range contributions one uses non-central interactions such as those between point dipoles.

1.3.2 Second-order Chapman–Enskog Theory

Although at each step one encounters greater and greater complexity, there is no conceptual difficulty in determining the second, third and higher order terms in the series (1.230). At each stage one evaluates the inhomogeneous portion of the appropriate linearized Boltzmann equation (cf 1.232, 1.233, ...) and then resolves it into a sum of terms each of which is the product of an irreducible tensor, $\mathscr{T}_i$, constructed from derivatives of the variables ρ, and a tensor (with the same rank and parity as $\mathscr{T}_i$) function, $\mathscr{S}_i$, of $\mathbf{W}$ and $\mathbf{\Omega}$. Then, because $\hat{\Omega}$ is a linear operator the solution of $\Sigma\mathscr{S}_i \, . \, \mathscr{T}_i = \hat{\Omega}(\phi^{(k)})$ can be written in the form $\phi^{(k)} = \sum_i \sigma_i^{(k)} \cdot \mathscr{T}_i$ where $\sigma_i^{(k)}$ must satisfy the equation $\mathscr{S}_i = \hat{\Omega}(\sigma_i^{(k)})$. From the solution of this equation one then can construct the corresponding contribution

$$J_\psi^{(k)} = \int \mathrm{d}\bar{1} f_1^{(k)} \mathbf{C}_1 \psi_1 = \sum_i n^0 \langle \mathbf{C}\psi\sigma_i^{(k)} \rangle \, . \, \mathscr{T}_i \tag{1.259}$$

with

$$n^0 \langle \alpha \rangle = \int \mathrm{d}\bar{1} f_1^{(0)} \alpha_1 \tag{1.260}$$

to the diffusive flux of the property ψ.

We previously identified the 'marker' ε in the Chapman–Enskog series $f = f^{(0)} + \varepsilon f^{(1)} + \cdots$ with $\varepsilon \approx \lambda_{f_p} |\nabla P|/P = (\mathbf{k}T/\sigma^2)(|\nabla P|/P)p^{-1}$ where P denotes some hydrodynamic variable $(n, T, \mathbf{u})$ and where $p = n\mathbf{k}T$ is the gas pressure. Therefore, the first-order transport coefficients $n^0\langle \mathbf{C}\psi\sigma_i^{(1)} \rangle$ are independent of pressure (as is found experimentally), the second-order coefficients vary inversely with pressure, and so on, with the succeeding terms in this asymptotic series becoming more and more important as the gas pressure diminishes.

The second-order flux contributions $J_\psi^{(2)}$ for structureless molecules were first investigated by Burnett[56]. A corresponding study for rough spheres, loaded spheres and spherocylinders was reported recently[77] and the third-order theory for monatomic species has also been investigated[78]. Those consequences of the second and higher order contributions to the fluxes which have been studied most thoroughly are the modifications they effect upon the structure of weak shocks and upon sonic propagation and absorption. For details the reader is referred to the papers mentioned above and to the references cited therein.

Two years ago Scott, Sturner and Williamson[79] discovered a most remarkable phenomenon. While studying the Einstein–de Haas effect they found that a heated cylinder which was suspended in a polyatomic gas and then immersed in an axially directed magnetic field experienced a torque. Because this torque was not due to spins associated with the cylinder material it was called 'anomalous'.

Since the original discovery, experiments have been performed on a number of gases[80–83], sometimes with modulated magnetic fields[81] and sometimes with variations of the experimental geometry[83]. In all experiments performed to date it is found that the torque varies linearly with the temperature gradient and that it is proportional to the reciprocal of the pressure (at constant H) except at pressures so low that there is evidence of Knudsen effects. The direction of the torque is reversed if the direction of the magnetic field or that of the temperature gradient is reversed. There is a field strength, characteristic of the gas, at which the torque reaches a maximum value. No effect has been observed with the noble gases and the torque does not appear to depend upon the character of the cylinder surface.

It soon was discovered that this 'Scott effect' could not be explained as a simple transfer of molecular rotational angular momentum[68]. Levi and Beenakker[84] then suggested that it arose from a stress which was proportional to $\mathbf{\nabla\nabla}T$ and that this stress could be computed from second-order Chapman–Enskog theory. They also argued that the phenomenon was closely related to the well-studied Senftleben–Beenakker effects. Recently, the Leiden group[85] has extended the theory of Levi and Beenakker and presented a more general and elegant theory based upon the Waldmann–Snider kinetic equation. By exploiting the connexion between this theory and that of Senftleben–Beenakker effects they have been able to demonstrate a considerable degree of internal consistency, e.g. agreement with experiment to within a factor of two. Results of a similar nature were reported in the Levi–Beenakker paper and, independently, by Klein, Hoffman and Dahler[68]. In these theories stress contributions proportional to $\mathbf{\nabla}T\mathbf{\nabla}T$ have been ignored, partly because of the great difficulty of performing such a calculation.

Waldmann[86] has suggested other mechanisms for the Scott effect, all of which depend rather strongly upon the condition and composition of the cylinder surface. However, these contributions to the torque apparently do not dominate since no evidence of such a dependence has been observed. Indeed, the torque is not altered when a metallic cylinder surface is covered with Teflon tape[79]. At the time of writing an unequivocal explanation of the anomolous torque experiment is lacking: possibly the explanation lies in a combination of the Levi–Beenakker mechanism and free-molecule flow or Knudsen effects.

A recent investigation by Park, Dahler, Cooper and Hoffman[87] has been designed to provide a quantitative test of the Levi–Beenakker mechanism. The theory developed by these authors is applicable to a rather broad class of molecular models (both kinematically and with regard to interactions) but the emphasis is upon the two models of rough spheres and loaded spherocylinders for which numerical calculations are presently practical. Since this theory is rather complex and since the numerical results derived from it are not yet in suitable form for presentation we describe it here only in barest detail.

If Knudsen effects are ignored, one can then show that the boundary conditions and equations of motion which are appropriate to the torque experiment are satisfied by a uniform gas pressure (there are negligibly small variations proportional to $(\nabla T)^2$) and by the steady-state velocity and temperature distributions $\mathbf{u} = 0$ and $(T-T_0)/(T-T_i) = [\log(r/r_0)]/[\log(r_i/r_0)]$. Here, T_i is the surface temperature of the suspended cylindrical rod (of radius r_i) and T_0 is the temperature of the outer concentric cylinder (with radius r_0). To be certain that these actually are the distributions encountered in the experiment one must examine their stability relative to other possible solutions of the equations of fluid mechanics. We have not yet completed our theoretical study of this stability but there already exists some experimental evidence that fluid flow begins to occur as the pressure drops into the Knudsen region.

For the conditions we have assumed $\phi^{(1)} = -n(2\mathbf{k}T/m)^{1/2}\mathbf{A}\,.\,\nabla\log T$ and the solubility condition for equation (1.233) reduces to $\partial_1 T = -2(3+r)^{-1}(\mathbf{k}n)^{-1}\nabla\,.\,\mathbf{Q}^{(1)}$. Therefore,

$$(\partial_0 + c_1\,.\,\nabla)f_1^{(1)} = -\frac{2k}{\rho}f_1^{(0)}\left\{\mathbf{A}_1\mathbf{W}_1 : \nabla\nabla T + \left[\left\{\partial_T A_1 + \frac{1}{T}\left(W_1{}^2 + \Omega_1{}^2 - \frac{6+r}{2}\right)A_1\right\}W_1\right] : \nabla T\nabla T\right\} \tag{1.261}$$

$$\partial_1 f_1^{(0)} = \frac{1}{3p} f_1^{(0)} \left(W_1{}^2 + \Omega_1{}^2 - \frac{3+r}{2} \right) [\lambda : \nabla\nabla T + \partial_T \lambda : \nabla T \nabla T] \tag{1.262}$$

and $I_1[f^{(1)}, f^{(1)}] = \mathsf{K}_1 : \nabla T \nabla T$ where

$$\mathsf{K}_1 = (2\mathbf{k}/n^2 mT) I_1[f^{(0)}\mathbf{A}, f^{(0)}\mathbf{A}] \tag{1.263}$$

Then, since the symmetric tensors $\nabla\nabla T$ and $\nabla T \nabla T$ are linearly independent we can express the solution of equation (1.233) in the form

$$\phi^{(2)} = -(nT)^{-1}[\mathsf{E}^{(2)} : \overset{\circ}{\overline{\overline{\nabla\nabla T}}} + \tfrac{1}{3} E^{(0)} \nabla^2 T + \mathsf{F}^{(2)} : \overset{\circ}{\overline{\overline{\nabla T \nabla T}}} + \tfrac{1}{3} F^{(0)} (\nabla T)^2] \tag{1.264}$$

where the scalers $E^{(0)}$ and $F^{(0)}$ and the two symmetric and traceless second rank tensors $\mathsf{E}^{(2)}$ and $\mathsf{F}^{(2)}$ are the solutions of soluble, linear integro-differential equations similar to those encountered in the first-order theory.

In terms of the solutions of these equations the second-order contributions to the fluxes are given by $\mathbf{Q}^{(2)} = 0$ and

$$\mathsf{p}^{(2)} = \int \mathrm{d}\Gamma f_1^{(2)} m \mathbf{C}_1 \mathbf{C}_1 = \mathsf{p}_1^{(2)} + \mathsf{p}_2^{(2)} \tag{1.265}$$

where

$$\begin{aligned} \mathsf{p}_1^{(2)} &= -2\mathbf{k}^0 \langle \mathbf{W}\mathbf{W}\mathsf{E}^{(2)} \rangle : \nabla\nabla T - (2\mathbf{k}/3)^0 \langle \mathbf{W}\mathbf{W} E^{(0)} \rangle \nabla^2 T \\ \mathsf{p}_2^{(2)} &= -2\mathbf{k}^0 \langle \mathbf{W}\mathbf{W}\mathsf{F}^{(2)} \rangle : \nabla T \nabla T - (2\mathbf{k}/3)^0 \langle \mathbf{W}\mathbf{W} F^{(0)} \rangle (\nabla T)^2 \end{aligned} \tag{1.266}$$

Let us now adopt cylindrical coordinates with an axis which is coincident with that of the suspended cylinder. The torque upon the surface of the cylinder is then related to the $r\theta$ component of the pressure by the formula $\tau = -2\pi r_i{}^2 h p_{r\theta}$ where h denotes the length of the cylinder. Since the temperature distribution in the gas is dependent only upon the radial coordinate, $\nabla\nabla T = \hat{r}\hat{r}(\mathrm{d}^2 T/\mathrm{d}r^2) + \hat{\theta}\hat{\theta} r^{-1}(\mathrm{d}T/\mathrm{d}r)$ and $\nabla T \nabla T = \hat{r}\hat{r}(\mathrm{d}T/\mathrm{d}r)^2$. Therefore, according to equations (1.265) and (1.266) the torque should equal the sum of

$$\tau_1 = 4\pi \mathbf{k} h \mathscr{T} [{}^0\langle \mathbf{W}\mathbf{W}\mathsf{E}^{(2)} \rangle_{\theta rrr} - {}^0\langle \mathbf{W}\mathbf{W}\mathsf{E}^{(2)} \rangle_{\theta r\theta\theta}]$$

and

$$\tau_2 = 4\pi \mathbf{k} h \mathscr{T}^2 [{}^0\langle \mathbf{W}\mathbf{W}\mathsf{F}^{(2)} \rangle_{\theta rrr} + \tfrac{1}{3}{}^0\langle \mathbf{W}\mathbf{W} F^{(0)} \rangle_{\theta r}] \tag{1.267}$$

where $\mathscr{T} = (T_i - T_0)/[\log(r_0/r_i)]$ is a constant characteristic of the experimental set-up. As mentioned previously, the experiments performed to date indicate a linear dependence of the torque upon the temperature gradient. From this one would presume τ_2 to be significantly smaller than τ_1, at least for the gradients and experimental geometries which investigators have hitherto employed.

The three tensors which appear in equation (1.267) are invariant with respect to the group of rotations about the cylinder axis, i.e. about the direction of the external field. Therefore, the second rank symmetric tensor ${}^0\langle \mathbf{WW}F^{(0)} \rangle$ must be a linear combination of U and $\hat{h}\hat{h}$ and consequently can contribute nothing to τ_2. Furthermore, it is readily proved that ${}^0\langle \mathbf{WW}\mathsf{E}^{(2)} \rangle_{\theta r\theta\theta} = -{}^0\langle \mathbf{WW}\mathsf{E}^{(2)} \rangle_{\theta rrr}$.

To simplify matters Park, Dahler, Cooper and Hoffman have restricted their studies to trial functions

$$\mathsf{E}_i^{(2)} = (5p/\rho n) \sum_1^2 \phi_i{}^p : \mathsf{E}^p$$
$$\mathsf{F}_i^{(2)} = (5p/\rho n) \sum_1^2 \phi_i{}^p : \mathsf{F}^p \tag{1.268}$$

which include on the two second-order homogeneous polynomials $\phi^1 = \overset{\circ}{\overline{\overline{\mathbf{WW}}}}$ and $\phi^2 = \overset{\circ}{\overline{\overline{\mathbf{\Omega\Omega}}}}$. The fourth rank tensors E^p and F^p are traceless and symmetric in their first and last index pairs.

The computational procedure for $\mathsf{E}^{(2)}$ and $\mathsf{F}^{(2)}$ is then very much like that encountered in the calculation of the thermal conductivity tensor $\boldsymbol{\lambda}$. As we have already stated, the numerical predictions of this theory are not yet ready for presentation. The only calculations which we can report now are estimates for rough spheres[88] obtained from the one-term trial function $\mathsf{E}^{(2)} = (5p/\rho n)\phi^1 : \mathsf{E}^1$. For CO and N_2 at 300°C and 0·05 Torr the computed values for the torque maxima were, respectively, $75{\cdot}3 \times 10^{-3}$ and $68{\cdot}4 \times 10^{-3}$ dyne cm. These are larger by a factor of three than the corresponding experimental values of 24×10^{-3} and 22×10^{-3} dyne cm.

1.3.3 Non-resonant Absorption and Dielectric Relaxation

As a next and relatively uncomplicated example let us apply the theory of Section 1.2 to microwave non-resonant absorption and to dielectric and magnetic relaxation. Our presentation will be patterned closely after that of Birnbaum[89]. The applied field is assumed to be spatially uniform and monochromatic. Therefore, $\mathbf{F}(t\mathbf{x}) = e^{i\omega t}\mathbf{F}_0$ where the vector $\mathbf{F}_0$ is independent of t and $\mathbf{x}$. The frequency ω is not resonant with any pair of rotational levels and so (in the approximation of the 'diagonal ensemble') the kinetic equation appropriate to the gas takes the simple form

$$\partial_t f(1) = \iiint \mathrm{d}1'\,\mathrm{d}2'\,\mathrm{d}2 w_F(12|1'2')[f(1')f(2') - f(1)f(2)] \quad (1.269)$$

We assume that the system is spatially uniform and that the translational degrees of freedom are in equilibrium with the surroundings. Therefore, the distribution function can be written in the form

$$f(1) = g(1)(m\beta/2\pi)^{3/2} \exp\left(-\tfrac{1}{2}\beta mc^2\right)$$

where $g(1)$ is a function of time and of the variables ($\mathscr{J}_1$ of Section 1.2) or quantum numbers which characterize the internal state of molecule 1. We then obtain for g the equation of motion

$$\partial_t g(1) = \sum_{1'} [g(1')A_{11'} - g(1)A_{1'1}] \quad (1.270)$$

with

$$A_{1'1} = \sum_{22'} g(2)B_F(12|1'2') \quad (1.271)$$

$$A_{11'} = \sum_{22'} g(2')B_F(12|1'2')\{{}^0g_F(1){}^0g_F(2)/{}^0g_F(1'){}^0g_F(2')\} \quad (1.272)$$

and where ${}^0g_F(\mathrm{i}) = z^{-1}(\beta, \beta\mathbf{F}) \exp[-\beta(\varepsilon_\mathrm{i} - \boldsymbol{\mu}_\mathrm{i} . \mathbf{F})]$, $z(\beta, \beta\mathbf{F}) = \mathrm{Tr} \exp[-\beta(\hat{\varepsilon} - \hat{\boldsymbol{\mu}} . \mathbf{F})]$ and

$$B_F(12|1'2') = \int \cdots \int \mathrm{d}^3c_1\,\mathrm{d}^3c_2\,\mathrm{d}^3c'_1\,\mathrm{d}^3c'_2 w_F(12|1'2') \quad (1.273)$$

From these formulas it can be verified that at equilibrium (in the presence of the field) the conditions of detailed balance $A_{11'} \exp[-\beta \times (\varepsilon'_1 - \boldsymbol{\mu}'_1 . \mathbf{F})] = A_{1'1} \exp[-\beta(\varepsilon_1 - \boldsymbol{\mu}_1 . \mathbf{F})]$ are satisfied. We now define $\mathscr{A}_{11'} = A_{11'} \exp[-\beta\varepsilon'_1]$ and $\mathscr{A}_{1'1} = A_{1'1} \exp[-\beta\varepsilon_1]$ and assume that these functions of the field strength can be expanded in the series $\mathscr{A}_{11'} = \mathscr{A}^{(0)}_{11'} + F\mathscr{A}^{(1)}_{11'} + \cdots$ and $\mathscr{A}_{1'1} = \mathscr{A}^{(0)}_{1'1} + F\mathscr{A}^{(1)}_{1'1} + \cdots$. Then, from the conditions of detailed balance it follows that $\mathscr{A}^{(0)}_{11'} = \mathscr{A}^{(0)}_{1'1}$ and $\bar{\mathscr{A}}^{(1)}_{11'} = \bar{\mathscr{A}}^{(1)}_{1'1}$ where $\bar{\mathscr{A}}_{11'} = \mathscr{A}^{(1)}_{11'} + \beta\boldsymbol{\mu}'_1 . \mathbf{F}\mathscr{A}^{(0)}_{1'1}$. We conclude that

$$A_{11'} = A^{(0)}_{11'}(1 - \beta\mathbf{F} . \boldsymbol{\mu}_1) + A^{(1)}_{11'}F + 0(F^2) \quad (1.274)$$

where

$$A^{(k)}_{11'}\,\mathrm{e}^{-\beta\varepsilon'_1} = A^{(k)}_{1'1}\,\mathrm{e}^{-\beta\varepsilon_1};\ k = 0, 1 \quad (1.275)$$

These relationships first were derived by van Vleck and Weisskopf[90].

Next we write the distribution function in the form $g(1) = {}^0g(1) + \boldsymbol{\sigma}_1 \,.\, \mathbf{F}_0 + 0(F^2)$ with ${}^0g(i) = ({}^0g_{\mathbf{F}}(i))_{\mathbf{F}=0}$. Then, by neglecting higher-order terms in the field strength we obtain Birnbaum's equation,

$$\partial_t \boldsymbol{\sigma}_1 = \sum_{1'} [(\boldsymbol{\sigma}_{1'} A_{11'} - \boldsymbol{\sigma}_1 A_{1'1}) + \beta\, \mathrm{e}^{\mathrm{i}\omega t}\{{}^0g(1)\boldsymbol{\mu}_1 A_{1'1} - {}^0g(1')\boldsymbol{\mu}_{1'} A_{11'}\}] \quad (1.276)$$

for the 'polarization' $\boldsymbol{\sigma}_1 = \boldsymbol{\sigma}_{\beta_1}(t)$. (Here, and henceforth the symbol $A_{ij}^{(0)}$ is replaced with A_{ij}.) If we pick the direction of the field as the axis of space quantization, then $\boldsymbol{\sigma} \to \sigma_{JM}$, $\boldsymbol{\mu} \to \boldsymbol{\mu}_{JM}$, and equation (1.276) becomes

$$\partial_t \sigma_{JM} = \sum_{J'M'} [(\sigma_{J'M'} A_{JMJ'M'} - \sigma_{JM} A_{J'M'JM}) + \beta\, \mathrm{e}^{\mathrm{i}\omega t}(n_J \mu_{JM} A_{J'M'JM} - n_{J'} \mu_{J'M'} A_{JMJ'M'})] \quad (1.277)$$

with $n_J\ (\equiv {}^0g_J) = n \exp(-\beta\varepsilon_J)/\sum_{J'} (2J'+1)\exp(-\beta\varepsilon_{J'})$.

The equation of change for the polarization $P_J = F_0 \sum_M \mu_{JM}\sigma_{JM}$ now can be manipulated into the form

$$\mathrm{d}P_J/\mathrm{d}t = \sum_{J'} P_J \left(\frac{P_{J'}}{P_J} R_{JJ'} - \sum_{M'} A_{J'M'JM} \right) + \beta F(t)(\Delta\varpi_J) \sum_M n_J \mu_{JM}^2 \quad (1.278)$$

where

$$\Delta\varpi_J = \sum_{J'M'} A_{J'M'JM}(1 - \mu_{J'M'}/\mu_{JM}) = \sum_{J'} (S_{J'J} - R_{J'J}) \quad (1.279)$$

$$R_{JJ'} = I_{J'}^{-1} \sum_{MM'} \mu_{JM}\mu_{J'M'} A_{J'M'JM} = (I_J n_J / I_{J'} n_{J'}) R_{J'J} \quad (1.280)$$

$I_J = \sum_M \mu_{JM}^2$, and $S_{J'J} = \sum_{M'} A_{J'M'JM}$. With $\mathrm{d}P_J/\mathrm{d}t = \mathrm{i}\omega P_J$ we obtain the set of algebraic equations

$$P_J = \beta F n_J (\Delta\varpi_J) I_J / \mathrm{i}\omega + \Delta'\varpi_J) \quad (1.281)$$

where

$$\Delta'\varpi_J = \sum_{J'M'} A_{J'M'JM} - \sum_{J'} (P_{J'}/P_J) R_{JJ'} = \sum_{J'M'} A_{J'M'JM}(1 - G_{J'}\mu_{J'M'}/G_J \mu_{JM}) \quad (1.282)$$

and $G_J(\omega) = (\Delta\varpi_J)/(\mathrm{i}\omega + \Delta'\varpi_J)$.

If the dipoles are magnetic then P_J is the magnetization per unit volume associated with molecules of angular momentum J. If the molecular dipoles are electric then P_J is the corresponding contribution to the polarization $\sum_J P_J$. The dielectric constant $\varepsilon(\omega)$ is related to this polarization by the formula $\varepsilon - 1 = (4\pi/F)\sum_J P_J$. In the limit $\omega \to 0$, $\Delta'\varpi_J = \Delta\varpi_J$ and we obtain the usual expression, $\varepsilon(0) = 1 + 4\pi n\beta \sum_{JM} n_J \mu_{JM}^2$, for the static dielectric constant. In many cases the experimental relaxation spectra are consistent with the assumption that $G_{J'}/G_J \approx 1$. To this approximation $\Delta'\varpi_J$ is equal to $\Delta\varpi_J$ even when the frequency differs from zero. The equations (1.281) are then simply formulas for computing the polarization in terms of the transition probabilities $A_{J'M'JM}$ and the dipole moments μ_{JM}.

It often happens that the experimental data can be represented by a single Debye term. Under these circumstances it is reasonable to introduce an average Debye frequency

$$\langle \Delta\omega \rangle \equiv \tau_d^{-1} = \sum_{JM} \sum_{J'M'} n_J \mu_{JM}^2 A_{J'M'JM}(1 - \mu_{J'M'}/\mu_{JM}) \Big/ \sum_{JM} n_J \mu_{JM}^2$$

$$= \left(\sum_{JM} n_J \mu_{JM}^2 \right)^{-1} \left[\sum_{J_1 M_1} \cdots \sum_{J'_2 M'_2} n_{J_1} n_{J_2} \mu_{J_1 M_1} (\mu_{J_1 M_1} - \mu_{J'_1 M'_1}) \times B_{\mathrm{F}}(J_1 M_1 J_2 M_2 | J'_1 M'_1 J'_2 M'_2) \right] \tag{1.283}$$

which is so defined that

$$P = \sum_J P_J = \beta F \sum_{JM} n_J \mu_{JM}^2 \frac{\Delta\varpi_J}{i\omega + \Delta'\varpi_J} \approx \beta F \left(\sum_{JM} n_J \mu_{JM}^2 \right) \frac{1}{1 + i\omega\tau_d} \tag{1.284}$$

We soon shall encounter this same 'dielectric relaxation time' τ_d in another, rather different context.

Birnbaum has estimated the cross-sections σ in the 'sudden approximation'. His studies indicate that the observed rates of dielectric relaxation cannot be explained by dipole–dipole scattering alone. It would appear that the situation can be set right only if one takes account of the close encounters which involve the short-ranged, strongly repulsive molecular interactions. We postpone a discussion of other exploratory calculations until the following section.

The classical theory of dielectric relaxation differs scarcely at all from its quantal counterpart. One has only to interpret $g(i)$ as a function $g(\mathbf{L}_i, t)$ of the angular momentum rather than of the corresponding quantum operator and/or quantum numbers, to replace the summation over states in the fashion $\sum_1 \to \int \mathrm{d}\mathscr{I}_1$ (that is, $\int \mathrm{d}^3L_1$ or $\int \mathrm{d}^3L_1/L_1$ in the case of spherical tops or rigid rotors, respectively) and substitute in place of the quantum cross-section the appropriate phase-averaged classical quantity. In particular, the classical expression for the Debye relaxation time of equation (1.283) is

$$\tau_d^{-1} = n\langle \boldsymbol{\mu} ; \boldsymbol{\mu}\rangle /{}^0\langle \mu^2\rangle \tag{1.285}$$

where the integral $\langle \boldsymbol{\mu} ; \boldsymbol{\mu}\rangle = \mathrm{Tr}\,\langle \boldsymbol{\mu}, \boldsymbol{\mu}\rangle$ is one of the 'matrix elements',

$$\langle \psi, \phi\rangle \equiv \int \mathrm{d}1 \psi_1 \hat{\Xi}_1(\phi) \tag{1.286}$$

of a linear operator,

$$n^2\hat{\Xi}_1(\phi) = \iiint \mathrm{d}1'\,\mathrm{d}2'\,\mathrm{d}2 f_1^{(0)} f_2^{(0)} w(12|1'2')(\phi_1 - \phi_{1'}) \tag{1.287}$$

which is very closely related to the operator $\hat{\Gamma}$ defined by equation (1.65). Thus, it can be proved that $\hat{\Xi}$ shares all six of the characteristics which we previously have attributed to $\hat{\Gamma}$ and that the integrals $\langle \psi, \phi\rangle$ and $\langle \psi ; \phi\rangle$ exhibit all properties of the corresponding integrals $[\psi, \phi]$ and $[\psi ; \phi]$ except those which specifically involve collisional invariants.

Here, just as in the case of gas transport coefficients, one can invoke the rough sphere, loaded sphere and spherocylinder models to obtain numerical estimates of the pertinent phenomenological coefficient, that is, the dielectric relaxation time. Although these models are obviously unable to simulate the long-range dipolar interactions of real polar molecules they should be adequate for estimating the contribution to the relaxation time due to 'core collisions'.

The integrals $\langle a, b\rangle$ also arise in the theory of

$$\mathbf{F}_{\mathrm{fr}}(1) = \iiint \mathrm{d}1'\,\mathrm{d}2'\,\mathrm{d}2 w(12|1'2') f(2)(\mathbf{p}_1 - \mathbf{p}'_1) \tag{1.288}$$

and

$$\mathbf{G}_{\mathrm{fr}}(1) = \iiint \mathrm{d}1'\,\mathrm{d}2'\,\mathrm{d}2 w(12|1'2') f(2)(\mathbf{L}_1 - \mathbf{L}'_1)$$

which are, respectively, the frictional force and torque acting upon a molecule in the state $(\mathbf{c}_1, \mathscr{J}_1)$[91]. From the properties of $\hat{\Xi}$ it is easily proved that $\mathbf{F}_{\text{fr}}$ will contain a 'Magnus force' contribution, proportional to $\mathbf{L} \times \mathbf{p}$, if and only if $\hat{\Xi}$ is non-selfadjoint[92].

1.3.4 The Method of Moments

Experimental conditions may be such that the Chapman–Enskog method is inapplicable. This is particularly likely to be so when the studies involve relaxation processes with characteristic times which are far less than those encountered in typical hydrodynamic or heat-transfer experiments. It will also be the case when the experiment involves two or more processes with significantly different time scales, as for example with spin diffusion and spin relaxation. Although it is certainly possible to cite many other examples these few clearly demonstrate that there do exist physically interesting and experimentally accessible situations which are not included in the quasi steady-state description of gas dynamics associated with the hydrodynamic time scale, $t_{\text{hy}} \approx L_{\text{hy}}/\bar{c} \approx P/|\nabla P|\bar{c}$, and to which the Chapman–Enskog method is appropriate.

A practical procedure for dealing with problems of this sort is the so-called method of moments. The general idea is to expand the distribution function in some complete set of functions and then to consider each member of this basis as a 'normal mode' with which one can hope to associate a characteristic relaxation time and between which there will be characteristic coupling coefficients. According to this picture the behaviour of the gas is to be described in terms of the amplitudes of the basis modes; the focus is shifted from the distribution function to its Fourier coefficients. From this there arises very naturally a language and a corresponding way of thinking. Thus, an experiment can be conceived (and perhaps even realized in the laboratory) which would involve the initial excitation of a single mode and a response which then would be measured in terms of the rate of decay of the amplitude of this mode, the strength and rate of propagation of it and of other coupled modes to distant parts of the fluid, and the steady or equilibrium state to which the system ultimately tends. Of course, one expects that when and if such a steady state is reached, its description as provided by the moment method will coincide with that derived from the normal solution of the Chapman–Enskog theory.

Since the collision term of the Boltzmann equation is a quadratic functional of $f(\mathbf{txc}\mathscr{J})$, the equations of motion for the amplitudes of the normal modes are generally non-linear. However, if the system lies sufficiently near some state of reference, we can replace this collision term with a linear functional of the difference. Specifically, if $f = {}^0f(1+\phi)$ where ${}^0f$ (of equations 1.162 and 1.163) refers to a reference state which closely approximates that of the actual system, then $\partial_c f \approx -n^2\hat{\Gamma}(\phi)$. Suppose now that we select for our basis a set of orthonormal scalar-valued functions $\psi_k = \psi_k(\mathbf{c}, \mathscr{J})$ which are so defined that ${}^0\langle\psi_k\psi_{k'}\rangle = \delta_{kk'}$*. Next, we substitute the expansion

$$\phi = \sum_j a_j(\mathbf{tx})\psi_j \tag{1.289}$$

into the linearized Boltzmann equation and obtain for the amplitudes a_j the equations of motion

$$n\partial_t a_m + \sum_j [\nabla . \{n^0\langle\mathbf{c}\psi_m\psi_j\rangle a_j\} + \{n^0\langle\psi_m[\overline{\psi_j, H^e}]\rangle + n^2[\psi_m, \psi_j]\}a_j] = 0 \tag{1.290}$$

From these one immediately identifies $\tau_m = (n_0[\psi_m, \psi_m])^{-1}$ to be the relaxation time characteristic of the mth normal mode. Furthermore, from the properties of $\hat{\Gamma}$ it follows that the coupling coefficients $[\psi_m, \psi_n]$ can differ from zero only if the product $\psi_m\psi_n$ contains a piece which transforms according to the totally symmetric representation of the rotation group $0_3{}^+$: if the functions ψ_m are chosen to transform as bases of irreducible representations of this group, then $[\psi_m, \psi_n]$ vanishes unless ψ_m and ψ_n belong to the same irreducible representation. We see also that $\sum_j n^0\langle\mathbf{c}\psi_m\psi_j\rangle a_j$, the flux of ψ_m, depends exclusively upon the amplitudes of modes which have the opposite parity.

Waldmann and his coworkers, Hess in particular, have developed this approach to a high degree and have used it to excellent advantage as a means of analysing the types of couplings which govern relaxation and transport in polyatomic gases[25,39]. This method seems to have originated with Clerk Maxwell[47] who used it to investigate a variety of processes occurring in dilute monatomic gases. The procedure then was refined

* We tacitly assume the basis to be discrete. There is something to be gained from using the eigenfunctions of $\hat{\Gamma}$ as a basis. However, these are almost never available in closed form and, even if they were, the non-selfadjointness of $\hat{\Gamma}$ would work somewhat to our disadvantage since the basis functions belonging to different eigenvalues of $\hat{\Gamma}$ are then not mutually orthogonal.

and extended by H. Grad[57,93] for the purpose (among others) of providing an alternative to the second- and higher-order Chapman–Enskog theory for very low pressure gases. Other modifications and applications of the moment method have arisen from research based upon the Krook and Fokker–Planck equations.

Here, we shall not reproduce the systematic studies of the moment equations which already have been reported in the literature[25,39,94]. It is our intention, instead, simply to illustrate the method by applying it to a few representative situations. The first problem to be faced is that of selecting the basis set $\{\psi_k\}$ of equation (1.289). In this we follow Hess and Waldmann[25] and choose the complete set of orthogonal functions which are obtained by combining Cartesian components of the two sets of irreducible tensors $\Phi^n_{lm}(\mathbf{W}) \propto S^{(n)}_{l+1/2}(W^2)W^l Y_{lm}(\hat{W})$ and $\Psi^n_{lm}(\boldsymbol{\Omega}) \propto S^{(n)}_{l-1+r/2}(\Omega^2)$ $\Omega^l Y_{lm}(\hat{\Omega})$ which are appropriate to the spaces of $\mathbf{W}$ and $\boldsymbol{\Omega}$, respectively. Actually, we shall be dealing only with a very few members of this set and so there is no reason to concern ourselves with formulas for the general term.*

One significant advantage of this basis is that the expansion coefficients of the first few terms can be identified with quantities which have obvious and familiar physical interpretations. To illustrate this let us select the quantities n_0, $\mathbf{u}_0$ and T_0 of ${}^0f$ equal to the local values of concentration, fluid velocity and temperature, respectively. One can then show that

$$f_1 = {}^0f_1\left[1+\frac{3}{r}\left(\frac{2}{\mathbf{k}T\Gamma}\right)^{1/2}\mathbf{m}\,.\,\boldsymbol{\Omega}_1+\frac{6}{rp}\left(\frac{m}{\Gamma}\right)^{1/2}\mathsf{C}:\boldsymbol{\Omega}_1\mathbf{W}_1+\frac{4}{5p}(\mathbf{k}T/m)\right.$$
$$\times\,\mathbf{Q}_{\mathrm{tr}}\,.\,\mathbf{C}_1\left(W_1{}^2-\frac{5}{2}\right)+\frac{4}{rp}(\mathbf{k}T/m)\mathbf{Q}_{\mathrm{rot}}\,.\,\mathbf{C}_1\left(\Omega_1{}^2-\frac{r}{2}\right)+\frac{1}{p}\mathsf{P}:\overline{\overline{\mathbf{W}_1\mathbf{W}_1}}^{\circ}$$
$$\left.+(\theta_{\mathrm{tr}}-1)\left(W_1{}^2-\frac{3}{2}\right)+(\theta_{\mathrm{rot}}-1)\left(\Omega_1{}^2-\frac{r}{2}\right)+\cdots\right] \tag{1.291}$$

where $\mathbf{m}$ and C denote the density and flux of spin, where $\mathsf{P} \equiv \mathsf{p}-\delta p$ is the viscous portion of the pressure tensor, and where $\mathbf{Q}_{\mathrm{tr}}$ ($\equiv \mathbf{Q}_K$) and $\mathbf{Q}_{\mathrm{rot}}$ ($\equiv \mathbf{Q}_I$) are the energy fluxes defined in Section 1.1. Finally $\theta_{\mathrm{tr}} = T_{\mathrm{tr}}/T$ and $\theta_{\mathrm{rot}} = T_{\mathrm{rot}}/T$ with $\frac{3}{2}\mathbf{k}T_{\mathrm{tr}} = \langle \frac{1}{2}mC^2\rangle$ and $\frac{1}{2}r\mathbf{k}T_{\mathrm{rot}} = \langle L^2/2\Gamma\rangle$.

The quantities $\mathbf{m}$, C, $\mathbf{Q}_{\mathrm{tr}}$, $\mathbf{Q}_{\mathrm{rot}}$, P, $\theta_{\mathrm{tr}}-1$ and $\theta_{\mathrm{rot}}-1$ are 'moments' of the exact distribution function which play well-established rôles in the phenomenological description of fluid mechanics: the coefficients of the

* Quite a few of these functions are displayed in References 25 and 92.

higher-order terms in equation (1.291) are less familiar. Now, one of the most important goals of transport theory is to derive approximate equations which can be solved for the low-order moments which are experimentally observable. However, from a glance at equation (1.153) we see that the equation of motion for a particular moment is invariably coupled to higher order moments. For example, the viscous stress P enters into the equation of motion

$$\rho\, \mathrm{d}_t \mathbf{u} = -\nabla p - \nabla . \mathsf{P}$$

for the fluid velocity. And when one goes a step further to the equation,

$$\partial_t \mathsf{P} + \nabla . [\mathbf{u}\mathsf{P} + n\langle m\mathbf{C}\overset{\circ}{\overline{\overline{\mathbf{CC}}}}\rangle] + 2\overset{\circ}{\overline{\overline{\mathsf{P} . \nabla\mathbf{u}}}} + 2p\overset{\circ}{\overline{\overline{\nabla\mathbf{u}}}} = n\partial_c\langle m\overset{\circ}{\overline{\overline{\mathbf{CC}}}}\rangle \quad (1.292)$$

for P one encounters an additional moment, the 'flux of viscous stress', $n\langle m\mathbf{C}\overset{\circ}{\overline{\overline{\mathbf{CC}}}}\rangle$. What one would like to have is a closed pair of equations for the velocity and pressure tensor and this is precisely what the moment method can provide. The first step is to truncate the expansion of the distribution function after the terms which are given explicitly in equation (1.291). To this approximation the stress flux is given by

$$n\langle m\mathbf{C}\overset{\circ}{\overline{\overline{\mathbf{CC}}}}\rangle \approx (4\rho/5p)\mathbf{Q}_{\mathrm{tr}} . {}^{0}\langle \mathbf{C}\mathbf{C}\overset{\circ}{\overline{\overline{\mathbf{CC}}}}(W^2 - 5/2)\rangle = (8p/5\rho)\mathbf{Q}_{\mathrm{tr}} . \Delta \quad (1.293)$$

with $\Delta_{ijkl} = \frac{1}{2}\delta_{ik}\delta_{jl} + \frac{1}{2}\delta_{il}\delta_{jk} - \frac{1}{3}\delta_{ij}\delta_{kl}$. To make things as uncomplicated as possible we shall assume that there is no flux of energy so that $n\langle m\mathbf{C}\overset{\circ}{\overline{\overline{\mathbf{CC}}}}\rangle$ vanishes.

Next, we linearize the right-hand side of equation (1.292) to obtain

$$n\partial_c\langle m\overset{\circ}{\overline{\overline{\mathbf{CC}}}}\rangle \approx -n^2[m\overset{\circ}{\overline{\overline{\mathbf{CC}}}}, \phi] \approx -n^2[m\overset{\circ}{\overline{\overline{\mathbf{CC}}}}, \overset{\circ}{\overline{\overline{\mathbf{WW}}}}] : \mathsf{P}/p = -\tau_p^{-1}\mathsf{P} \quad (1.294)$$

where

$$\tau_p = (5/2n)[\overset{\circ}{\overline{\overline{\mathbf{WW}}}}; \overset{\circ}{\overline{\overline{\mathbf{WW}}}}]^{-1} \quad (1.295)$$

As a result of these steps the formal identity (1.292) reduces to the differential equation

$$\partial_t \mathsf{P} + \nabla . (\mathbf{u}\mathsf{P}) + 2\overset{\circ}{\overline{\overline{\mathsf{P} . \nabla\mathbf{u}}}} + 2p\overset{\circ}{\overline{\overline{\nabla\mathbf{u}}}} = -\tau_p^{-1}\mathsf{P} \quad (1.296)$$

which involves only P and $\mathbf{u}$. Since this is the goal proclaimed earlier, we could pass on immediately to some other example. However, the value of the result just obtained will become more apparent if we carry the analysis a bit further. To first order in the gradient of velocity the second and third terms of equation (1.296) can be neglected. Then, by integrating the resulting equation one obtains the 'linear constitutive equation',

$$\mathsf{P}^{(1)}(t) = \mathsf{P}(0)\,\mathrm{e}^{-t/\tau_p} - 2\eta \int_0^t \frac{\mathrm{d}s}{\tau_p}\,\mathrm{e}^{-(t-s)/\tau_p}\,\overline{\overline{\nabla \mathbf{u}}}^{\circ}(s) \tag{1.297}$$

where $\eta = p\tau_p$ is precisely equal to the coefficient of shear viscosity which results from the Chapman–Enskog theory.

From this we conclude that the viscous stress relaxes to its 'normal form'—that given by the (first-order) Chapman–Enskog theory—within a time of the order of τ_p. After this brief period of adjustment the pressure tensor becomes a functional of the lower moments $\mathbf{u}$, n and T. A very much longer 'hydrodynamic' interval of time must pass before these lower moments adjust to the values characteristic of mechanical and thermodynamic equilibrium. Finally, we learn from equation (1.297) that the flux of momentum is determined by the entire mechanical (and thermal) history of the system although, to be sure, the effective retention of the system's memory is really only of the order of the relaxation time τ_p.

Now it might seem that to obtain the non-linear analogue of equation (1.297) one would need only to include the second and third terms of equation (1.296) which we previously omitted. However, this would be inconsistent since our derivation of equation (1.296) included a neglect of quadratic terms in ϕ. Although most investigations based upon the moment method have been restricted to the linear approximation for the collision term, there is no real need to do so. Thus, we retain the quadratic terms and, to simplify matters, concern ourselves only with the simple distortion $\phi = \overline{\overline{\mathbf{WW}}}^{\circ}:\mathsf{P}/p$. In place of equation (1.294) we then obtain

$$n\partial_c\langle m\overline{\overline{\mathbf{CC}}}^{\circ}\rangle = -\mathsf{P}/\tau_p - \overline{\overline{\mathsf{P}\,.\,\mathsf{P}}}^{\circ}/2p\tau'_p \tag{1.298}$$

where the algebraic sign of the coefficient

$$\frac{1}{\tau'_p} = \frac{16}{5n^3}\iiiint \mathrm{d}1\,\mathrm{d}2\,\mathrm{d}1'\,\mathrm{d}2' w(1'2'|12)\,{}^0\!f_1\,{}^0\!f_2$$

$$\times\,\boldsymbol{\delta}:[(\overline{\overline{\mathbf{W}_1\mathbf{W}_1}}^{\circ} - \overline{\overline{\mathbf{W}'_1\mathbf{W}'_1}}^{\circ})\,.\,(\overline{\overline{\mathbf{W}_1\mathbf{W}_1}}^{\circ})\,.\,(\overline{\overline{\mathbf{W}_2\mathbf{W}_2}}^{\circ})] \tag{1.299}$$

can be either positive or negative. When this is substituted into equation (1.292) and the result integrated one finds that

$$\mathsf{P}(t) = \mathsf{P}^{(1)}(t) - \tau_p \int_0^t \frac{ds}{\tau_p} e^{-(t-s)\tau_p} [\nabla . \{\mathbf{u}(s)P(s)\} + 2\overline{\overline{\overset{\circ}{\mathsf{P}(s) . \nabla u(s)}}}]$$

$$- \frac{\tau_p}{2p\tau'_p} \int_0^t \frac{ds}{\tau_p} e^{-(t-2s)/\tau_p} \overline{\overline{\overset{\circ}{\mathsf{P}(s) . \mathsf{P}(s)}}}$$

$$= \mathsf{P}^{(1)}(t) + \mathsf{P}^{(2)}(t) + \cdots \tag{1.300}$$

where the second-order term is given by

$$\mathsf{P}^{(2)}(t) = 2\eta\tau_p \int_0^t \frac{ds'}{\tau_p} e^{-(t-s)/\tau_p} \int_0^s \frac{ds'}{\tau_p} e^{-(s-s')/\tau_p}$$

$$\times [\nabla . \{\mathbf{u}(s) \overline{\overline{\overset{\circ}{\nabla \mathbf{u}(s')}}}\} + 2 \overline{\overline{\overset{\circ}{\overline{\overline{\overset{\circ}{\nabla \mathbf{u}(s')}}} . \nabla \mathbf{u}(s)}}}$$

$$- \frac{\tau_p}{\tau'_p} \int_0^s \frac{ds''}{\tau_p} e^{-(s-s'')/\tau_p} \overline{\overline{\overset{\circ}{\overline{\overline{\overset{\circ}{\nabla \mathbf{u}(s')}}} . \overline{\overline{\overset{\circ}{\nabla \mathbf{u}(s'')}}}}}}] \tag{1.301}$$

Since τ_p is very much smaller than the usual time scale of **u**, equations (1.297) and (1.301) can be replaced with $\mathsf{P}^{(1)} = -2\eta \overline{\overline{\overset{\circ}{\nabla \mathbf{u}}}}$ and

$$\mathsf{P}^{(2)} = 2\eta\tau_p [\mathbf{u} . \nabla (\overline{\overline{\overset{\circ}{\nabla \mathbf{u}}}}) + \tfrac{4}{3} (\nabla . \mathbf{u}) \overline{\overline{\overset{\circ}{\nabla \mathbf{u}}}} + (1 - \tau_p/\tau'_p) \overline{\overline{\overset{\circ}{\overline{\overline{\overset{\circ}{\nabla \mathbf{u}}}} . \overline{\overline{\overset{\circ}{\nabla \mathbf{u}}}}}}}] \tag{1.302}$$

respectively. Although there is no difficulty in extending the calculation to higher order in the derivatives of **u**, there is little point in doing so here.

To the same approximation used to obtain equation (1.294) we find that

$$\partial_c \langle \tfrac{1}{2} m C^2 \rangle = -\partial_c \langle L^2 / 2\Gamma \rangle \approx -n[\tfrac{1}{2} m C^2, \phi]$$

$$\approx -p\{[W^2, W^2 - 3/2](\theta_{tr} - 1)$$

$$+ [W^2, \Omega^2 - r/2](\theta_{rot} - 1)\} = -\frac{1}{\tau_T} \frac{3\mathbf{k}}{2} (T_{tr} - T_{rot}) \tag{1.303}$$

and

$$\partial_c\langle \mathbf{L}\rangle \approx -n[\mathbf{L}, \phi] \approx -\frac{3}{r}\left(\frac{2}{\mathbf{k}T\Gamma}\right)^{1/2}[\mathbf{L}, \mathbf{\Omega}]\,.\,\mathbf{m} = -\frac{1}{\tau_{sp}}\mathbf{m} \quad (1.304)$$

where the two relaxation times are given by

$$\tau_T = (2/3n)[W^2; W^2]^{-1}; \qquad \tau_{sp} = (r/2n)[\mathbf{\Omega}; \mathbf{\Omega}]^{-1} \quad (1.305)$$

Therefore, in the event that $\mathbf{Q}_{tr} = \mathbf{Q}_{rot} = \overset{\circ}{\overline{\overline{\nabla \mathbf{u}}}} = 0$ we obtain from equations (1.36) and (1.41) the equations of change $\dot{T}_{rot} = -(T_{rot} - T_{tr})/\tau_{rot}$ and

$$\partial_t(T_{tr} - T_{rot}) + (p/nc_{tr})\nabla\,.\,\mathbf{u} = -(T_{tr} - T_{rot})/\tau_T(c_{rot}/c_v) \quad (1.306)$$

where $c_{tr} = 3\mathbf{k}/2$, $c_{rot} = r\mathbf{k}/2$, $c_v = c_{tr} + c_{rot}$ and $\tau_{rot} = \tau_T(c_{rot}/c_{tr})$. The steady-state solution of equation (1.306) is $T_{tr} - T_{rot} = -T(\mathbf{k}\tau_{rot}/c_v)(\nabla\,.\,\mathbf{u})$. Then, since $c_{tr}T_{tr} + c_{rot}T_{rot} = c_v T$ it follows that

$$\theta_{tr} = 1 - (\mathbf{k}\tau_{rot}c_{rot}/c_v^2)\nabla\,.\,\mathbf{u}$$

and so, in this steady state the isotropic part of the pressure tensor is given by

$$\mathsf{p}^{(iso)} = \boldsymbol{\delta}\tfrac{1}{3}n\langle mC^2\rangle = \boldsymbol{\delta}p\theta_{tr} = \boldsymbol{\delta}(p - \eta_b\nabla\,.\,\mathbf{u}) \quad (1.307)$$

where the formula $\eta_b = p\tau_{rot}(\mathbf{k}c_{rot}/c_v^2)$ for the coefficient of bulk viscosity is precisely that of the Chapman–Enskog theory (cf. equation 42 of Reference 28).

Although there is more that can be wrung from this specific example, we shall leave it now. The results it has yielded are constitutive relationships with far less restrictive ranges of applicability than those of the quasi-steady-state Chapman–Enskog theory, the same formulas for phenomenological coefficients as can be obtained from that theory, and information about dynamical couplings among the various macrovariables which is not at all obvious from the steady-state approach.

1.3.5 Spin Diffusion and Relaxation

The next example we consider is that of spin diffusion and relaxation. We study a dilute gas composed of molecules or atoms which have magnetic moments. This system is immersed in a uniform, steady field

H and displaced from equilibrium by some means, say for example by an intermittent radiofrequency signal. One then records a response which consists either of the relaxation of the excess magnetization which has been induced by the disturbance or of the diffusion of spins which follows upon the imposition of the rf signal. The rate of attenuation of spin echoes provides one means for studying these processes. We are concerned with the behaviour of the system only after the perturbing signal has ceased and so the external field which appears in the kinetic equation not only will be spatially uniform but also independent of time. To deal with this particular situation and with the remaining examples of this section we shall rely upon the Boltzmann (or Wang Chang–Uhlenbeck) equation instead of the equation of Waldmann and Snider. Actually, it makes very little difference which equation one chooses since it is usually so easy to transcribe results from one theory to the other*.

The magnetization and its flux are related to the molecular magnetic moment $\bar{\boldsymbol{\mu}} = \bar{\gamma}\mathbf{L}$ by the definitions $\mathbf{M} = n\langle\bar{\boldsymbol{\mu}}\rangle$ and $\mathsf{J}_\mathrm{M} = n\langle\mathbf{C}\bar{\boldsymbol{\mu}}\rangle$, respectively, and satisfy the equations of change[95]

$$\partial_t\mathbf{M}+\boldsymbol{\nabla}\,.\,[\mathbf{uM}+\mathsf{J}_\mathrm{M}]-n\langle\bar{\gamma}\bar{\boldsymbol{\mu}}\rangle\times\mathbf{H} = n\partial_\mathrm{c}\langle\bar{\boldsymbol{\mu}}\rangle \tag{1.308}$$

$$\partial_t\mathsf{J}_\mathrm{M}+\boldsymbol{\nabla}\,.\,[\mathbf{u}\mathsf{J}_\mathrm{M}+n\langle\mathbf{CC}\bar{\boldsymbol{\mu}}\rangle]-n\langle\bar{\gamma}\mathbf{C}\bar{\boldsymbol{\mu}}\rangle\times\mathbf{H}$$
$$+(\boldsymbol{\nabla}\mathbf{u})^\dagger\,.\,\mathsf{J}_\mathrm{M}-\rho^{-1}(\boldsymbol{\nabla}\,.\,\mathsf{p})\mathbf{M} = n\partial_\mathrm{c}\langle\mathbf{C}\bar{\boldsymbol{\mu}}\rangle \tag{1.309}$$

We next write the distribution function in the form $f = {}^0f_\mathrm{F}(1+\phi)$ and then neglect quadratic terms in ϕ to obtain the approximation

$$n\partial_\mathrm{c}\langle\psi\rangle \approx -n^2[\psi,\phi]_\mathrm{F} \equiv \iiiint \mathrm{d}1\,\mathrm{d}2\,\mathrm{d}1'\,\mathrm{d}2'\,{}^0f_\mathrm{F}(1)\,{}^0f_\mathrm{F}(2)$$
$$\times\, w(12|\,1'2')\psi_1(\phi_{1'}+\phi_{2'}-\phi_1-\phi_2) \tag{1.310}$$

For present purposes the simple two-term distortion

$$\phi_i = \mathbf{a}\,.\,\bar{\boldsymbol{\mu}}_i+\mathbf{b}\,.\,\bar{\boldsymbol{\mu}}_i\times\mathbf{C}_i \tag{1.311}$$

* In the case of spin relaxation there is a significant argument favouring use of the Waldmann–Snider equation over that of Wang Chang and Uhlenbeck. For in the approximation of the diagonal ensemble, to which the latter of these equations applies, there is no transverse component of magnetization. Consequently, one cannot even hope to compute the corresponding spin relaxation time. In practice the distinction between the two is not nearly so momentous, for in the high-temperature limit to which we confine ourselves here the longitudinal and transverse relaxation times are equal. Both theories lead to the same phenomenological equations of change for the magnetization and spin flux and to identical formulas for the relaxation time and for the coefficient of spin diffusion.

will suffice. Thus, one finds immediately that $\mathbf{a} = 3\mathscr{M}/n^0\langle\bar{\boldsymbol{\mu}}^2\rangle$ and $\mathbf{b} = (m/2\mathbf{k}T)(\boldsymbol{\varepsilon}:\mathbf{J}_{\mathrm{M}})/n^0\langle\bar{\boldsymbol{\mu}}^2\rangle$ where $\mathscr{M} = \mathbf{M}-\mathbf{M}_0$ is the deviation of the magnetization from its stationary value of $\mathbf{M}_0 = n^0\langle\bar{\boldsymbol{\mu}}\rangle$. And furthermore, to this approximation $n\langle\mathbf{CC}\bar{\boldsymbol{\mu}}\rangle \approx (\mathbf{k}T/m)\boldsymbol{\delta}\mathbf{M}$, $n\langle\bar{\gamma}\mathbf{C}\bar{\boldsymbol{\mu}}\rangle = \bar{\bar{\gamma}}\mathbf{J}_{\mathrm{M}}$ with $\bar{\bar{\gamma}} = {}^0\langle\bar{\gamma}\bar{\boldsymbol{\mu}}^2\rangle/{}^0\langle\bar{\boldsymbol{\mu}}^2\rangle$ and $n\langle\bar{\gamma}\bar{\boldsymbol{\mu}}\rangle = n^0\langle\bar{\gamma}\boldsymbol{\mu}\rangle + \bar{\bar{\gamma}}\mathscr{M}$ where ${}^0\langle\bar{\gamma}\bar{\boldsymbol{\mu}}\rangle$ is parallel to H. Finally,

$$n\partial_{\mathrm{c}}\langle\bar{\boldsymbol{\mu}}\rangle \approx -n^2[\bar{\boldsymbol{\mu}},\bar{\boldsymbol{\mu}}]\,.\,\mathbf{a} - n^2[\bar{\boldsymbol{\mu}},\bar{\boldsymbol{\mu}}\times\mathbf{C}]\,.\,\mathbf{b}$$
$$= -\mathbf{a}\tfrac{1}{3}n^2[\bar{\boldsymbol{\mu}};\bar{\boldsymbol{\mu}}] = -\mathscr{M}/\tau_{\mathscr{M}} \tag{1.312}$$

and

$$n\partial_{\mathrm{c}}\langle\mathbf{C}\bar{\boldsymbol{\mu}}\rangle \approx \boldsymbol{\varepsilon}\,.\,\mathbf{b}\tfrac{1}{6}n^2[\bar{\boldsymbol{\mu}}\times\mathbf{C};\bar{\boldsymbol{\mu}}\times\mathbf{C}] = -\mathbf{J}_{\mathrm{M}}/\tau'_{\mathscr{M}} \tag{1.313}$$

where

$$\tau_{\mathscr{M}} = {}^0\langle\bar{\boldsymbol{\mu}}^2\rangle/(n[\bar{\boldsymbol{\mu}};\bar{\boldsymbol{\mu}}]); \qquad \tau'_{\mathscr{M}} = {}^0\langle\bar{\boldsymbol{\mu}}^2\rangle/(2n[\bar{\boldsymbol{\mu}}\times\mathbf{W};\bar{\boldsymbol{\mu}}\times\mathbf{W}]) \tag{1.314}$$

When the gyromagnetic ratio $\bar{\gamma}$ is independent of $\mathbf{L}$, $\tau_{\mathscr{M}}$ reduces to the spin–relaxation time τ_{sp} defined in equation (1.305). Chen and Snider[96] have used a variant of the first Born approximation in order to evaluate the cross-sections for the spin transitions caused by dipole–dipole interactions. For an atom with a hard core of diameter d and a nuclear spin I they obtained the same relaxation time, $\tau_{\mathrm{sp}}^{-1} = 2I(I+1)\times \gamma^4\hbar^2 d^{-2} n(\pi m\beta)^{1/2}$, as had been obtained earlier by the correlation function method. Just as in the case of dielectric relaxation, this dipole–dipole mechanism is found to be too inefficient to account for the measured values of the spin–relaxation time. To correct this deficiency Torrey proposed a spin–rotation interaction of the form $(-\gamma\hbar/\mu r^2)f(r) \times(\mathbf{I}+\mathbf{I}_1)\,.\,(\mathbf{r}\times\mathbf{p})$. Chen and Snider have confirmed Torrey's estimate of the corresponding relaxation time by using precisely the same kinetic theory and scattering approximations as they employed in the case of dipole–dipole interactions.

At very low temperatures particle statistics play a significant rôle in determining the correct forms of the quantum scattering amplitudes. We have made no attempt to include these effects in our considerations. The interested reader is referred to a paper by Emery[97] which deals with the effects of particle statistics upon the coefficient of spin diffusion.

To apply our results let us assume the fluid is at rest ($\mathbf{u} = 0$) and that the concentration and temperature fields are spatially uniform. Then, from equations (1.308), (1.309), (1.312) and (1.313) we obtain the coupled differential equations.

$$\partial_t\mathscr{M} + \omega_{\mathrm{H}}\hat{h}\times\mathscr{M} + \mathbf{j} + \tau_{\mathscr{M}}^{-1}\mathscr{M} = 0 \tag{1.315}$$

and

$$\partial_t \mathbf{j} + \omega_H \hat{h} \times \mathbf{j} + \omega_{sp} \mathscr{D}_{sp} \Delta \mathscr{M} + \omega_{sp} \mathbf{j} = 0 \tag{1.316}$$

where, in Hess' notation[98], $\omega_H = \bar{\bar{\gamma}} H$, $\omega_{sp} = (\tau'_{\mathscr{M}})^{-1}$, $\mathscr{D}_{sp} = \mathbf{k}T/m\omega_{sp}$ and $\mathbf{j} = \nabla . \mathsf{J}_M$.

As a first approximation we neglect the inertial and field-dependent terms in equation (1.315). This yields the 'constitutive relation' $\mathbf{j} = -\mathscr{D}_{sp} \Delta \mathscr{M}$ (or $\mathsf{J}_M = -\mathscr{D}_{sp} \nabla \mathscr{M}$) between the spin flux and the gradient of the spin deviation. By substituting this relationship into equation (1.315) we arrive at the equation,

$$\partial_t \mathscr{M} + \omega_H \hat{h} \times \mathscr{M} - \mathscr{D}_{sp} \Delta \mathscr{M} + \mathscr{M}/\tau_{\mathscr{M}} = 0 \tag{1.317}$$

of Bloch[99] and Torrey[100] and, consequently, are able to identify $\mathscr{D}_{sp}$ and $\tau_{\mathscr{M}}$, respectively, with the coefficient of spin diffusion (which is *not* equal to the ordinary diffusion coefficient) and the spin relaxation time.

To this approximation the spin flux depends solely upon the gradient of the excess magnetization $\mathscr{M}$. The conditions to which this simple description of spin diffusion is applicable have recently been established by Hess[98]. He has also derived an equation of change for the magnetization which is valid under conditions where that of Bloch and Torrey is not. To derive Hess' equation one neglects only the inertial term $\partial_t \mathbf{j}$. The solution of the resulting, linear algebraic equation can be written in the form

$$\mathbf{j} = -\mathscr{D}^{\|} \hat{h}\hat{h} . \Delta \mathscr{M} - D^{\perp}(\Delta \mathscr{M} - \hat{h}\hat{h} . \Delta \mathscr{M}) + D^{tr} \hat{h} \times \Delta \mathscr{M} \tag{1.318}$$

where $D^{\|} = \mathscr{D}_{sp}$, $D^{\perp} = (1+\varphi^2)^{-1} \mathscr{D}_{sp}$, and $D^{tr} = \varphi(1+\varphi^2)^{-1} \mathscr{D}_{sp}$. The quantity $\varphi = \omega_H/\omega_{sp} = \bar{\bar{\gamma}} H m \beta \mathscr{D}_{sp}$ which appears in these expressions can be identified with the number of times the magnetic moment precesses during the interval between two successive molecular collisions. With this constitutive relation one obtains Hess' modification,

$$\partial_t \mathscr{M} + \left(\omega_H + \frac{\varphi}{1+\varphi^2} \mathscr{D}_{sp} \Delta \right) \hat{h} \times \mathscr{M} + (\tau_{sp}^{-1} - \mathscr{D}_{sp} \Delta) \hat{h}\hat{h} . \mathscr{M}$$

$$+ \left(\tau_{sp}^{-1} - \frac{1}{1+\varphi^2} \mathscr{D}_{sp} \Delta \right) (\mathscr{M} - \hat{h}\hat{h} . \mathscr{M}) = 0 \tag{1.319}$$

of the Bloch–Torrey equation.

This equation reduces to the Bloch–Torrey equation, (1.317), whenever φ is very much less than unity. Thus, if $D^{tr} = \varphi(1+\varphi^2)^{-1} \mathscr{D}_{sp}$ and $D^{\|} - D^{\perp} = \varphi^2(1+\varphi^2)^{-1} \mathscr{D}_{sp}$ are to differ significantly from zero, φ must be of the order of unity. For example, this will be so provided that $\omega_H \approx 10^8 \text{ sec}^{-1}$, $p \approx 100$ Torr, and that $\mathscr{D}_{sp}$ is assumed to be approximately equal to the coefficient of self-diffusion. As yet no n.m.r. or spin diffusion experiments have been performed under these or other conditions which would permit a direct test of Hess' theory. Nevertheless, there is little reason to doubt its correctness.

Actually, it is a trivial matter to obtain from equations (1.315) and (1.316) uncoupled partial differential equations for $\mathscr{M}$ and $\mathbf{j}$. We differentiate the first of these equations with respect to time and eliminate $\mathbf{j}$ and $\partial_t \mathbf{j}$ from the resulting expression by use of equations (1.315) and (1.316). This leads directly to the differential equation

$$\ddot{\mathscr{M}} + (\omega_{sp} + \tau_{\mathscr{M}}^{-1})\dot{\mathscr{M}} + \omega_{sp}\tau_{\mathscr{M}}^{-1}\mathscr{M} - \omega_{sp}\mathscr{D}_{sp}\Delta\mathscr{M} + 2\omega_H \hat{h} \times \dot{\mathscr{M}}$$
$$+ \omega_H(\omega_{sp} + \tau_{\mathscr{M}}^{-1})\hat{h} \times \mathscr{M} + \omega_H{}^2 \hat{h} \times (\hat{h} \times \mathscr{M}) = 0 \tag{1.320}$$

To simplify this equation we transform to a frame of reference which rotates with the Larmor frequency $\boldsymbol{\omega} = -\hat{h}\omega_H = -\bar{\gamma}\mathbf{H}$. Thus, $\dot{\mathscr{M}} \to \dot{\mathscr{M}} - \omega_H \hat{h} \times \mathscr{M}$, $\ddot{\mathscr{M}} \to \ddot{\mathscr{M}} - 2\omega_H \hat{h} \times \dot{\mathscr{M}} + \omega_H{}^2 \hat{h} \times (\hat{h} \times \mathscr{M})$, and equation (1.320) reduces to

$$\dot{\mathscr{M}} - \mathscr{D}\Delta\mathscr{M} + \frac{1}{\tau}\mathscr{M} + \frac{\mathscr{D}}{c^2}\ddot{\mathscr{M}} = 0 \tag{1.321}$$

wherein $\mathscr{D} = \mathscr{D}_{sp}\omega_{sp}/(\omega_{sp} + \tau_{\mathscr{M}}^{-1})$, $\tau = \tau_{\mathscr{M}}(\omega_{sp} + \tau_{\mathscr{M}}^{-1})/\omega_{sp}$ and $c^2 = \mathbf{k}T/m$.

To appreciate the significance of the terms $\ddot{\mathscr{M}}$ which appear in equations (1.320) and (1.321) it is important to note that a parabolic differential equation such as (1.317) does not describe the propagation of spin 'signals' from one part of the gas to another but instead a collective motion of the entire medium, a motion which tends to 'smooth out' irregularities of the spin deviation. According to this equation an excess magnetization in one region stimulates an *immediate* response throughout the entire system! However, according to the 'telegrapher's equations', (1.320) and (1.321), spin deviations are propagated in a wavelike fashion and, like all disturbances with finite velocities of propagation, are confined to the interior of a characteristic signal cone. Telegrapher's equations of this sort are standard products of the moment method and have been discussed elsewhere[101].

1.3.6 Electric Polarization

We have now become acquainted with a number of situations in which gradients of temperature, velocity and other macroscopic variables give rise to a partial alignment of molecular angular momentum or spin. The resulting anisotropy of the gas can be observed in a variety of ways, many of which have already been discussed in connexion with the Senftleben–Beenakker–Scott phenomena. Another and rather exotic consequence of this anisotropy (conjectured by Waldmann[102] several years ago and recently formalized by Waldmann and Hess[103]) is the induction of electric polarization by a temperature gradient. To see how this can happen we recall that the axis of a freely rotating (polar) symmetric top molecule precesses very rapidly about the direction of the rotational angular momentum, **L**. This motion produces a phase-averaged dipole moment with the magnitude $\bar{\gamma}L = \mu_0 K/L$ and the direction of **L**. Therefore, if the spin distribution can be suitably distorted, it is possible for a net electric polarization to result.

The analysis of this phenomenon differs scarcely at all from that which we have just conducted for the magnetization. Unlike its magnetic counterpart, the electric moment $\bar{\boldsymbol{\mu}} = \bar{\gamma}\mathbf{L}(\bar{\gamma} = \mu_0 K/L^2$, with K a constant, for symmetric tops) is a polar vector; in place of **M**, $\mathscr{M}$ and J_M we use the symbols **P**, $\mathscr{P}$ and J_P, respectively. Then, to account for the polarizing action of a temperature gradient we add to the truncated moment expansion (1.311) the term $\mathbf{C}(W^2+\Omega^2-7/2).(4\mathbf{Q}/7p)$ where $\mathbf{Q} = \mathbf{Q}_K+\mathbf{Q}_\iota$ is the conductive flux of energy. This term contributes nothing to **P**, J_P, $n\langle\mathbf{CC}\bar{\boldsymbol{\mu}}\rangle$, or $n\partial_c\langle\mathbf{C}\boldsymbol{\mu}\rangle$, but in place of equation (1.312) we now obtain

$$
\begin{aligned}
n\partial_c\langle\bar{\boldsymbol{\mu}}\rangle &\approx -n^2[\bar{\boldsymbol{\mu}}, \bar{\boldsymbol{\mu}}_1].\mathbf{a} - n^2[\bar{\boldsymbol{\mu}}, \mathbf{C}(W^2+\Omega^2-7/2)].(4\mathbf{Q}/7p) \\
&= -\mathscr{P}/\tau_{\mathscr{P}} - ({}^0\langle\bar{\boldsymbol{\mu}}^2\rangle^{1/2}/\tau_{\mathscr{Q}})\mathbf{Q}/(\mathbf{k}T)(2\mathbf{k}T/m)^{1/2} \qquad (1.322)
\end{aligned}
$$

where $\tau_{\mathscr{P}}$ is the electric analogue of $\tau_{\mathscr{M}}$ and where

$$\tau_{\mathscr{Q}} = 21\,{}^0\langle\bar{\boldsymbol{\mu}}^2\rangle^{1/2}/(2n[\bar{\boldsymbol{\mu}};\mathbf{W}(W^2+\Omega^2-7/2)]) \qquad (1.323)$$

is a phenomenological coefficient whose algebraic sign can be either positive or negative, depending upon the nature of the molecular interactions.

In the steady state $n\partial_c\langle\bar{\boldsymbol{\mu}}\rangle$ will vanish and so there is established a (density independent) polarization

$$\mathscr{P} = -\frac{\tau_{\mathscr{P}}}{\tau_{\mathscr{Q}}}{}^0\langle\bar{\boldsymbol{\mu}}^2\rangle^{1/2}\mathbf{Q}/(\mathbf{k}T)(2\mathbf{k}T/m)^{1/2} \approx v\nabla\mathrm{T} \qquad (1.324)$$

with

$$\nu = (2\lambda/21)\{{}^{0}\langle\bar{\boldsymbol{\mu}}^2\rangle/[\bar{\boldsymbol{\mu}};\bar{\boldsymbol{\mu}}]\}$$
$$\times\{[\bar{\boldsymbol{\mu}};\mathbf{W}(W^2+\Omega^2-7/2)]/(\mathbf{k}T)(2\mathbf{k}T/m)^{1/2}\} \qquad (1.325)$$

and where λ is the thermal conductivity.

Although we shall pursue this topic no further, it is clearly possible to generate for electric polarization a precise analogue of the spin relaxation and diffusion theory given above.

1.3.7 Electric Permeability

As a final example we consider the dynamical behaviour of the symmetric, traceless tensor, $\mathscr{A} = n\langle\overset{\circ}{\overline{\overline{ll}}}\rangle$, which previously was shown to be proportional to the anisotropic portion of the electric permeability. The sources of interest in this quantity, which plays a central rôle in the theory of flow birefringence and the spectrum of depolarized Rayleigh scattering, will become apparent as we proceed. The equations of change for $\mathscr{A}$ and its flux, $\mathsf{J}_{\mathscr{A}} = n\langle\mathbf{C}\overset{\circ}{\overline{\overline{ll}}}\rangle$, are given by

$$\partial_t\mathscr{A}+\boldsymbol{\nabla}.[\mathbf{u}\mathscr{A}+\mathsf{J}_{\mathscr{A}}] = \overline{n\langle[\overset{\circ}{\overline{\overline{ll}}}, H^{\mathrm{e}}]\rangle}+n\partial_{\mathrm{c}}\langle\overset{\circ}{\overline{\overline{ll}}}\rangle \qquad (1.326)$$

and

$$\partial_t\mathsf{J}_{\mathscr{A}}+\boldsymbol{\nabla}.[\mathbf{u}\mathsf{J}_{\mathscr{A}}+n\langle\mathbf{C}\mathbf{C}\overset{\circ}{\overline{\overline{ll}}}\rangle]+(\boldsymbol{\nabla}\mathbf{u})^{\dagger}.\mathsf{J}_{\mathscr{A}}$$
$$-\rho^{-1}(\boldsymbol{\nabla}.\mathsf{p})\mathscr{A} = \overline{n\langle[\mathbf{C}\overset{\circ}{\overline{\overline{ll}}}, H^{\mathrm{e}}]\rangle}+n\partial_{\mathrm{c}}\langle\mathbf{C}\overset{\circ}{\overline{\overline{ll}}}\rangle \qquad (1.327)$$

and it is readily verified that the external field terms in both vanish identically. Since we shall be interested in coupling between $\mathscr{A}$ and gradients of temperature and velocity, we select for the distortion ϕ the four-term moment expansion

$$\phi = (15/2n)\overset{\circ}{\overline{\overline{ll}}}:\mathscr{A}+(15/n)(m/2\mathbf{k}T)^{1/2}\overset{\circ}{\overline{\overline{ll}}}\mathbf{W}\vdots\mathsf{J}_{\mathscr{A}}+\overset{\circ}{\overline{\overline{\mathbf{W}\mathbf{W}}}}:\mathsf{P}/p$$
$$+\mathbf{C}(W^2+\Omega^2-7/2).(4\mathbf{Q}/7p)(m/\mathbf{k}T) \qquad (1.328)$$

From this one finds that $n\langle CC\overset{\circ}{\overline{\overline{ll}}}\rangle \approx (\mathbf{k}T/m)\boldsymbol{\delta}\mathscr{A}$

$$n\partial_{\mathrm{c}}\langle\overset{\circ}{\overline{\overline{ll}}}\rangle \approx -\mathscr{A}/\tau_{\mathscr{A}}-(\mathsf{P}/p)/\tau_{\mathsf{P}} \qquad (1.329)$$

and

$$n\partial_c\langle \mathbf{C}\overset{\circ}{\overline{\overline{ll}}}\rangle \approx -15n[\mathbf{W}\overset{\circ}{\overline{\overline{ll}}}, \overset{\circ}{\overline{\overline{ll}}}\mathbf{W}]:\mathsf{J}_{\mathscr{A}} \approx -\frac{1}{\tau'_{\mathscr{A}}}\mathsf{J}_{\mathscr{A}} \tag{1.330}$$

where the three characteristic relaxation times are given by the formulas

$$\begin{aligned} \tau_{\mathscr{A}} &= (2/3n)[\overset{\circ}{\overline{\overline{ll}}}; \overset{\circ}{\overline{\overline{ll}}}]^{-1} \\ \tau_{\mathrm{P}} &= (5/n)[\overset{\circ}{\overline{\overline{ll}}}; \overset{\circ}{\overline{\overline{\mathbf{WW}}}}]^{-1} \end{aligned} \tag{1.331}$$

and

$$\tau'_{\mathscr{A}} = (1/n)[\mathbf{W}\overset{\circ}{\overline{\overline{ll}}}; \overset{\circ}{\overline{\overline{ll}}}\mathbf{W}]^{-1}$$

Therefore, to the approximation defined by equation (1.328), the dynamical behaviour of n, $\mathbf{u}$, T, $\mathscr{A}$, $\mathsf{J}_{\mathscr{A}}$, $\mathbf{Q}$ and P are governed by the equation of continuity, $\mathrm{d}_t n - n\nabla\cdot\mathbf{u} = 0$, the fluid mechanical equation of motion (1.32) of Section 1.1, the energy balance $nc_V\,\mathrm{d}_t T - \nabla\cdot\mathbf{Q} - \mathsf{p}:\nabla\mathbf{u} = 0$, equation (1.296) of this section and a similar equation for $\partial_t\mathbf{Q}$,

$$\partial_t\mathscr{A} + \nabla\cdot[\mathbf{u}\mathscr{A} + \mathsf{J}_{\mathscr{A}}] + \mathscr{A}/\tau_{\mathscr{A}} + (\mathsf{P}/p)/\tau_{\mathrm{P}} = 0 \tag{1.332}$$

and

$$\partial_t\mathsf{J}_{\mathscr{A}} + \nabla\cdot\left[\mathbf{u}\mathsf{J}_{\mathscr{A}} + \frac{p}{\rho}\boldsymbol{\delta}\mathscr{A}\right] + (\nabla\mathbf{u})^{\dagger}\cdot\mathsf{J}_{\mathscr{A}} - \rho^{-1}(\nabla\cdot\mathsf{p})\mathscr{A} + \mathsf{J}_{\mathscr{A}}/\tau'_{\mathscr{A}} = 0 \tag{1.333}$$

There are many applications of these equations but we shall consider only two. Specifically, we neglect $\mathsf{J}_{\mathscr{A}}$ in equation (1.332) and so conclude that to first order in P,

$$\mathscr{A}(t, \mathbf{x}) = \mathscr{A}(0, \mathbf{x})\,\mathrm{e}^{-t/\tau_{\mathscr{A}}} - (\tau_{\mathscr{A}}/\tau_{\mathrm{P}})\int_0^t \frac{\mathrm{d}s}{\tau_{\mathscr{A}}}\,\mathrm{e}^{-(t-s)/\tau}\,\{\mathsf{P}(s)/p\} \tag{1.334}$$

In the steady state we then obtain from equation (1.334) [or directly from 1.332] the relationship (first derived by Hess[104]),

$$\mathscr{A} = -\frac{\tau_{\mathscr{A}}}{\tau_{\mathrm{P}}}\left(\frac{\mathsf{P}}{p}\right) \approx \frac{\tau_{\mathscr{A}}}{\tau_{\mathrm{P}}}\frac{2\eta}{p}\overset{\circ}{\overline{\overline{\nabla\mathbf{u}}}} \tag{1.335}$$

between the polarization $\mathscr{A}$ and the shear field $\overset{\circ}{\overline{\overline{\nabla\mathbf{u}}}}$. From this one concludes that a steady shear field gives rise to an anisotropic electric permeability $\overset{\circ}{\overline{\overline{\boldsymbol{\varepsilon}}}} = -2\beta\overset{\circ}{\overline{\overline{\nabla\mathbf{u}}}}$ where β is the pressure independent quantity

$$\beta = 4\pi n(\alpha_{\|} - \alpha_{\perp})(\tau_{\mathscr{A}}/\tau_{\mathrm{P}})(\eta/p) \tag{1.336}$$

This optical anisotropy is to be associated with an alignment of molecular rotational angular momentum whereas the 'streaming birefringence' observed in liquids and colloidal suspensions is due to an alignment of molecular axes.

Here, as with electric dipolarization it is only a matter of a few minutes' work to derive analogues of the Bloch–Torrey, Hess and telegrapher's equations: from there one then can proceed to an examination of the propagation, diffraction and attenuation of magnetization (or spin), dipolarization or electric permeability disturbances. A relatively uncomplicated but very important example of this is provided by Hess' recent study[105] of depolarized Rayleigh scattering. The spectrum of this light is described by the function

$$S(\omega|\mathbf{k}) = \frac{1}{\pi}\,\mathrm{Re}\int_0^\infty \mathrm{d}t\, \mathrm{e}^{\mathrm{i}\omega t} A(t, \mathbf{k}) \tag{1.337}$$

where A, defined by $\qquad \tilde{\mathscr{A}}(t, \mathbf{k}) = A(t|\mathbf{k})\tilde{\mathscr{A}}(0, \mathbf{k})$

contains the time dependence of the Fourier transform,

$$\tilde{\mathscr{A}}(t, \mathbf{k}) = \int \mathrm{d}^3x\, \mathrm{e}^{-\mathrm{i}\mathbf{k}.\mathbf{x}} \mathscr{A}(t, \mathbf{x})$$

of the electric permeability.

To determine $\mathscr{A}(t, \mathbf{x})$, and hence $A(t|\mathbf{k})$ and $S(\omega|\mathbf{k})$, Hess proceeded from equations (1.332) and (1.333) to the Bloch–Purcell equation (without external field terms)

$$\partial_t \mathscr{A} - D_{\mathrm{T}}\Delta\mathscr{A} + \omega_{\mathrm{T}}\mathscr{A} = 0 \tag{1.338}$$

where $D_{\mathrm{T}} = (\mathbf{k}T/m)\tau'_{\mathscr{A}}$ and $\omega_{\mathrm{T}} = \tau_{\mathscr{A}}^{-1}$. To this approximation, $A(t|\mathbf{k}) = \exp\{-(\omega_{\mathrm{T}} + \mathbf{k}^2 D_{\mathrm{T}})t\}$ and the spectrum of the depolarized Rayleigh light is given by the Lorentzian $S_{\mathrm{L}}(\omega|\mathbf{k}) = (\omega_{\mathrm{T}} + \mathbf{k}^2 D_{\mathrm{T}})/\{\omega^2 + (\omega_{\mathrm{T}} + \mathbf{k}^2 D_{\mathrm{T}})^2\}\pi$. Hess has also examined the consequences of using more accurate approximation than equation (1.338) for the equation of motion of the permeability. (Some comments added in proof, see appendix 3).

ACKNOWLEDGMENTS

Our thanks are extended to Dr Siegfried Hess for providing us with the results of his recent research prior to their publication. We also gratefully acknowledge financial support by the National Science Foundation for a portion of the studies reported here.

REFERENCES

1. B. J. McCoy, S. I. Sandler and J. S. Dahler, *J. Chem. Phys.*, **45**, 3485 (1966); S. I. Sandler and J. S. Dahler, *J. Chem. Phys.*, **46**, 3520 (1967).
2. A. C. Eringen, *J. Math. Mech.*, **16**, 1 (1966).
3. D. W. Condiff and J. S. Dahler, *Phys. Fluids*, **7**, 842 (1964).
4. J. S. Dahler in *Research Frontiers in Fluid Dynamics*, (Ed. George Temple and Raymond Seeger), Interscience, New York, 1965, Chap. 15.
5. H. S. Green, *J. Math. Phys.*, **2**, 344 (1961).
6. M. S. Green, *J. Chem. Phys.*, **20**, 1281; **22**, 398 (1954).
7. J. G. Kirkwood and D. D. Fitts, *J. Chem. Phys.*, **33**, 1317 (1960).
8. Robert Zwanzig, *J. Chem. Phys.*, **33**, 1338 (1960); in *Lectures in Theoretical Physics*, Interscience, N.Y., Vol. 3, 1961.
9. B. Robertson, *Phys. Rev.*, **144**, 151 (1966).
10. R. A. Piccirelli, *Phys. Rev.*, **175**, 77 (1968).
11. J. C. McLennan, Jr., *Phys. Rev.*, **115**, 1405 (1959); *Phys. Fluids*, **3**, 493 (1960); *Phys. Fluids*, **4**, 1319 (1961).
12. D. N. Zubarev, *Dokl. Akad. Nauk SSSR*, **6**, 776 (1962); **10**, 452 (1965); **10**, 526 (1965); **10**, 850 (1966).
13. J. S. Dahler, *Phys. Rev.*, **129**, 1464 (1963).
14. B. Robertson, *Phys. Rev.*, **153**, 391 (1967).
15. C. Swartz and J. S. Dahler (to be published).
16. D. K. Hoffman and J. S. Dahler, *J. Stat. Physics*, **1**, 521 (1970).
17. S. Watanabe, *Rev. Mod. Phys.*, **27**, 26 (1955).
18. C. S. Wang Chang, G. E. Uhlenbeck and J. de Boer, in *Studies in Statistical Mechanics*, (Ed. J. de Boer and G. E. Uhlenbeck), Vol. 2, Interscience, N.Y., 1964.
19. L. Waldmann, *Z. Naturforsch.*, **12a**, 660 (1957); **13a**, 609 (1958).
20. J. S. Dahler, *J. Chem. Phys.*, **30**, 1447 (1959).
21. R. F. Snider, *J. Chem. Phys.*, **32**, 1051 (1960).
22. S. Hess, *Z. Naturforsch.*, **22a**, 1871 (1967).
23. R. F. Snider, *J. Math. Phys.*, **5**, 1580 (1964).
24. F. R. McCourt and R. F. Snider, *J. Chem. Phys.*, **43**, 2276 (1965); **46**, 2387 (1967); **47**, 4117 (1967).
25. L. Waldmann, *Z. Naturforsch.*, **15a**, 19 (1960); **18a**, 1033 (1963); S. Hess and L. Waldmann, *Z. Naturforsch.*, **21a**, 1529 (1966).
26. L. Waldmann, *Handbuch d. Physik*, Springer-Verlag, Berlin, 1955, Bd. XII.
27. E. A. Mason and L. Monchick, *J. Chem. Phys.*, **36**, 1622 (1962).

28. L. Monchick, K. S. Yun and E. A. Mason, *J. Chem. Phys.*, **39**, 654 (1963).
28. L. Monchick, A. N. G. Pereira and E. A. Mason, *J. Chem. Phys.*, **42**, 3241 (1965); see also L. Monchick, R. J. Munn and E. A. Mason, *J. Chem. Phys.*, **45**, 3051 (1966).
30. C. F. Curtiss, *J. Chem. Phys.*, **21**, 2045 (1953); G. Gioumousis and C. F. Curtiss, *J. Chem. Phys.*, **29**, 996 (1958); *J. Math. Phys.*, **2**, 96 (1961); *J. Math. Phys.*, **3**, 1059 (1962); C. F. Curtiss, *J. Chem. Phys.*, **48**, 1725 (1968); Louis Biolsi and C. F. Curtiss, *J. Chem. Phys.*, **48**, 4508 (1968); C. F. Curtiss, *J. Chem. Phys.*, **49**, 1952 (1968); C. F. Curtiss and R. B. Bernstein, *J. Chem. Phys.*, **50**, 1168 (1969); R. W. Fenstermaker, C. F. Curtiss and R. B. Bernstein, *J. Chem. Phys.*, **51**, 2439 (1969).
31. J. J. Mueller and C. F. Curtiss, *J. Chem. Phys.*, **46**, 298 (1967); **46**, 1252 (1967).
32. C. F. Curtiss and A. Hardisson, *J. Chem. Phys.*, **46**, 2618 (1967).
33. C. C. Rankin and J. C. Light, *J. Chem. Phys.*, **51**, 1701 (1969); D. Russell and J. C. Light, *J. Chem. Phys.*, **51**, 1720 (1969).
34. R. C. Gordon, *J. Chem. Phys.*, **51**, 14 (1969).
35. W. A. Lester, Jr. and R. B. Bernstein, *J. Chem. Phys.*, **48**, 4896 (1968).
36. B. R. Johnson and D. Secrest, *J. Chem. Phys.*, **48**, 4682 (1968).
37. R. J. Cross, Jr., *J. Chem. Phys.*, **46**, 609 (1967); **47**, 3724 (1967); H. K. Shin, *J. Chem. Phys.*, **41**, 2864 (1964); **42**, 59 (1965); **46**, 744 (1967); **47**, 3302 (1967); **48**, 3644 (1968); *Intern. J. Quantum Chem.*, **2**, 265 (1968); R. T. Pack and J. S. Dahler, *J. Chem. Phys.*, **50**, 2397 (1969).
38. L. Monchick, S. I. Sandler, E. A. Mason, *J. Chem. Phys.*, **49**, 1178 (1968); S. I. Sandler, *Phys. Fluids*, **11**, 2549 (1968).
39. D. W. Condiff and J. S. Dahler, *J. Chem. Phys.*, **44**, 3988 (1966); B. J. McCoy and J. S. Dahler (unpublished); S. Hess, *Z. Naturforsch.*, **23a**, 597 (1968); **23a**, 1095 (1968).
40. T. F. Morse, *Phys. Fluids*, **7**, 159 (1964); F. B. Hanson and T. F. Morse, *Phys. Fluids*, **10**, 345 (1967); see also C. A. Brau, *Phys. Fluids*, **10**, 48 (1967).
41. R. J. Cross and D. R. Herschbach, *J. Chem. Phys.*, **43**, 3530 (1965).
42. L. Monchick and E. A. Mason, *J. Chem. Phys.*, **35**, 1676 (1961).
43. J. G. Parker, *Phys. Fluids*, **2**, 449 (1959).
44. J. van der Ree, *Physica*, **37**, 553 (1967).
45. P. Resibois, *Physica*, **25**, 725 (1959); J. R. N. Miles and J. S. Dahler, *J. Chem. Phys.*, (manuscript in press).
46. J. R. N. Miles and J. S. Dahler, *J. Chem. Phys.*, (manuscript in press).
47. J. H. Jeans, *Dynamical Theory of Gases*, Cambridge University Press, New York, 1904.
48. E. W. Becker and E. Doernenburg, *Naturwissenschaften*, **37**, 165 (1950); E. W. Becker and W. Heyrich, *J. Chem. Phys.*, **26**, 911 (1952).
49. A. E. de Vries, A. Haring and W. Slots, *Physica*, **22**, 247 (1956).
50. J. Schirdewahn, A. Klemm and L. Waldmann, *Z. Naturforsch.*, **16a**, 133 (1961); see also C. J. G. Slieker, *thesis*, Amsterdam, 1964; *Physica*, **31**, 1388 (1965).
51. C. F. Curtiss, *J. Chem. Phys.*, **24**, 225 (1956).
52. D. K. Hoffman, *J. Chem. Phys.*, **50**, 4823 (1969).

53. V. D. Borman, B. I. Nikolaev and N. I. Nikolaev, *Zh. Eksperim i Teor. Fiz.*, **50**, 821 (1966) [English transl., *Soviet Phys. JETP*, **23**, 544 (1966); see also A. C. Levi, F. R. McCourt and A. Tip, *Physica*, **39**, 165 (1968).
54. For a very excellent bibliography of work in this area see J. J. M. Beenakker, *Festkörperprobleme VIII*, Friedr. Vieweg, Braunschweig, Germany, 1968.
55. H. Grad, *Third Intern. Symp. Rarified Gas Dynamics*, Vol. I, Academic Press, New York (1963).
56. S. Chapman and T. G. Cowling, *The Mathematical Theory of Non-Uniform Gases*, Cambridge University Press, New York (1953).
57. H. Grad, in *Handbuch d. Physik*, (Ed. S. Flügge), Springer-Verlag, Berlin, 1958, Bd. XII.
58. H. Grad, *Phys. Fluids*, **6**, 147 (1963).
59. L. L. Gorelik and V. V. Sinitsin, *Physica*, **41**, 486 (1969).
60. J. Korving, *Thesis*, University of Leiden, 1967; see also Reference (54).
61. H. Senftleben and J. Pietzner, *Z. Physik*, **35**, 986 (1934); *Ann. Physik.*, **27**, 108 (1936); **27**, 117 (1936); **30**, 541 (1937).
62. L. L. Gorelik and V. V. Sinitsyn, *JETP*, **46**, 401 (1964); L. L. Gorelik, Yu. N. Redkoborody and V. V. Sinitsyn, *JETP*, **48**, 761 (1965) and additional papers cited in Reference 54.
63. D. W. Condiff, W.-K. Lu and J. S. Dahler, *J. Chem. Phys.*, **42**, 3445 (1965).
64. S. I. Sandler and J. S. Dahler, *J. Chem. Phys.*, **43**, 1750 (1965).
65. S. I. Sandler and J. S. Dahler, *J. Chem. Phys.*, **47**, 2621 (1967).
66. S. I. Sandler and J. S. Dahler, *J. Chem. Phys.*, **44**, 1229 (1966).
67. W. M. Klein, J. S. Dahler, E. Cooper and D. K. Hoffman, *J. Chem. Phys.*, **52**, 4752 (1970).
68. W. M. Klein, D. K. Hoffman and J. S. Dahler, *J. Chem. Phys.*, **49**, 2321 (1968).
69. F. R. McCourt, H. F. P. Knaap, H. Moraal, *Physica*, **43**, 485 (1969).
70. E. Cooper and D. K. Hoffman, *J. Chem. Phys.*, (to be published).
71. E. Trübenbacher, *Z. Naturforsch.*, **17a**, 539, 936 (1962).
72. D. K. Hoffman and J. S. Dahler, (unpublished).
73. N. F. Sather and J. S. Dahler, *Phys. Fluids*, **5**, 754 (1962).
74. B. J. McCoy and J. S. Dahler, *J. Chem. Phys.*, **50**, 2411 (1969).
75. S. A. Rice and A. R. Allnatt, *J. Chem. Phys.*, **34**, 2144 (1961); **34**, 2156 (1961).
76. K. Luks, H. T. Davis, J. S. Dahler, (to be published).
77. B. J. McCoy and J. S. Dahler, *Phys. Fluids*, **12**, 1392 (1969).
78. C. Y. Cha and B. J. McCoy, (to be published).
79. G. G. Scott, H. W. Sturner and R. M. Williamson, *Phys. Rev.*, **158**, 117 (1967).
80. G. G. Scott, H. W. Sturner and R. M. Williamson, *Phys. Lett.*, **25A**, 573 (1967).
81. G. W. Smith and G. G. Scott, *Phys. Rev. Letters*, **20**, 1469 (1968).
82. T. W. Adair, III, D. A. Avery, C. F. Squire and S. Wolfson, *Phys. Rev. Letters*, **20**, 142 (1968).
83. W. van Dael, *Phys. Lett.*, **26A**, 523 (1968).
84. A. C. Levi and J. J. M. Beenakker, *Phys. Lett.*, **25A**, 350 (1967).
85. A. C. Levi, F. R. McCourt and J. Hajdu, *Physica*, **42**, 347 (1969); A. C. Levi, F. R. McCourt and J. J. M. Beenakker, *Physica*, **42**, 363 (1969).

86. L. Waldmann, *Z. Naturforsch.*, **22a**, 1678 (1967).
87. W. H. Park, J. S. Dahler, E. Cooper and D. K. Hoffman, (to be published).
88. W. M. Klein and J. S. Dahler: these and other similar results were alluded to in Reference (68) and reported by one of us (J.S.D.) at the Deutsche Physiker Tagung, Berlin (1968).
89. George Birnbaum, *Phys. Rev.*, **150**, 101 (1966).
90. J. H. van Vleck and V. F. Weisskopf, *Rev. Mod. Phys.*, **17**, 227 (1945).
91. J. T. O'Toole and J. S. Dahler, *J. Chem. Phys.*, **33**, 1496 (1960); N. F. Sather and J. S. Dahler, *J. Chem. Phys.*, **35**, 2029 (1961); **37**, 1947 (1962).
92. Unpublished correspondence between S. Hess and J. S. Dahler.
93. H. Grad, *Commun. Pure Appl. Math.*, **2**, 331 (1949).
94. I. L. McLaughlin and J. S. Dahler, *J. Chem. Phys.*, **44**, 4453 (1966).
95. For the generalization to high density see Reference 13.
96. F. M. Chen and R. F. Snider, *J. Chem. Phys.*, **46**, 3937 (1967).
97. V. J. Emery, *Phys. Rev.*, **133A**, 661 (1964).
98. S. Hess, *Z. Naturforsch.*, **23a**, 898 (1968).
99. F. Bloch, *Phys. Rev.*, **70**, 460 (1946); **105**, 1206 (1957).
100. H. C. Torrey, *Phys. Rev.*, **104**, 563 (1956).
101. S. I. Sandler and J. S. Dahler, *Phys. Fluids*, **7**, 1743 (1964); J. S. Dahler, unpublished lecture notes in kinetic theory.
102. Private conversation with J.S.D. (1964).
103. L. Waldmann and S. Hess, (to be published).
104. S. Hess, *Z. Naturforsch.*, **24a**, 1675, 1852 (1969).
105. S. Hess, *Phys. Lett.*, **29A**, 108 (1969); *Phys. Lett.*, **30A**, 239 (1969).

2

Studies of Molecular Rotational Processes by Dielectric Methods

Mansel Davies

2.1 GENERAL CONSIDERATIONS. FACTORS MEASURED AND UNITS

Dielectric studies can be taken to embrace the interaction of material systems with electromagnetic fields from the electrostatic (frequency zero Hz) to those of frequencies associated with millimetre wavelengths. Over this range of the electromagnetic spectrum the interaction with atoms and molecules is almost exclusively of a non-quantized character and its intensity can be expressed in terms of mean atomic or molecular parameters such as polarizability, dipole, quadrupole or higher electric moments. These factors are intimately dependent upon the conditions pertaining in the electron-charge distribution of the atomic or molecular particle which ensures their relevance to molecular structural studies. In turn, the molecular polarizability and electric moment vectors give rise to molecular interactions which play a predominant role in the macroscopic behaviour of atomic and molecular systems, interactions whose energies frequently determine the possibility of physical or chemical change. It is a simplification to write 'the molecular electric vectors give rise to interactions': the situation is that the latter are currently represented and are best understood in terms of the former models of the molecular constitution, and this is likely to persist for a decade or longer.

In practice, dielectric studies can now be made over the whole frequency range from the equivalent of 10^{-4} Hz to 3×10^{11} Hz (i.e. $\lambda = 1$ mm). This continuous spectrum of more than fifty octaves presents a surprisingly uniform pattern of behaviour. The dispersion, i.e. frequency dependence, of the molecular responses encountered in it provides the earliest and one of the best developed models of relaxation methods (Debye, 1913). The recurrent feature is the orientation of a dipole in the applied electric field: the rate of reorientation determines the frequency range characteristic of the process, and the amplitude of the dispersion and the temperature dependence of its frequency provide important measures of the molecular energies involved in the reorientation. As the reorientation is very frequently (but not exclusively) a molecular or an intra-molecular rotation, the relevance of those studies to molecular energy behaviour is clear.

It will be helpful to summarize some aspects of dimensions and units†. Physicists and engineers find it desirable to use the *rationalized M.K.S. system* of units which leads to a dielectric constant (or permittivity as it will be called here) for a vacuum, $\varepsilon\,(\text{vacuum}) = (36\pi)^{-1} \times 10^9$ farad metre^{-1}. These units lead to electric current density in ampere metre^{-2} and molecular moments of inertia in kg metre^2. We shall throughout use the e.s.u.-c.g.s. system which gives the permittivity relative to that of a vacuum, and so makes it a dimensionless factor. In the dispersion regions the permittivity is complex,

$$\varepsilon^* = \varepsilon' - \mathrm{j}\varepsilon''$$

the real component, ε', being accompanied by a factor ε'' measuring the effective ohmic conductance which is known as the dielectric loss; j is the operator algebraically equivalent to $\sqrt{-1}$; $\varepsilon''/\varepsilon' = \tan\delta$ where δ is the loss angle measuring the phase angle of the resultant displacement current in the medium with respect to the applied voltage. In a loss-less dielectric, this phase angle is $\pi/2$ but becomes $(\pi/2 - \delta)$ when the loss factor ε'' is finite: the latter is equivalent to a conductance in the medium: $(\varepsilon'' \times \nu) = 1{\cdot}80 \times 10^{12}\,\kappa$ [In the M.K.S. system $\kappa = \omega\varepsilon''$, where $\omega = 2\pi\nu$]. The frequency ν is in cycle sec^{-1} or Herz (Hz), the appropriate multiples being 10^3 Hz (kHz): 10^6 Hz (MHz): 10^9 Hz (GHz): 10^{12} Hz (THz). Similarly, time units are reduced to 10^{-3} sec (msec): 10^{-6} sec (μsec); 10^{-9} sec (nsec): 10^{-12} sec (psec). The wavelength is advisedly never used as an alternative to frequency although they are invariably related by $\nu \times \lambda = c/n$ where $c = 3 \times 10^{10}$ cm sec^{-1} and n is the refractive index of the medium. Measured in the absence of absorption *at the same frequency* $\varepsilon' = n^2$. To emphasize and regulate the transition through the microwave region to the far infrared, frequencies between 30 GHz ($\lambda_{\text{vacuum}} = 1$ cm) and 300 GHz will also be given in wavenumbers, i.e. $\bar{\nu} = \nu/c$ = waves per cm:

† Written in 1967 (author's insertion).

the frequencies mentioned are 1 cm^{-1} and 10 cm^{-1}. At higher frequencies, the wavenumber only will be used.

The symbol ε' emphasizes that the permittivity is frequency dependent and that, in itself, is adequate reason for abandoning the term 'dielectric constant'. The symbols ε_0 and ε_∞ indicate the specific values of ε' at frequencies below the onset of its dispersion and at frequencies in excess of its dispersion region.

2.2 ELECTROSTATIC RELATIONS

2.2.1 The Permittivity of Gases

A homogeneous medium in an electric field $\bar{E}_0$ becomes polarized, i.e. its component electric charges are displaced in a way such as to reduce the field strength to an effective value $\bar{E}$ where $(\bar{E}_0/\bar{E}) = \varepsilon$, the specific inductive capacity of Faraday, the 'dielectric constant' or better the (electric) permittivity. The conventional representation is of the medium as a slab of material between supposedly infinite parallel condenser plates The polarization can then be measured as the effective charge density per unit area which is established on the surface of the medium, or equivalently, as the effective dipole moment per unit volume in the medium. Thus

$$\bar{p} = \text{specific polarization} = s\bar{\mu}$$

$$\bar{P} = \text{molar polarization} = Vs\bar{\mu} = N\bar{\mu}$$

where s = number of molecules per cm^3; V = molar volume (cm^3); N = Avogadro's number; $\bar{\mu}$ is the mean effective moment contributed per molecule: and $\bar{\mu} = \alpha_t\bar{E}$ where α_t is a proportionality constant assumed independent of $\bar{E}$, which measures the total polarizability of the molecule. The molecule may consist of a single atom, a (neutral) complex of ions or a covalently-bonded grouping of atoms. It is usual to represent α_t in terms of components arising from various modes of charge displacement which can occur for individual molecules. When this charge displacement is merely that of the electron cloud of the molecule, the contribution is the electronic polarizability (α_e). This will be the only displacement in atomic systems (e.g. argon) and it will itself vary with direction within polyatomic molecules; but this anisotropy will not be apparent in most of our interests and we can adequately use a mean value, i.e. $\alpha_e = \frac{1}{3}(\alpha_x+\alpha_y+\alpha_z)$ where

α_x, α_y and α_z are the principal polarizabilities. Molecules have a total charge distribution made up of atomic nuclei located within a non-spherically symmetric electron cloud. The nuclei may now undergo slight displacements from their equilibrium positions when a field is applied: this contributes a term α_a conventionally called the atomic polarizability (also anisotropic). To such distortion polarizability $(\alpha_e + \alpha_a)$ the dipole component (α_{dip}) must be added. In the neighbourhood of the molecular multicharge system the electrostatic field will be a function of the distance (r) in terms of r^{-2}, r^{-3}, r^{-4} etc. The first term (r^{-2}) arises if the molecule has a net electric charge, i.e. is an *ion*: the second (r^{-3}) is present if the centres of its positive and negative charge distributions do not coincide, i.e. if an *electric dipole* is present: the third (r^{-4}) corresponds to the field from an opposed pair of dipoles, such as in carbon dioxide (O=C=O), which is an axial *quadrupole*. By its partial orientation in the applied field the dipole tends to neutralize the latter and so contributes the term α_{dip}. In this pattern, ionic molecules ($Na^+\bar{C}l$; $NH_4^+\bar{C}lO_4$, etc.) can be treated as ionic dipoles whose moment like those of other electric dipoles (HCl, SO_2, etc.) is measured by the electric couple needed to orient the dipole perpendicular to unit electric field: this is equivalent to $\bar{\mu} = e \times \bar{r}$ where the charge distribution is represented by opposite charges ($\pm e$) separated by distance r. To maintain consistency with the electric-field convention the vector of magnitude $|\mu|$ must point from the negative to the positive pole. Chemists invariably represent the dipole direction ($\leftrightarrow$) as from the positive to the negative centre ($\overset{+\!\!\longrightarrow}{\text{H—Cl}}$): little, if any, confusion arises.

In a dilute gas the effective field acting on any one molecule will be that applied externally ($\bar{E}_0$)—i.e. the orientation of one molecule will not be influenced by that of the others. Electrostatic theory* then shows that

$$(\varepsilon_0 - 1) = \frac{4\pi\bar{p}}{\bar{E}_0} = \frac{4\pi s\bar{\mu}}{\bar{E}_0}$$

For a non-polar molecule $\bar{\mu} = (\alpha_e + \alpha_a)\bar{E}_0$ where α_a is only some ten to twenty per cent of α_e. A dipolar molecule of scalar moment μ will con-

* For this and other relations not deduced here many standard texts may be consulted: *inter alia* C. J. F. Böttcher, *Theory of Electric Polarization*, Elsevier, Amsterdam, 1952; A. von Hippel, *Dielectrics and Waves*, John Wiley, New York, 1954; P. Debye, *Polar Molecules*, Chemical Catalog Co., New York, 1929. Nora Hill, W. E. Vaughan, A. H. Price and Mansel Davies, *Dielectric Properties and Molecular Behaviour*, Van Nostrand Reinhold Co., London, 1969.

tribute a component moment ($\mu \cos \theta$) in its tendency to neutralize the applied field in which it will have an energy ($-\mu \bar{E}_0 \cos \theta$) where θ is the angle between the dipole and the field. This energy, via the Boltzmann factor, determines the mean $\langle \cos \theta \rangle$ for the molecules. Debye (1912) following Langevin's treatment of the orientation of molecular magnets in paramagnetism (1904) showed for gases

$$(\varepsilon_0 - 1) = 4\pi s(\alpha_e + \alpha_a + \alpha_{dip})$$

where $\alpha_{dip} = (\mu^2/3\mathbf{k}T)$. This equation can be deduced for a notable variety of models, including the purely quantum-mechanical origin of the molecular parameters[1].

As soon as the dilute gaseous state is left behind dielectric theory has to deal with the departure of the local field acting on a molecule $\bar{E}_L$ from that $\bar{E}$ macroscopically measured in the medium. The relation of $\bar{E}_L$ to $\bar{E}$ has a precise and explicit form only in special cases, the general situation (illustrated by pure polar liquids) being approximated to by a number of functions (Onsager, Kirkwood, Fröhlich, etc.) which are reasonably adequate in practice[2]. By considering the effective field at the centre of a (molecularly large) spherical cavity in the medium, a result equivalent to that deduced by Mossotti and by Clausius is obtained (see Böttcher p. 52 *et seq.*)

$$\bar{E}_L = \frac{(\varepsilon_0 + 2)}{3\varepsilon_0} \bar{E} = \frac{(\varepsilon_0 + 2)}{3\varepsilon_0} \bar{E}_0$$

Using this as the effective field acting on the molecules leads to the equation often described as the 'Debye dipole moment equation':

$$\bar{P} = \left(\frac{\varepsilon_0 - 1}{\varepsilon_0 + 2} \right) V = \frac{4\pi}{3} N \left(\alpha_e + \alpha_a + \frac{\mu^2}{3\mathbf{k}T} \right)$$

This equation proves amply accurate for gases to beyond atmospheric pressure: in general, deviations are to be expected when $(C\mu^2) > 1$: where C = concentration in mole litre^{-1}, and μ is the electric moment in the practical Debye unit, 10^{-18} e.s.u.-c.g.s. If the interactions between the gaseous molecules are pronounced (e.g. formic acid, acetic acid, hydrogen fluoride, etc.) deviations will be obvious at much lower concentrations. One appropriate representation is in terms of a virial function for $\bar{P}$:

$$\bar{P} = \left(\frac{\varepsilon_0 - 1}{\varepsilon_0 + 2}\right) V = A' + B'\left(\frac{1}{V}\right) + C'\left(\frac{1}{V}\right)^2 + \cdots$$

Classical statistical mechanics provide the coefficients A', B', etc. for moderate gas densities and field strengths by expansion in terms of the distortion polarizability and dipole moment energies of interaction as a function of the gas density. Kirkwood[3] gave values A' = (Lorentz–Debye) $= (4\pi/3)N\alpha_{\text{total}}$; $B' = [8\pi N^2/3V]\alpha^3_{\text{total}}\langle r_{ij}^{-6}\rangle$, where r_{ij} is the separation of molecules i and j. Thus the B' term takes account of pair-wise interactions and, for typical α_{total} values, predicts a deviation of a few per cent from A' when $C \sim 5$ mole litre^{-1}. [This Kirkwood treatment was confined to non-polar gases, $\mu = 0$.] The C' term, arising from triple-molecule interactions will not usually be significant below 100 atm pressure. This evaluation only takes account of the non-ideal spatial distribution of the molecules and presupposes α_{total} independent of molecular interactions. There is evidence from studies of the inert gases above 100 atm of significant deviations from that condition (Cole): in sufficiently energetic collisions a reduction in α_e is implied.

For dipolar gases an explicit evaluation of the virial coefficients B' and C' becomes difficult. Under the conditions of concentration (i.e. close molecular approach) where they become significant the anisotropy of α_e and the departures of μ from a point-dipole become obvious. The theoretical treatments lead to complex calculations[4]. In the calculation of Buckingham and Pople the molecular-pair interaction energy is approximated by a Lennard-Jones 6–12 radial function. For ammonia it leads to $B' \sim 990\ \text{cm}^6\ \text{mole}^{-2}$; the observed increase of about 3% in $\bar{P}$ for ammonia as the pressure is raised to 100 atm corresponds to $B' = 440\ \text{cm}^6\ \text{mole}^{-2}$. This agreement is probably as good as the model will allow. A molecule of greater anisotropy, methyl fluoride (CH_3—F), shows a decrease of $\bar{P}$ with pressure, equivalent to B' (obs) $= -600\ \text{cm}^6\ \text{mole}^{-2}$. Buckingham and Pople calculate $B' = +1700$ to $+7500\ \text{cm}^6\ \text{mole}^{-2}$ if the dipole term only or dipole and polarizability contribute to the interaction energy. These great differences emphasize the problem presented by the anisotropy of the local interaction energy for even pairs of molecules. Buckingham and Pople proposed to explain B' (obs) for methyl fluoride in terms of a plausible prolate ellipsoidal molecular shape with the dipole parallel to the long axis: this tends to favour the antiparallel alignment of the dipoles which leads to the negative B' (obs). However, there are other features characteristic of the molar polarization in pure liquids which offer alternative interpretations.

2.2.2 The Permittivities of Liquids

That the Debye equation is completely inadequate for the typical polar liquid can be seen by writing it in the form giving the specific polarization:

$$\bar{p} = \left(\frac{\varepsilon_0 - 1}{\varepsilon_0 + 2}\right) = \frac{4\pi}{3}\frac{N}{V}\left(\alpha_e + \alpha_a + \frac{\mu^2}{3\mathbf{k}T}\right)$$

Neglecting α_e and α_a and taking $\mu = m \times 10^{-18}$ e.s.u. $= m$ debyes, $= m \times 3{\cdot}333 \times 10^{-30}$ SI, the order of magnitude of the right-hand term at 300 K is $15m^2/V$ where $V =$ molar volume in cm^3. For many polar liquids (water, acetone, chlorobenzene, etc.) this factor will exceed unity—which is the limit imposed by an infinite ε_0 value. The prediction of this ferroelectric catastrophe is a consequence of applying the Mossotti–Clausius cavity field to the orientation of the individual polar molecules. Onsager (1936) clarified the situation in concentrated polar media by showing that the total field acting in the cavity could be resolved into two components, $(\bar{F} + \bar{R})$. $\bar{R}$ is the reaction field arising from the induced moment produced in the medium by the polar molecule itself and, being invariably parallel to μ, it clearly cannot help orient the latter. With a uniform macroscopic field $\bar{E}$ outside the cavity containing the molecular dipole, $\bar{F}$ the orienting field within, which effectively influences the molecule is

$$\bar{F} = \frac{3\varepsilon_0}{2\varepsilon_0 + 1}\bar{E}$$

Onsager makes the molecular cavity a sphere of volume v such that $Nv = V =$ molar volume, and evaluates the contribution of the polarizabilities $(\alpha_e + \alpha_a)$ in terms of the permittivity which would be found with zero contribution from the dipole moment; this is ε_∞, the value measured in the microwave region (or far infrared) where the frequency is too high for any contribution by dipolar reorientation. The details of Onsager's calculation then result in:

$$\frac{(\varepsilon_0 - \varepsilon_\infty)(2\varepsilon_0 + \varepsilon_\infty)}{\varepsilon_0(\varepsilon_\infty + 2)^2} = \frac{4\pi N}{3V}\cdot\frac{\mu^2}{3\mathbf{k}T}$$

This Onsager equation shows a great improvement over Debye's when applied to concentrated polar media: ε_0 and ε_∞ values for pure liquids often give μ values within a few per cent of those found in the gaseous state (Böttcher). If ε_∞ has not been evaluated from the microwave data it can be estimated from refractive indices in the far infrared: $\varepsilon_\infty \approx n_{ir}^2$; or from ε_0 values for the polar compound in the solid state where, especially at sufficiently low temperatures, the dipoles will be frozen into immobility, i.e. ε_0 (low-temperature solid) $= \varepsilon_\infty$; or approximated from refractive indices in the visible, e.g. $\varepsilon_\infty = 1{\cdot}1 n_D{}^2$. The proper evaluation of ε_∞ is essential in any further assessment of condensed-phase permittivities, and the Onsager equation or various equivalents or developments of it must be used in examining polar molecule interactions at $C\mu^2 > 1$ in the liquid (including solution) or high-pressure gaseous states. Other sources[5] should be consulted for details of these considerations.

The treatment of dipole polarization in the solid state will frequently require an entirely different model from that so far considered[6] as the dipole character and the dipole motion can be very different from that envisaged above. Some pure crystalline solids behave dielectrically like liquids[7]: these are the 'rotator phase' solids whose molecules reorient as completely as they do in the liquid and with no increase in the rotational energy barrier. Guest molecules in clathrates simulate dilute-solution behaviour and the latter is also found for some spherical polar molecules in the unlikely medium of a solid polymer matrix[8]. These varied systems are amongst the many which show considerable molecular rotational mobility. They are mentioned here as examples in which the mere measurement of ε_0, i.e. a low-frequency permittivity, suffices to establish labile molecular reorientation. When the latter ceases completely, i.e. at sufficiently low temperatures, the permittivity falls to that of an essentially non-polar material, i.e. $\varepsilon_0 \approx n_{ir}^2$.

2.3 FREQUENCY DISPERSION STUDIES

2.3.1 Dielectric Absorption in Gaseous Systems

The static or low frequency dielectric properties are derived from the equilibrium conditions in the molecular system. Even so they provide valuable information on the freedom of reorientation of groups within and of whole rigid molecules provided these molecular units contain a dipole element. Much more intimate detail on molecular rotational processes becomes available when a study is made of the rate at which these dipole reorientations occur. This involves the study of the dielectric properties as a function of the frequency of the field used to measure them.

Elementary ideas suggest that when the frequency of the electric field approaches that of the dipolar motion a pronounced interaction with some absorption of the radiation is probable. This expectation is fully realized in practice and a pronounced variation of the dielectric parameters accompanies such interaction: this is seen as the dispersion (frequency dependence) of the permittivity and of the dielectric absorption.

Electromagnetic interactions in conservative, non-relativistic systems are fully summarized by Maxwell's equations. Such a general phenomenological description, however, cannot provide any indication except of a negative character concerning the ranges of behaviour commonly occurring in material systems. Only experimental investigation and the construction of specific models provide adequate representations of such molecular processes as actually occur. Accordingly, as did the meaningful approach to static or low-frequency conditions, the extension of dielectric studies in terms of the frequency variable requires the consideration of a number of model processes to provide a basis for understanding experimental observations.

The interpretation of the specific or molar polarization has invoked three molecular electric parameters: α_e: α_a: μ. It can be anticipated that each of these will contribute to the frequency dependence of the electric polarization. This is so.

2.3.2 Resonance Absorption and Anomalous Dispersion

The classical origin of the electronic polarizability presents it as an 'elastic displacement' of the electron-charge distribution by the applied electric field. This displacement gives rise to an induced dipole, $\bar{\mu}$ (induced) = $\bar{\alpha}_e \,.\, \bar{E}_0$. Here the vector character of $\bar{\alpha}_e$ for any molecular structure may be emphasized although that aspect of its character will not be considered further. Supposing a displacement in the x-direction, the electron-cloud distortion will invariably be small in any electric field encountered in normal measurement ($\bar{E}_0$ in measurement $\ll \bar{E}$ within the electron cloud). This justifies the simplest representation of the displacement in terms of an 'elastic' (Hooke's Law) force constant (k):

$$m\frac{d^2x}{dt^2}+kx = 0$$

Here m is an inertial factor appropriate to the electron-cloud displacement (x). As μ (induced) is proportional to x it is seen that the former has a natural angular frequency (radian $\sec^{-1}$) of undamped oscillation, $\omega_0 = 2\pi\nu_0$.

$$\nu_0 = \frac{1}{2\pi}\sqrt{\frac{k}{m}}$$

However, a damping coefficient is certain to limit the amplitude (and energy) of any such oscillation: one major source of damping being the radiation of energy from an oscillating dipole. The damped oscillator is accordingly represented with a term proportional to the displacement velocity:

$$m\frac{d^2x}{dt^2}+2a'\frac{dx}{dt}+kx = 0 \quad \text{or,} \quad \frac{d^2x}{dt^2}+2a\frac{dx}{dt}+\frac{k}{m}x = 0$$

The solution for this motion is

$$x = x_0 \,.\, e^{-at} \,.\, \cos \omega'_0 t$$

where $\omega'_0 = (\omega_0{}^2 - a^2)^{1/2}$. The frequency difference introduced by the damping coefficient, a, can be very small, its principal role being the reduction of the amplitude.

In an absorption process, such an oscillator is driven by the applied electric field: if the local value of the latter is $\bar{E}_L$ and it acts on a charge q of inertia m, the previous equation for free vibration is extended to

$$\frac{d^2x}{dt^2}+2a\frac{dx}{dt}+\omega_0{}^2 x = \frac{q}{m}\bar{E}_L$$

If there are s displaceable charges (i.e. polarizable molecules) per cm^3 then, by definition of the specific polarization

$$\bar{p} = s\bar{\mu} \text{ (induced)}$$

$$= sq\bar{x}$$

The simple vector displacement $\bar{x}$ is assumed to be coincident in direction with $\bar{E}_L$. The Mossotti–Clausius expression for $\bar{E}_L$ is equivalent to

$$\bar{E}_L = \frac{(\bar{\varepsilon}_0+2)}{3}\bar{E} = \bar{E}+\frac{4\pi}{3}\bar{p}$$

and the equation can be rewritten:

$$\frac{d^2\bar{p}}{dt^2}+2a\frac{d\bar{p}}{dt}+\left(\omega_0{}^2-\frac{sq^2}{3m}\right)4\pi\bar{p} = \frac{sq^2}{m}\,.\,\bar{E}$$

One effect of the medium is to reduce the frequency of the resonating system to $\omega_m^2 = (\omega_0^2 - sq^2/3m)$. The field in the medium $\bar{E}$ may be fully represented by $\bar{E}_m \, e^{j\omega_m t}$, i.e. of amplitude $\bar{E}_m$ and angular frequency ω_m, and the solution for $\bar{p}$ is

$$\bar{p} = \bar{p}_0 \, e^{j(\omega t + \phi)} = \frac{sq^2/m}{\omega_m^2 - \omega^2 + j \, . \, 2a\omega} \bar{E} + j \, . \, 2a\omega$$

A phase shift ϕ is involved between the applied field $\bar{E}$ and the response in the molecules owing to the presence of the damping factor (2a). The solution immediately gives the complex permittivity, ε^* from the relation

$$\varepsilon^* = 1 + \frac{4\pi\bar{p}}{\bar{E}}$$

$$\varepsilon^* = (n^*)^2 = 1 + \frac{4\pi sq^2/m}{\omega_m^2 - \omega^2 + j2a\omega}$$

The (complex) refractive index and the permittivity at the same frequency are always related for non-magnetic materials by the Maxwell condition, $\varepsilon = n^2$. This single-oscillator model has to be generalized for the real case of atoms or molecules capable of an extended number of 'natural' frequencies governed by the permissible electron-energy changes: Δ (electronic energy) $= h\omega_0/2\pi$: each of the supposed r such frequencies will make its own contribution to ε^* so that:

$$\varepsilon^* = 1 + \sum_r \frac{4\pi sq_r^2/m_r}{\omega_r^2 - \omega^2 + j2a_r\omega} \tag{2.1}$$

This equation is the classical representation of the dispersion associated with resonant frequencies $(\omega_r/2\pi)$ within the atom or molecule.

Considering the real part of equation (2.1), it is seen that far below its characteristic frequency ω_r, each oscillator makes a contribution $(4\pi sq_r^2/m_r\omega_r^2)$ to the (real) permittivity, ε': this is equivalent to its contribution to α_e. Well above the frequency ω_r this component disappears giving a fall in ε' as a result of traversing the frequency ω_r, the simple oscillator no longer responding for $\omega \gg \omega_r$.

For polyatomic structures, in addition to the frequencies associated with ultraviolet or visible absorptions due to electronic charges, there will be the infrared vibrational frequencies (fundamentals, overtones and combination tones): Δ (vibrational energy) $= h\omega_r/2\pi$. Each of these likewise contributes to the polarization and permittivity. The charge q associated with the vibration in the electronic process is now replaced by the ratio of a dipole moment matrix element appropriate to the vibrational transition to the value of the corresponding displacement coordinate: $q = \Delta\bar{\mu}/\Delta\bar{Q}$. In all these instances the amplitude of the contribution is proportional to $(q)^2$, which can be translated to $(\Delta\bar{\mu})^2$. Anticipation and observation shows that the infrared absorptions even when integrated are usually much smaller in intensity than the electronically-based visible and ultraviolet absorptions. This is the condition which makes $\alpha_e \approx 10\alpha_a$.

To consider the behaviour near resonance one such frequency may be singled out of those in equation (2.1), the contribution of the remainder being represented by a constant A, and in the neighbourhood of the one resonant frequency it is convenient (and adequate) to write $\Delta\omega = (\omega_r - \omega) \ll \omega$; $(\omega_r + \omega) \simeq 2\omega_r$. The equation is thus simplified to:

$$\varepsilon^* = \mathrm{A} + \frac{\mathrm{B}}{\Delta\omega + \mathrm{ja}}$$

$$\mathrm{B} = \frac{2\pi s_r q_r^{\,2}}{m_r \omega_r}$$

The definitions

$$\varepsilon^* = \varepsilon' - \mathrm{j}\varepsilon'' = (n^*)^2$$

$$n^* = n(1 - \mathrm{j}k)$$

$$\varepsilon' = n^2(1 - k^2); \qquad \varepsilon'' = 2n^2 k$$

relate the permittivity (ε'), the dielectric loss (ε''), the (real) refractive index (n) and the absorption index (k). The frequency dependence of ε' is then represented:

$$\varepsilon' = \mathrm{A} + \frac{\mathrm{B}\,.\,\omega}{(\Delta\omega)^2 + \mathrm{a}^2} \tag{2.2}$$

Below the resonance frequency value ω_r the permittivity rises from $(\mathrm{A} + 2\mathrm{B}/\omega_r)$ to a maximum, $\varepsilon'_{\mathrm{max}} = (\mathrm{A} + \mathrm{B}/2\mathrm{a})$ for $\Delta\omega = +\mathrm{a}$: falls with a slope $(-\mathrm{B}/\mathrm{a}^2)$ essentially linearly through $\omega = \omega_r$ to $\varepsilon'_{\mathrm{min}} = (\mathrm{A} - \mathrm{B}/2\mathrm{a})$ and then goes asymptotically to the constant value A. It is seen that the effective amplitude of the dispersion $(\varepsilon_0 - \varepsilon_\infty) = 2\mathrm{B}/\omega_r = (2\pi s_r q_r^{\,2}/m_r)$.

The dielectric loss factor

$$\varepsilon'' = 2n^2k = \frac{\mathrm{Ba}}{(\Delta\omega)^2 + \mathrm{a}^2} \tag{2.3}$$

follows a simple pattern which is symmetrical about $\Delta\omega = 0$ where it shows a maximum value, (B/a). The frequency width of the absorption at half-peak height is 2a. See Figure 2.1.

Early observations using visible radiation almost invariably showed ε' (or n^2) to increase with increasing frequency: this is so for all those transparent media whose electronic absorptions lie in the ultraviolet. This progression became known as *normal dispersion* and so the rapid fall of ε' with increasing frequency in passing through the resonance frequency became known as *anomalous dispersion*. The term is especially justified in that, near ω_r the value ε' can be less than unity, but it is important to emphasize that a fall of ε' with increasing frequency is the invariable normal condition in the absence of resonant absorptions and virtually all dielectric materials show such a decrease of ε' at frequencies up to 100 GHz (3 cm^{-1}), both through and on the high frequency side of their

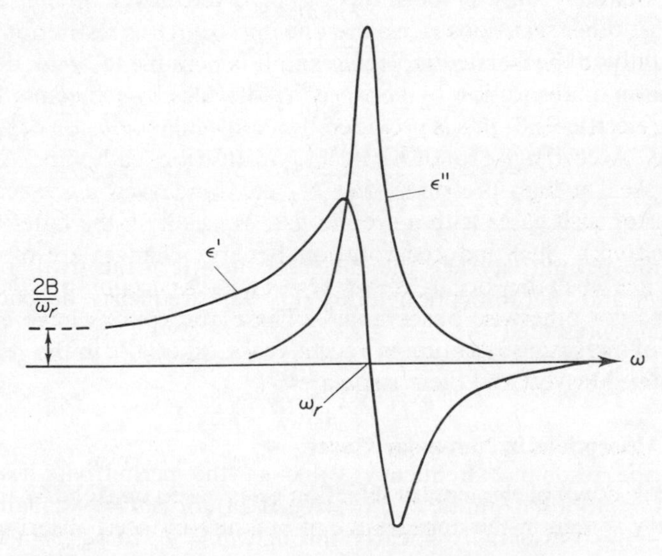

Figure 2.1 The characteristic form of ε' and ε'' in a resonance absorption region

absorptions: such resonant absorptions as are found relate to rotational-energy changes in the gaseous state. They will be considered later.

That the simple model envisaged in this classical representation of resonant absorptions and that the numerical approximations made suffice to account for the form of these processes, in ε' and ε'', is well known. Examples are detailed in texts on physical optics[9]. It is equally clear that the simple model is a heuristic one. Quantum-mechanical theory[10] is necessary to provide physically adequate representations of the amplitude factor (B) and the observed width parameter (a). In the form of a damping factor the latter is an 'omnium gatherum' term: a number of sources contribute to it not all of them giving the frequency behaviour imposed by 'a'. Accordingly, the observed contours of the absorption can show departures from the functional form quoted: some of the relevant factors will be mentioned later.

2.3.3 Microwave Absorptions by Gases

It is now appropriate to consider the form of absorptions found to be associated with the rotational motions of gaseous molecules. Classical and quantum-mechanical considerations show that such absorptions will occur essentially only if the molecule involved has a dipole moment. Like most other selections rules, the one imposing this restriction applies strictly only to the isolated molecule and it is possible to break down the prohibition of absorption by non-polar molecules by subjecting them to a strong electric field: this is provided by a molecular collision of sufficient violence. Accordingly, rotational-energy transitions in homopolar gases can be seen at high pressures (H_2, N_2, etc.) and even more readily in mixtures of such gases with a second gas, especially if the latter is polar (Kranendonk). Such induced rotational-energy changes are of especial importance when they occur, as they reveal details pertaining to molecular collisions not otherwise observable*. These absorptions have been the subject of increasing attention in recent years, especially in the studies by Birnbaum, Maryott and their associates[11].

2.3.4 Absorptions in Non-polar Gases

The break-down of elementary selection rules based on isolated-molecule symmetry is seen in the appearance of pressure-induced absorptions for

* For CO_2 see, J. E. Harries, in course of publication (1970).

non-polar gases in the microwave region[12]. In the case of the linear CO_2 molecule, the absorption in the pure gas is ascribed to the dipole induced on collision by the quadrupole field of adjacent molecules. Particularly significant is the very short relaxation time ($\tau = 3{\cdot}4 \times 10^{-13}$ sec) for the absorption which is independent of the pressure. As this is very much 'the time of collision', it is deduced that the moment appears only during that interval and is not derived from any long-lived bimolecular or other complex. A 'bimolecular complex' might have been invoked as the absorption is proportional to $(p_{CO_2})^2$: so are, of course, the number of collisions. Kranendonk has now reported similar absorptions in *mixtures* of inert gases. A temporary dipole can appear in an (A–B) collision: no such effect is found in pure inert gases i.e. for (A–A) collisions. These features are of particular significance in relation to the so-called 'charge transfer complexes' and also because related absorptions appear in pure non-polar liquids: there, it is clear, molecular polarizability plays an important role in the absorptions which appear.

The electric moments induced on collision of non-polar molecules have been ascribed to the electric quadrupole field which many of them possess (e.g. O=C=O, CH_2=CH_2, C_6H_6, etc.). It is a fact that molecules of the highest symmetry (e.g. CH_4, SF_6, etc.) which have zero quadrupole moments give extremely weak absorption in the gaseous state. However, liquid-phase studies (see later) suggest that such molecules can be sufficiently distorted on collision to give rise to temporary dipole moments. The quadrupole-moment molecules can, as a result of collisional interaction give observable resonance-type absorptions: such features will usually be at frequencies in the far infrared. This behaviour is illustrated by the absorption in compressed CO_2[13]: a very broad absorption is proportional to $(\text{pressure})^2$ and is centred at 50 cm^{-1}. Analysis of the width and contours of such absorptions will clarify the collisional processes*. The successful development of the Fourier interferometric spectrometer will markedly promote studies of these absorptions.

In the more strictly dielectric region up to 300 GHz ($\bar{\nu} = 10\ \text{cm}^{-1}$) the non-polar gases are found to exhibit a non-resonant dielectric loss which, on its low-frequency side, can be represented[14]

$$\varepsilon'' = (\varepsilon_0 - \varepsilon_\infty)_Q \frac{\omega\tau}{4}$$

* For CO_2 see, J. E. Harries, in course of publication (1970).

The amplitude $(\varepsilon_0 - \varepsilon_\infty)_Q$ is the total contribution of the quadrupole moment to the low frequency permittivity, ω is the angular frequency (radian sec^{-1}) and τ is the effective relaxation time for the collisional, or more accurately, the induced moment process. In gaseous mixtures of components A and B the total absorption is a composite factor for the various pair interactions:

$$\varepsilon''(\text{total}) = \varepsilon''_{AA} + \varepsilon''_{AB} + \varepsilon''_{BA} + \varepsilon''_{BB}$$

The individual contributions are functions of the quadrupole moments, the polarizabilities, the gas densities and the interaction energies of the molecular pairs. The latter are approximated in terms of the Lennard-Jones (12–6) potential. A typical study (Maryott and Kryder[14]) is based on the loss values (ε''_{total}) measured at 24 GHz for carbon dioxide at 25°C in presence of other non-polar gases.

2.3.5 Absorptions in Polar Gases

A minimum of two quite distinct processes pertain to the dispersion of the dipole contribution of a polar gas. Each of them, when fully studied, can contribute much insight into rotational-energy conditions and the interactions of gaseous molecules on collision. In a symmetrical-top molecule (CH_3—Cl) the dipole is necessarily along the symmetry axis. The total angular momentum of rotation will only exceptionally be along this axis and rotational momentum will be present about the axes perpendicular to the symmetrical element. Accordingly, the molecular dipole moment can be resolved into one component along the axis of total angular momentum and an orthogonal component perpendicular to that momental axis. It is the latter component which is associated with the quantized molecular rotation and which contributes the observed total amplitude of the resonance absorptions arising from the allowed changes in rotational energy. Each such microwave line that occurs has its intensity determined by the elements in the dipole-moment matrix appropriate to their rotational quantum states directly involved[15]. Their total contribution is that derived from the component of the molecular moment perpendicular to the rotational momental axis: under defined conditions (especially temperature) the Boltzmann factor, with appropriate weighting from spectroscopic multiplicities, will determine the population of each rotational state and hence the partition of the total spectroscopically active dipole absorption. Other accounts should be consulted for the definition of the quantized levels contributing to the microwave absorptions in the various molecular symmetry types[16].

In contrast, the dipole component along the axis of total momentum is not involved in the molecular rotation. In the presence of an applied electric field this component becomes partly oriented thanks to the maintenance of equilibrium with the field which is brought about by energy exchanges on molecular collision. The corresponding contribution to the gaseous permittivity is characterized by a relaxation time determined by the mean interval between molecular collisions. This latter mechanism will clearly be of a non-resonant character. That the relaxation time in this instance is dependent on the collisional frequency results from the randomization of the molecular orientations after collision.

A further complication arises in non-rigid molecules where a dipole component is able to assume alternative orientations with respect to the molecular framework. This is a very common occurrence in molecules based on ethane ($CH_3{-}CH_3$) or its higher homologues and in those many molecules containing ($O{-}H$), ($O{-}CH_3$), (NH_2), etc., groupings. In the free (gaseous) molecule such group rotations will occur in a fashion controlled by the intramolecular force field. The essential aspects of the latter are largely presented if the variation of the molecular energy as a continuous function of the group rotational angle is known. Statistical methods can be applied to evaluate the average molecular behaviour: interesting but difficult problems of quantization may be involved. The *locus classicus* of these conditions is that of methyl alcohol $CH_3{-}O{\diagup}H$, studied by Dennison and coworkers over a number of years[17]. Other microwave spectra have contributed much to the classification of these features[18]. In principle, the dipole element associated with the group rotation is treated as providing a third mode of relaxation added to the quantized and non-quantized molecular framework processes. In the limiting case of a polar group which is a 'free rotator' about its bond to the molecular skeleton, its contribution to the dipole relaxation (and corresponding absorption) will be governed by its moment of inertia about the bond ($I/\mathbf{k}T$)—as though it were an independent polar molecule. Essentially similar but practically simpler aspects arise for such non-rigid molecules in condensed-phase (e.g. solution) studies.

The ammonia molecule is unique in providing a fully-developed inversion process whose absorptions have been intensively studied[19].

2.3.6 Resonance-type Microwave Absorptions

The essential character of the function $\varepsilon^*(\nu)$ in the neighbourhood of the resonance line has been outlined. The form of the $\varepsilon''(\nu)$ function can be explored in particular detail for microwave gas-rotational absorptions thanks to the great precision with which frequency can be defined and swept over a restricted region from appropriately monitored klyston sources: at the same time the detector systems have considerable accuracy for and sensitivity to small changes in the absorption intensities[20].

A valuable feature is the assessment of alternative mechanisms contributing to the observed overall width of the rotational line. A review[21] of these factors includes: (i) the natural line width. This is determined by the perturbations of the local field around an isolated molecule in space and the finite life-time of the states between which the transition occurs. This ultimate limit (*ca* 10^{-4} Hz) is many orders of magnitude less than the best microwave spectrometer can define. (ii) A Doppler broadening will always be present: it can be calculated with some accuracy and can be effectively eliminated by observations for a molecular beam directed perpendicularly to the direction of observation: typically this factor may give a total width of 5×10^4 Hz to ν_r which is *ca* 3×10^{10} Hz. (iii) At the lowest pressures, irradiation with a considerable intensity can lead to a condition showing significant departures from equilibrium in the gas examined—an excessive proportion of molecules being in the excited level: this could amount to a term $\Delta\nu \sim 2 \times 10^5$ Hz but it is reduced at lower irradiation intensities and becomes negligible on increase of pressure. (iv) Collisions with the cell wall have a similar effect, normally far smaller (*ca* 2×10^4 Hz) than that of (v) Gas-collisional broadening. This is the essential and important experimental variable: with increase in collisional frequency, this effect can (and usually does) override all the others. (vi) Instrumental conditions, such as the electric field modulation imposed for the Stark effect, lead to a lack of definition in either the frequency or the rotational-energy levels: a figure of 10^5 Hz could be reached in $\Delta\nu$ in this way[22].

The Lorentz treatment (1906) provided the first of very many formal theories of collisional line broadening. Physically the essential feature is the interruption of the otherwise precisely defined (frequency and phase) interaction between the radiation and the (rotational) absorber. Each collision is assumed completely to disrupt this phasing. If the uninterrupted life-time of an energy level is t_E, the uncertainty in its energy (Δu) is given by $\Delta u \times t_E \gtrsim h/2\pi$. If t_E is reduced to the period ($1/\nu_C$) between collisions of average frequency ν_C sec^{-1}, $\Delta u \sim h\nu_C/2\pi$: and the absorption breadth is markedly dependent upon ν_C. A further possibility considered in treatments such as Anderson's[23] is that collision leads to energy exchange and so perturbs the molecular population participating in a given absorption: collisions must then be divided into those which are diabatic (with energy change) and adiabatic (no energy change). Specifically, the lines of the ammonia inversion spectrum which, at lowest pressures give a width ($\Delta\nu_{1/2} \sim 20$ kHz, cf. $\nu_0 \sim 3 \times 10^{10}$ Hz) limited by experimental conditions, are found to be broadened by energy change on each collision[24].

In the van Vleck–Weisskopf treatment[25] it is assumed that interruption of the phase condition between molecule and radiation is the essential mechanism but that the energy distribution (Boltzmann) is at its equilibrium immediately after the disturbing collision. For a symmetrical top whose rotational lines are defined by the quantum numbers (J, K) the absorption at frequency ν near the resonance value $\nu_0 = [(J, K)$ transition energy$]/h$ is

$$\varepsilon''(\nu) = \frac{2\pi \,.\, f_{JK} \,.\, s}{3\mathbf{k}T}|m_{JK}|^2\left\{\frac{\nu \,.\, \Delta\nu}{\Delta\nu^2 + (\nu_0 - \nu)^2} + \frac{\nu\,\Delta\nu}{\Delta\nu^2 + (\nu_0 + \nu)^2}\right\}$$

Here f_{JK} is the fraction of the s molecules per cm^3 in the appropriate initial energy state for the absorption; $|m_{JK}|$ is the dipole-moment matrix element determining the transition intensity, and $\Delta\nu$ is the line width factor, related to the average time (τ_C) between collisions by $\Delta\nu = \frac{1}{2}\pi\tau_C$. For a symmetrical top molecule $|m_{JK}|^2 = \mu^2_{(\text{total})} \,.\, K^2/(J^2 + J)$. Corresponding expressions for other molecular rotator types are well-established[15]. In this expression ν_0 is constant, and the half-width is clearly proportional to pressure. Up to medium pressures (<1 atm) a very good fit to this analytical contour is found. Eventually the absorptions tend to acquire markedly dissymmetric features which cannot be fitted. Such departures are expected when the time between collisions is approximating to the time interval of collision.

Special interest pertains to the $\varepsilon''(\nu)$ function when $\nu_0 \to 0$. It can be anticipated that this corresponds to the condition for a non-resonant absorption.

$$\varepsilon''_{\nu_0 \to 0}(\nu) = \frac{4\pi s}{3\mathbf{k}T} \cdot f_{JK}|m_{JK}|^2 \cdot \frac{\nu \,.\, \Delta\nu}{\nu^2 + \Delta\nu^2}$$

If it be envisaged that under high pressure conditions the individual rotational resonance frequencies ν_0 will all degenerate to zero and that simultaneously they acquire the same half-width factors $(\Delta\nu)$, the total absorption will acquire this simple form: the intensity factor involving $\Sigma f_{JK}|m_{JK}|^2$ and $\varepsilon''(\nu)$ becomes characterized by a single relaxation time, $\tau = \frac{1}{2}\pi\Delta\nu$. Precisely this condition is found as a more or less constant feature for the non-resonant gaseous polar molecule absorptions and the special interest of the van Vleck–Weisskopf function is that it is able to link the two types of gas rotational absorptions. Nevertheless, doubts have been expressed whether genuine physical (as opposed to convenient functional) continuity exists between the two processes[26]. It may be that the degeneracy of the form of the absorption from the strictly quantized condition is not sensitive to the means (e.g. excessive collisional interaction) by which the quantized states are destroyed. It is a certain result that the limiting $(\nu_0 \to 0)$ form of the van Vleck–Weisskopf condition is essentially identical with that adduced by Debye for molecular dipolar relaxation occurring in a classical fashion governed by Brownian motion and Boltzmann considerations. Von Hippel's doubts may perhaps be quelled on the basis of the Correspondence Principle, that when quantization is sufficiently perturbed, classical conditions are reinstated.

The behaviour envisaged in the van Vleck–Weisskopf relations has been fully substantiated in a number of detailed studies of microwave absorptions. Especially is this so for the ammonia inversion absorption which, although not a rotational process, will clearly have the same line-shape behaviour[27]. The contours are very well fitted although a shift in ν_0 is necessarily involved even at low pressure ($p < 1$ torr): Table 2.1. The quantization in the inversion has been completely blurred at 2 atm pressure and the absorption has now acquired an enormous half-width

Table 2.1

Wavenumber ($\bar{\nu}_0$ cm^{-1}) of the resonance absorption frequency characterizing the ammonia inversion process in the van Vleck–Weisskopf representation

p (atm)	0	0·5	1·0	1·5	2·0
$\bar{\nu}_0$ (NH_3 inversion) cm^{-1}:	0·78	0·71	0·54	0·30	0·00

($\nu_{-1/2}/\nu_{+1/2} \sim 14$) relative to the dilute gas. For ND_3, $\nu_0 \sim 0$ is already found at 10 cm Hg. It is important to note that measurements of refractive indices in the microwave region[28], equivalent to determining $\varepsilon'(\nu)$, confirm the correctness of the van Vleck–Weisskopf function for $\varepsilon''(\nu)$.

2.3.7 Non-resonance Microwave Absorptions

Birnbaum and Maryott[29] have been responsible for definitive studies of the non-resonant absorptions in polar gases. Using resonant cavity techniques from 1 to 24 GHz and pressures up to 2 atm, systematic study of many 'rigid' polar molecules derived from methane (CH_3Cl, CHF_3, $CClF_3$, etc.) has been made. The line width $\Delta\nu$ is usually linearly dependent upon pressure, a result which establishes that only binary collisions are important in the relaxation mechanism: the absorption peak is a little lower and the absorption width somewhat greater in most cases than corresponds to a single $\tau(\frac{1}{2}\pi\Delta\nu)$ value for $\nu_0 = 0$. The small departures from the minimum width for a non-resonant absorption (Debye case—see later) are adequately represented by a factor which defines a distribution function for τ which is symmetrical around a mean value (the τ deduced). Many alternative empirical forms for such a distribution function exist: Birnbaum has used the convenient Cole–Cole version[30]. For two molecules with low dipole moments ($CClF_3$, 0·50D and $(CH_3)_3N$, 0·60D) no departure from the single-valued (Debye) case was found. This correlation with the dipole moment (μ) extends in a quantitative form to the broadening of the absorption measured by the constant ($\Delta\nu/p$):

$$\Delta\nu/p = a \,.\, \mu^2 \langle J^{-1} \rangle$$

'a' is a constant and $\langle J^{-1}\rangle$ a weighted (Boltzmann) average of (J^{-1}), where J is the principle rotational quantum number. This means that the first-order dipole–dipole interaction between the colliding molecules causes a large molecular reorientation. The theoretical interpretation of 'a' by Birnbaum and Maryott indicates it is (at a given pressure) proportional to the number of molecules per cm^3: this implies $a \propto (T^{-1})$: and as $\langle J^{-1}\rangle \propto (T^{-0\cdot5})$, it is expected that $(\Delta\nu/p) \propto (T^{-m})$ where $m = 1\cdot5$. Experimentally, for CHF_3 ($\mu = 1\cdot81$D), $m = 1\cdot59 \pm 0\cdot03$; CH_3F ($\mu = 1\cdot64$D), $m = 1\cdot60 \pm 0\cdot02$; $CClF_3$ ($\mu = 0\cdot50$D), $m = 1\cdot27 \pm 0\cdot02$. According to Anderson's theory, the collision broadening is due to an intermolecular energy which can be approximated by the form (A/r^n): on this basis

$m = (n+1)/(2n-2)$. Dipole–dipole forces give $n = 3$, whilst incompressible molecular spheres give $n \to \infty$. This gives m values in the range 1·0 to 0·5. The experimental data clearly eliminate the Anderson mechanism and favour that of Birnbaum–Maryott which is based on a torque–impulse model of collisions. This detail is quoted as a typical instance of the intimate features of molecular collisional exchanges which appropriate spectroscopic studies provide.

From their measurements of absorptions and relaxation times in single component gases, Maryott and Birnbaum[31] have evaluated the corresponding collision diameters, and also observed that the transition from resonant to non-resonant absorption is not complete in some cases even at 30 atm pressure. The studies of the collisional process between a polar molecule and a non-polar one has emphasized the inadequacy of the concept of 'collisional diameters'. In many cases the 'collisional diameter' is less a reflection of the size of the molecule than of its polarizability and the presence of anisotropy in this regard is also detectable. This establishes, as can be readily accepted, that the interaction energy is characterized structurally by the dipole-induced dipole term. Quantitative considerations suggest that other factors are involved—or that simple concepts of molecular polarizability need extension. It is certain that the extension of these studies will add a new insight into gas-collision conditions.

Other studies confirming the deductions based on $\varepsilon''(\nu)$ have been made by detailed evaluation of $\varepsilon'(\nu)$ for polar gases[32]. Where the total molecular moment is associated with quantized rotation (e.g. linear polar molecules) no variation of $\varepsilon'(\nu)$ is expected until the rotational resonance lines are encountered. This is found to be so, as also $\varepsilon'(\nu) < \varepsilon'(0)$ for many symmetric (CH_3Cl, etc.) and asymmetric top molecules ($CHCl_2F$, etc.) due to non-resonant dipolar dispersion. One particular feature of symmetry considerations arises in describing part of the non-resonant absorptions of such molecules as due to 'an inversion transition'. The rotational states of a symmetric top can be divided into those which are (wave-functionally) symmetric with respect to inversion about the molecular centre $[\psi_s(J, K)]$ and those which are antisymmetric $[\psi_a(J, K')]$. These can appear for the same J but different K values and they are then essentially degenerate in energy. A transition between such states $\psi_s(J, K) \leftrightarrow \psi_a(J, K')$ within the molecule will therefore be 'a resonance absorption of zero frequency ($\nu_0 = 0$)', i.e. a non-resonant absorption, and it will contribute to the latter feature. There is no implication that the molecule inverts its structure: it is an angular momentum vector which changes. In view of the special significance of structural inversion, it is unfortunate that the term 'inversion absorption' has been used without the appropriate qualification, symmetry-inversion absorption.

2.4 GENERAL ASPECTS OF POLAR MOLECULE RELAXATIONS IN CONDENSED PHASES

Molecular rotation of usually a non-quantized and frequency hindered character occurs in dense gases, in liquids, in solutions and in some solid phase. The loss of rotational energy quantization as gas density increases has already been emphasized. The resulting situation has usually been treated by reference to the liquid state: this is appropriate in the practical sense that much more extensive data are available for liquids but, in general there is likely to be little difference between the liquid and the highly compressed gas. Whilst dielectric relaxation studies necessarily require the presence of a dipole in the molecular species studied, the molecular reorientation is not primarily influenced by that feature. Appropriately detailed study can be made with molecules of such small polarity that the extrapolation to non-polar behaviour can in general be made with confidence.

The rotational behaviour of a rigid spherical molecule having a centrally located dipole was the model treated by Debye (1912) to evaluate the frequency dispersion of dipole polarization. A brief consideration of the general relations will be helpful.

When an electric field is applied to measure the permittivity of a polar liquid elementary considerations show that the observed polarization, or ε' value, corresponds to only a minute displacement from the complete spatial randomization of the dipoles. (In water, $\varepsilon' \approx 80$, the specific polarization is equivalent to 1 in 10^7 molecules being completely aligned by the field). This small bias produced by the field will clearly decay on its removal. It is this time dependence of the polarization which gives rise to the dipole relaxation.

It is appropriate to refer to the two components of the total specific polarization:

$$\bar{p}_{\text{total}} = \bar{p} + \bar{p}_{\text{dip}}$$

where $\bar{p}$ is the combined $(\bar{p}_{\text{el}} + \bar{p}_{\text{at}})$ and $\bar{p}_{\text{dip}}$ is the contribution due only to the dipoles. In a static field the latter will attain its equilibrium value $\bar{p}_0(\text{dip})$. Imposing the simplest, i.e. first-order, rate equation the decay of

$\bar{p}$(dip) when a field is instantaneously removed at time $t = 0$ can be written

$$\bar{p}_t(\text{dip}) = \bar{p}_0(\text{dip}) \,.\, e^{-kt}$$

or, for the growth when the field is applied at $t = 0$:

$$\bar{p}_t(\text{dip}) = \bar{p}_0(\text{dip})[1 - e^{-kt}]$$

where k is the rate coefficient. These relations each derive from

$$\frac{d\bar{p}_t(\text{dip})}{dt} = k[\Delta\bar{p}(\text{dip})] = k[\bar{p}_e(\text{dip}) - \bar{p}_t(\text{dip})]$$

where $\bar{p}_e$(dip) is the equilibrium value at $t = \infty$. No formal justification can be provided *ab initio* for the use of the dynamic equivalent of Hooke's Law for this process†, but its assumption can be tested and it is found to be a close approximation to general behaviour. In dielectric studies k is usually replaced by its reciprocal $\tau(= 1/k)$, the relaxation time—which measures the interval for the decay of p(dip) to (1/e)th of its initial value.

The definition of ε' in terms of the dielectric displacement $\bar{D}$, i.e. $\bar{D}_0 = \varepsilon_0 \bar{E}$ where $\bar{E}$ is the applied electric field, leads to:

$$\bar{p}_0 = \bar{p} + \bar{p}(\text{dip}) = \frac{\bar{D}_0 - \bar{E}}{4\pi} = \frac{(\varepsilon_0 - 1)}{4\pi} \,.\, \bar{E}$$

and

$$\bar{p}_0(\text{dip}) = \frac{(\varepsilon_0 - 1)}{4\pi}\bar{E} - \frac{(\varepsilon_\infty - 1)}{4\pi}\bar{E}$$

The sinusoidal variation of $\bar{E}$ can be written:

$$\bar{E}_t = \bar{E}_0 \, e^{j\omega t}$$

where $\bar{E}_0$ is the amplitude and $\omega = 2\pi\nu$ is the angular frequency of the applied field.

$$\frac{d\bar{p}_t(\text{dip})}{dt} = k[\bar{p}_0(\text{dip}) - \bar{p}_t(\text{dip})]$$

$$= k\left[\frac{(\varepsilon_0 - \varepsilon_\infty)}{4\pi}\bar{E}_0 \, e^{j\omega t} p_t(\text{dip})\right]$$

The general solution of this relation is

$$\bar{p}_t(\text{dip}) = C\, e^{-kt} + \frac{1}{4\pi}(\varepsilon_0 - \varepsilon_\infty)\frac{\bar{E}_0 \, e^{j\omega t}}{1 + (j\omega)/k}$$

† Debye's justification was based on randomization of orientation obeying a rotational Brownian motion pattern.

The first term represents a transient which decays with t and so the steady-state condition in the oscillating field is given by the second term:

$$\bar{p}_t(\text{total}) = \bar{p} + \bar{p}(\text{dip}) = \left[\frac{\varepsilon_\infty - 1}{4\pi} + \frac{1}{4\pi}\frac{(\varepsilon_0 - \varepsilon_\infty)}{1 + j\omega\tau}\right]\bar{E}_0\, e^{j\omega t}$$

Defining the complex permittivity ε^* by

$$\begin{aligned}\bar{D}_t &= \varepsilon^* \bar{E}_t \\ &= \bar{E}_t + 4\pi\bar{p}_t(\text{total}) \\ &= \bar{E}_0\, e^{j\omega t}\left[\varepsilon_\infty + \frac{(\varepsilon_0 - \varepsilon_\infty)}{1 + j\omega\tau}\right] \\ \varepsilon^* &= \varepsilon' - j\varepsilon'' = \varepsilon_\infty + \frac{(\varepsilon_0 - \varepsilon_\infty)}{1 + j\omega\tau}\end{aligned}$$

or

$$\varepsilon' = \varepsilon_\infty + \frac{\varepsilon_0 - \varepsilon_\infty}{1 + \omega^2\tau^2} \tag{2.4}$$

$$\varepsilon'' = (\varepsilon_0 - \varepsilon_\infty)\frac{\omega\tau}{1 + \omega^2\tau^2} \tag{2.5}$$

The relations for the frequency dependence of ε' and ε'' are clearly arrived at on a very general basis (Pellat 1899), and they became part of dielectric studies with Debye's use of them (1912). They offer a representation in terms of a single rate coefficient (k) or corresponding time constant (τ). It is a surprising fact that, despite the complexity of many molecular systems (e.g. in liquids), these equations often suffice quantitatively to represent the experimental data. They presuppose that the dipole contribution to the polarization can be completely separated on the frequency scale from the polarizability terms, the latter contributing ε_∞, the former $(\varepsilon_0 - \varepsilon_\infty)$ to the total permittivity (ε_0). In condensed phase this separation is usually reasonably clear: the dipoles are too sluggish to respond to fields of frequency beyond say 10^{12} Hz ($\bar{\nu} = 33\ \text{cm}^{-1}$) whilst lattice and proper mode vibrations are usually beyond $\bar{\nu} = 50\ \text{cm}^{-1}$.

A number of useful and important relations follow from the equations for ε' and ε''. These include $\omega(\text{maximum}) \times \tau = 1$: where $\omega(\text{max})$ is for

ε''(max); the Cole–Cole semi-circular arc for $\varepsilon'' = f(\varepsilon')$:

$$\left(\varepsilon' - \frac{\varepsilon_0 + \varepsilon_\infty}{2}\right) + \varepsilon'' = \left(\frac{\varepsilon_0 - \varepsilon_\infty}{2}\right)^2 = \text{constant}$$

and

$$\int_0^\infty \frac{\varepsilon''}{\omega}\,\mathrm{d}\omega = \frac{\pi}{2}(\varepsilon_0 - \varepsilon_\infty)$$

The $(\omega_{max} \times \tau) = 1$ relation is retained in defining an 'effective relaxation time, when departures from the Pellat–Debye equations occur. Such departures are very frequently adequately expressed by assuming a distribution of relaxation times (symmetrically) around the τ (effective) value. The extent of the distribution on a reduced time scale (τ/τ effective) is assessed by a number of alternative empirical factors.

One particularly useful function is due to Fuoss and Kirkwood[33]: they wrote

$$\varepsilon'' = \varepsilon''_{max} \cdot \text{sech}\,[\beta \ln(\omega/\omega_{max})]$$

The Pellat–Debye function corresponds to $\beta = 1{\cdot}0$ and this factor provides a convenient (and generally effective) means of extending the width of the loss-function, and this on the basis of ε'' values only: $(1-\beta)$ measures the spread of relaxation times demanded by the observations.

The Cole–Davidson function

$$\varepsilon^* - \varepsilon_\infty = \frac{\varepsilon_0 - \varepsilon_\infty}{(1 + \mathrm{j}\omega\tau)^\alpha}$$

provides a corresponding factor[34] and, at the same time adequately represents many loss processes that show in the ε'' vs ε' plot, not a circular arc but one which is skewed to a linear form at the high-frequency side. This form commonly occurs in such liquids as the glycols (dihydroxylic compounds, $HOCH_2{\cdot}(CH_2)_n{\cdot}CH_2OH$) and others, especially at low temperatures.

In addition to those many systems showing an apparently continuous distribution of values around a mean τ, a number of instances occur of non-rigid molecules showing two or more distinct relaxation processes. Usually such instances are revealing in terms of intra-molecular group mobility. Examples will be mentioned later in this account.

The τ values deduced from these relations clearly refer to the process of dipole relaxation occurring in the bulk medium. Numerous attempts from Debye (1913) onwards have been made to relate τ(medium) with τ(molecular) by consideration of the appropriate internal field correction applying between them. It is now established that for nearly all practical considerations, the difference between the terms may be neglected, i.e. $\tau(\text{medium}) \approx \tau(\text{molecular})$[35].

The interpretation of these relaxation-time distribution functions involves detailed analysis of the relevant molecular mechanisms. Their current understanding is largely based on a model emphasizing the co-operative nature of the molecular reorientation process in condensed phases and the role played by a (pseudo-lattice) defect diffusing through the medium[36].

Another aspect is concerned not so much with the existence of a range of τ-values in some system as with the absolute τ-values encountered. As these can range from 10^{-11} sec to 10^{+3} sec a variety of mechanisms is involved. For liquids and solutions Debye provided a molecular interpretation based on a rigid sphere reorienting with a Brownian-motion form of momental distribution: for a molecular volume v and (effective) medium viscosity η this gave

$$\tau = \frac{3v}{\mathbf{k}T}\eta \tag{2.6}$$

More satisfactory versions of this relation have been advanced[37] allowing both for the general shapes of molecules (approximated as ellipsoids of revolution) and for the relevant 'molecular viscosity'. One aspect of the equation, that $\mathrm{d}\ln(T\tau)/\mathrm{d}(1/T) = \mathrm{d}\ln\eta/\mathrm{d}(1/T)$ is equivalent to the coincidence of the corresponding activation enthalpies, $\Delta H^*(T\tau) = \Delta H^*(\eta)$. The two factors are, for many liquids, within ten per cent of one another.

The well-known Eyring rate equation has also been much used in dielectric relaxation studies, usually in the form:

$$\tau = \frac{1}{k} = \frac{h}{\mathbf{k}T}\exp\left(\frac{\Delta H_{\mathrm{E}}^*}{RT}\right)\exp\left(-\frac{\Delta S_{\mathrm{E}}^*}{R}\right) \tag{2.7}$$

ΔH_{E}^*, the Eyring activation enthalpy is derived from the $\log(T\tau)$ *vs* $(1/T)$ plot. The associated ΔS_{E}^* activation entropy term must be regarded as a measure of the departure of the observed relaxation frequency factor $(1/\tau)$ from an arbitrary reference frequency $(\mathbf{k}T/h)$ implied in the Eyring relation. It is surprising how small the correction imposed by ΔS_{E}^* is when ΔH_E^* is less than 4 kcal mole^{-1}, which it usually is for most liquids or polar molecules studied in solution[38]. Again, a more rational form of absolute rate equation than Eyring's is that due to Bauer[39] incorporating the moment of inertia (I) of the polar molecule.

$$\tau = \left(\frac{2\pi T}{\mathbf{k}T}\right)^{1/2} \exp\left(\frac{\Delta H_B^*}{RT}\right) \exp\left(-\frac{\Delta S_B^*}{R}\right) \tag{2.8}$$

Although there is preference for the Bauer equation as of greater theoretical and practical validity than Eyring's, over the usual temperature range of measurement (*ca* 100°C) the normal small variation in ln τ does not give significantly better constancy for ΔH_B^* than for ΔH_E^*: with the Arrhenius factor ΔH_A^* evaluated from the same data, they are related†:

$$\Delta H_A^* = \Delta H_E^* + RT = \Delta H_B^* + \tfrac{1}{2}RT$$

Far the most detailed and successful calculations of dipole relaxation rates on an *a priori* molecular model have been made by Hoffman and his collaborators[40] in analysing the behaviour of systems involving long-chain polar aliphatic compounds. These compounds provide from their lattice-site symmetry a local force-field on the dipole which can be calculated with sufficient precision to give a good fit with the observed data. This is clearly the desirable end in all dielectric relaxation studies—that the actual rates should be understood in terms of the dipole and molecular interaction energies. It is only exceptionally that a quantitative appreciation can be achieved‡.

The remainder of this chapter will be concerned with the results and interpretation of dielectric relaxations in condensed phases. In choosing the systems presented an attempt has been made to illustrate the range of molecular environments which has already been explored by dielectric methods. It can be said at once that these methods have revealed a considerable range of molecular rotational processes which were either unsuspected or of which only scanty appreciation existed. It will become obvious that only in a small fraction of the systems have the dielectric aspects been at all fully delineated. And it will be equally fully appreciated that the study of these systems by methods other than the dielectric (e.g. calorimetric, ultrasonic, n.m.r., X-ray diffraction, infrared, spectrophotometric, etc.) could, in particular instances, add greatly to the under-

† See C. Brot, *Chem. Phys. Lett.*, **3**, 319 (1969).

‡ See C. Brot and I. Darmon, *J. Chem. Physics*: in course of publication (1970).

standing of the molecular behaviour. Further advances will be facilitated by such cooperative studies.

2.4.1 Dielectric Relaxation in Simple Liquids

The use of 'simple' in the context of a liquid is itself a simplification, but it can be accepted that some liquids are simpler than others. A norm can be taken as formed by rigid polar molecules not spherical in shape and lacking tendencies to interact specifically with their neighbours. Chlorobenzene and many related molecules conform to this pattern.

Debye (1913) early provided a treatment of the relaxation times of rigid (and spherical) polar molecules in a medium whose retardation of the molecular rotation was expressed in terms of a viscous drag. He showed that the dispersion could be represented by a single rate coefficient or its reciprocal, the single-valued relaxation time. The frequency dependence of the loss

$$\varepsilon'' = (\varepsilon_0 - \varepsilon_\infty)\frac{\omega\tau}{1+\omega^2\tau^2}$$

and the interpretations of τ already quoted

$$\tau = \frac{3v}{\mathbf{k}T}\eta$$

were part of Debye's treatment. The first of these relations has had very considerable success both for the 'simple' polar liquids and for polar solutions. One essential qualification of it has recently been pursued. In deducing this form for a molecular model, Debye neglected the inertia of the molecule—i.e. he assumed its angular rotation would instantly attain the steady value appropriate to the applied field and the viscous drag. [Formally this may be regarded as equivalent to the neglect of $C\,e^{-kt}$ on p. 160.] For the motion under the torque arising from the field acting on the dipole

$$I\frac{d^2\theta}{dt^2} + \rho\frac{d\theta}{dt} = -\mu F_0\, e^{i\omega t}\sin\theta$$

Debye omitted the first term on the practically adequate basis that the acceleration of angular reorientation would require almost negligible time (*ca* 10^{-13} sec) compared with the period of the highest frequency fields he was contemplating (10^{11} Hz). Accordingly, the loss value at the frequencies well beyond $\omega\tau = 1$ becomes from equation (2.5),

$$\varepsilon''(\text{residue}) \times \omega\tau = (\varepsilon_0 - \varepsilon_\infty)$$

When this absorption coefficient (ε'') which is essentially in terms of the wavelength of the radiation is converted to an absorption (α) per cm:

$$\alpha = \frac{1}{\text{length}} \ln \frac{I_0}{I} = \text{neper cm}^{-1}$$

$$= \frac{\omega}{c} \frac{\varepsilon''(\text{residual})}{(\varepsilon_\infty)^{1/2}}$$

$$= \frac{1}{c\tau} \frac{(\varepsilon_0 - \varepsilon_\infty)}{(\varepsilon_\infty)^{1/2}}$$

This α(residual) is independent of frequency and sufficiently large to make chlorobenzene or water opaque in the visible region. This is the anomaly caused by Debye's conscious neglect of the inertial term. When it is included[41], ε''(residual) falls off from the Debye value and reaches zero at 10^3–10^4 times the frequency for its maximum value. This fall-off, which commences at a frequency about $10^2\ \omega(\max)$, largely governed by the factor $(I/2\mathbf{k}T\tau^2)$ for the polar molecule, has only recently been plotted as a very far infrared contour[42(i)]: it is found to occur, when not complicated by other features, from 20 cm^{-1} to 100 cm^{-1}. As yet there has been little precise assessment of the contour in relation to deductions from the fuller equation of motion. This aspect will certainly be pursued, as it provides a direct representation of the relaxation of molecular angular velocity. Whilst other absorptions to be mentioned later will frequently interfere with the contour of the inertial relaxation in pure liquids, it is possible that more favourable conditions will exist in dilute solutions of 'spherical' polar molecules in spherical molecule non-polar solvents (CCl_4:C_6H_{12}: *neo* C_5H_{12}). It is reasonable to anticipate that new features of molecular rotation in the liquid phase will be observed in this way*.

* Birnbaum has now observed rotational line structure in the absorptions of HCl dissolved in SF_6.

2.4.2 Non-polar Liquids

All early measurements emphasized that the u.h.f. dielectric absorption of non-polar liquids (CCl_4, C_6H_6, C_6H_{12}, etc) was very small and perhaps due only to impurities or experimental errors. However, careful measurements by Bleaney[43], Whiffen[44], Parry[45] and others at frequencies up to 100 GHz indicated that small absorptions appeared to be real, e.g. in benzene. Whiffen, in particular, emphasized their probable significance as arising from dipole moments of small value appearing for collisionally distorted molecules. These features have now been fully confirmed. In liquid benzene a weak-absorption has a flat maximum near 70 cm^{-1} where $\alpha = 5{\cdot}0$ neper cm^{-1} [42(ii)], Figure 2.2. The authors first describing these absorptions imply a relation between those observed in the liquid and the crystalline phases: this is a very doubtful proposition. In crystalline benzene four absorption centres are found—at 53, 64, 72 and 97 cm^{-1}, the latter being of half-width *ca* 20 cm^{-1} and markedly the strongest: but the integrated intensity is larger in the liquid where only a featureless absorption appears. There is very much less 'lattice structure' in the liquid than in the crystal and the increased intensity can be understood if the liquid absorption arises from a collisionally induced dipole moment. For benzene this means the quadrupole induced dipole whose strength must be markedly dependent upon the anisotropy of polarizability in the molecule.

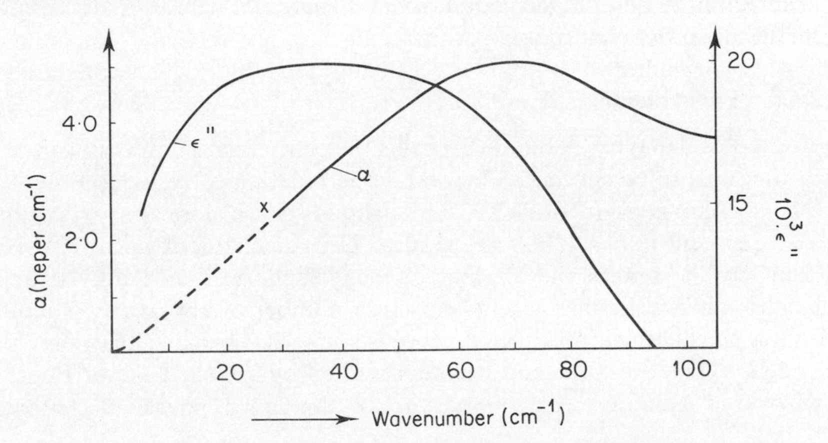

Figure 2.2. Absorption of liquid benzene in the very far infrared: (i) α (neper cm^{-1}) (after Chantry, Gebbie, Lassier and Wyllie, Reference 42(ii)); (ii) ε'' (after G. W. F. Pardoe)

In liquid CCl_4 and CS_2 at room temperature, similar broad peaks appear centred near 50 cm^{-1} and 70 cm^{-1} respectively which are, again, more intense than the lattice absorptions of the crystalline phases. The integrated liquid phase absorptions are in the sequence CCl_4 (140 cm^{-2}) < CS_2 (230 cm^{-2}) < C_6H_6 (450 cm^{-2}): these do not correlate with the mean molecular polarizabilities for which $\alpha(CCl_4) \approx \alpha(C_6H_6)$. Furthermore, CCl_4 in its symmetrical tetrahedral form has no quadrupole moment. Whilst it is possible to invoke the higher moments as responsible for the absorption in the liquid, a more plausible origin is the distortion of the CCl_4 tetrahedron on close collisional contact*. If this assignment is correct, the CCl_4 absorptions should be far more dependent on pressure and especially temperature than the feature in C_6H_6. In CS_2 it is probable that both collisional distortion and the quadrupole moment play a part. That cyclohexane (C_6H_{12}) has only a very small monotonically rising absorption from 20 cm^{-1} to beyond 120 cm^{-1} is a significant finding: it suggests the component (C—H) moments in this molecule are close to zero value. Whilst these absorptions in non-polar liquids were only suspected from dielectric studies in the microwave region, there is little doubt now that they are a general feature. It remains to be disclosed to what extent the absorption sources giving rise to them are due to translational or to hindered rotational motion. Such correlation as can be suggested with similar frequency absorptions in polar liquids points to a translational–vibrational (rattling) mode as a probable origin†.

2.4.3 Polar Liquids

The molecular type treated in Debye's dielectric relaxation is illustrated by the substituted benzenes. Typical modern data mostly due to Poley[45a] (21 ± 1°C) are given in Table 2.2. These liquids give a close approximation (*ca* 1 per cent in $\varepsilon'' = f(\omega)$) to the ideal Debye behaviour up to 10^{11} Hz (3 cm^{-1}). The same is true of a large range of aliphatic liquids: alkyl or aryl halides, nitriles, ketones, esters, etc. In a number of apparently 'simple' liquids the dielectric absorption curve is somewhat broader than that for a single relaxation time and it is represented by a Cole–Cole or Fuoss–Kirkwood function. The amplitude of this liquid phase dispersion,

* See G. W. F. Pardoe, *Ph.D. thesis*, University of Wales, 1969.

† See Mansel Davies, G. W. F. Pardoe, John Chamberlain and H. A. Gebbie, *Trans. Faraday Soc.*, **66**, 273–292, 1970.

Table 2.2

Parameters of Debye-type dipole relaxations in pure liquids (21°C)

Liquid:	C_6H_5F	C_6H_5Cl	C_6H_5Br	C_6H_5I	$C_6H_5NO_2$	C_6H_5CN
ε_0		5·69	5·43		35·7	25·6
ε_∞	2·33	2·56	2·62	2·76	4·07	3·85
n_{ir}^2	2·15	2·33	2·43	2·63	2·43	2·34
μ^2(gas)D^2	2·6	2·9	3·0	2·8	17·6	17·1
$(\varepsilon_\infty - n_{ir}^2)/\mu^2$	0·07	0·08	0·06	0·05	0·09	0·09

$(\varepsilon_0 - \varepsilon_\infty)$ on substitution into the Onsager or Kirkwood equations for specific polarization, gives dipole moments which are within a few per cent of those deduced from studies of the gaseous state[46]. This result and the approximation in many cases of ε_∞ in the microwave region to $n_D{}^2$ for such liquids led to the acceptable conclusion that the total molecular rotational relaxation had been passed in the liquid at frequencies of 10^{11} Hz or a little beyond. Such small values as were established for the difference $(\varepsilon_\infty - n_D{}^2)$ could usually be regarded as arising from the atomic polarization terms—i.e. from the totality of the infrared vibrational-mode absorptions. It was Poley, on the basis of the data in Table 2.2, who first insisted that the differences $(\varepsilon_\infty - n_D{}^2)$ were probably too large for this to be true and that the use of Errera and Cartwright's far infrared (*ca* 120 cm^{-1}) refractive indices[47] quoted as n_{ir} in Table 2.2 showed a residual discrepancy itself proportional to μ^2. This implied that a further small dipole rotational dispersion should be found between say 5 cm^{-1} and 100 cm^{-1}. This deduction has been fully confirmed and it is now clear that most polar liquids give an absorption in the region of 50 cm^{-1}, which immediately reflects the individual molecular behaviour in the liquid state. Furthermore, it appears to be derived from a residual angular rotational or librational motion.

As these absorptions have only recently been established by the work of Gebbie and collaborators[48] and of Leroy and others[49] much further clarification of their character must be awaited. Nevertheless the general situation is perhaps simply represented: Figure 2.3. At any instant the local field around one molecule in a pure polar liquid has an energy value varying with the orientational angle and showing minor peaks and troughs. The major (Debye) rotational relaxation involves the small activation energy ($\Delta H^*(\tau) \sim 3$ kcal mole^{-1}) needed for the encaged

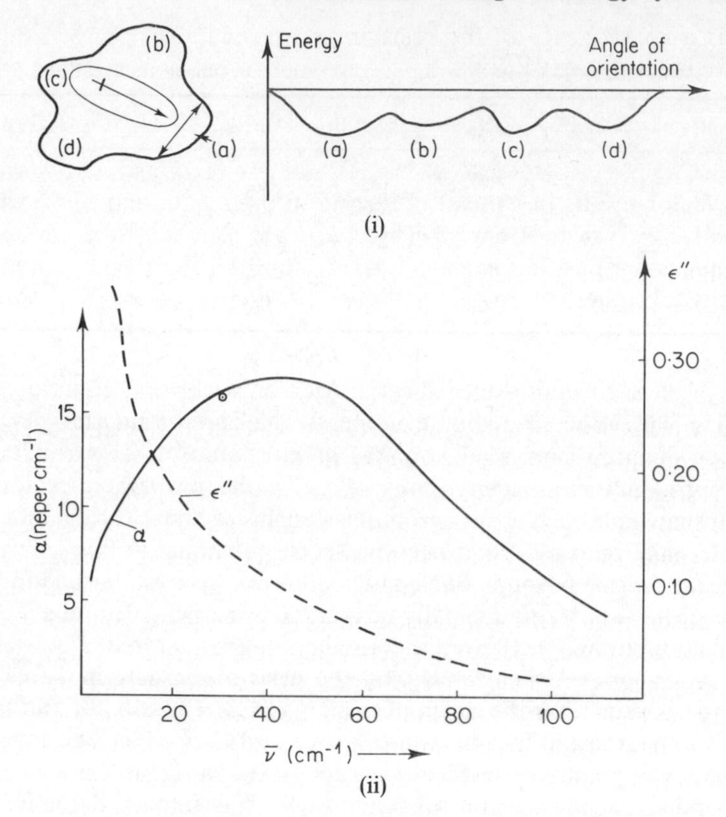

Figure 2.3. (i) Schematic instantaneous energy contour in neighbourhood of single rigid polar molecule; (ii) absorption of liquid chlorobenzene in the very far infrared (after G. W. F. Pardoe)

molecule and its neighbours to reorient cooperatively and gives a relaxation of Debye character of the order τ *ca* 10^{-11} sec. However, a higher frequency librational oscillation can occur within the 'instantaneous' energy troughs at a frequency of the order of 10^{12} Hz (30 cm^{-1}). This fragmented angular oscillatory reorientation is what now appears as an important absorption in polar liquids near 50 cm^{-1}.

It must be emphasized that the absorption centre near 50 cm^{-1} in polar liquids is seen as a separate (flat) peak only in terms of the α (neper cm^{-1}) plot: in the plot of ε'' against $\bar{\nu}$ or $\log \bar{\nu}$ it forms only a small increase in the level of the high-frequency tail of the Debye absorption.

This is a consequence of the relation, at wavenumber $\bar{\nu}$ cm^{-1} and refractive index $n = (\varepsilon')^{1/2}$,

$$\alpha = \frac{2\pi\varepsilon''\bar{\nu}}{n}$$

The $\bar{\nu}$ factor results in ε''(max) occurring in the region of 1 cm^{-1} whilst the peak of α is near 50 cm^{-1}. This frequency factor has the important consequence of locating much the larger contribution to the integrated intensity taken in the form

$$\int_0^\infty \alpha(\bar{\nu})\, \mathrm{d}\bar{\nu}$$

in the α peak near 50 cm^{-1}. Kroon and van der Elsken[50] have concluded from this that the major fraction of the total spectroscopic intensity to be associated with the rotational mode of polar molecules is found in the 50 ± 30 cm^{-1} α-absorption in the liquid state. They correlate the $\bar{\nu}$ cm^{-1} values for α(max) in the liquids with the moments of inertia of the molecules involved. The implication is that this very far infrared liquid phase absorption is the condensed-phase feature corresponding to the gas-phase rotational spectrum. Leroy has independently come to the same conclusion[49]. This interpretation, as is emphasized by Leroy, regards the Debye feature peaking in ε'' near 1 cm^{-1} as the cooperative reorientational process in the liquid state and, as such, it is far less closely related to the behaviour of the individual isolated molecule.

The Poley absorption near 50 cm^{-1} is of a degenerate (i.e. broadened) resonance type. The broadening presumably reflects the physical situation in that a wide range of 'resonant frequencies' characterize the rotational motions at any instant: and, of course, an exceptional degree of perturbation broadening can be envisaged. The absorptions are in fact so broad[50a] that only the careful delineation of the refractive index function (Leroy: Pardoe) suffices to establish its anomalous (i.e. resonance) character. These aspects are in general agreement with Dr Nora Hill's predictive treatment[51].

A very much larger body of data is concerned with the temperature coefficient of the major dipole relaxation (i.e. Debye) process. An Arrhenius type representation

$$\tau = \mathrm{A}' \exp \frac{\Delta H_{\mathrm{A}}{}^*}{RT}$$

often suffices over the small temperature ranges ($\Delta t°$ 40°C) for which many measurements extend. There are good reasons for using (as is frequently done) an absolute rate equation: e.g. that of Eyring or of Bauer already quoted, equations (2.7) and (2.8). As $\Delta H_A{}^*$ is itself often small (<3 kcal mole^{-1}) unequivocal choice between these alternatives is sometimes difficult: Eyring's equation has been much used although Bauer's is certainly of the more correct form for the pseudo-rotational process involved. The entropy terms ($\Delta S_E{}^*$ and $\Delta S_B{}^*$) are to be regarded as essentially corrections to the over-simplified evaluation of the absolute frequency term, and it is significant that generally $|\Delta S_B{}^*| < |\Delta S_E{}^*|$ and that this is especially so for large rigid polar molecules[52].

One of Debye's equations is conveniently written:

$$\tau = \frac{3v}{\mathbf{k}T} s\eta_{\text{macroscopic}}$$

The factor 's' is an empirical expression of the ratio of the molecularly effective viscosity term to the immediately measurable macroscopic term. Debye himself found by experiment s < 1; it is often 0·2±0·1 and it can be *ca* 10^{-2}, for the most viscous solvents. The general reason for this is well understood: the rotational reorientation of individual molecules can often occur without too great a displacement of neighbouring molecules. Only when the latter have to move translation-wise will the full viscosity factor (which necessarily involves relative translational motion) come into play. Large non-spherical molecules give $s \approx 1$: small, and especially spherical molecules in viscous solvents, give $s \ll 1$ (Hase, Smyth, Meakins). One deduction, $(T\tau) = \text{constant} \times \eta$(macroscopic) suggests $\Delta H_E{}^* = \Delta H^*(\eta)$: i.e. $[T\tau/\eta_{\text{macroscopic}}]$ should be independent of temperature. Some data[53] illustrating this condition are given in Table 2.3. In all four of these liquids $\Delta H_E{}^*(\tau) \approx \Delta H^*(\eta) \simeq 2{\cdot}8$ kcal mole^{-1}: this value is typical of many simple organic liquids near room temperatures. For many polar liquids these relations are essentially correct, i.e. $\Delta H_E{}^*$ and $\Delta H^*(\eta)$ differ by less than 10%, and although the use of the macroscopic viscosity to measure the barrier to molecular reorientation is an over-simplification this correlation is a possible approximation to the norm. Distinct departures occur when specific interactions (e.g. hydrogen bonding) interfere with the dipole or molecular mobility: for non-rigid polar molecules, and also for molecules which are close to spherical in form. The latter two categories will be considered later: the variations in $(T\tau/\eta)$ or in the corresponding ΔH^* terms are

Table 2.3

Dipole relaxation times and viscosities of pure liquids†

	Toluene			*o*-Xylene		
t°C	20	40	60	20	40	60
$10^{12} \times \tau$(sec)	6·5	5·1	4·1	8·2	6·4	5·3
$10^{12} \times (\tau/\eta)$	11·0	10·8	10·8	10·1	10·1	10·6
	p-Chlorotoluene			α-Picoline		
t°C	20	40	60	20	40	60
$10^{12} \times \tau$(sec)	25·7	20·4	16·0	12·7	9·8	8·0
$10^{12} \times (\tau/\eta)$	28·9	29·2	28·6	15·7	15·5	15·7

illustrated in Table 2.4 for hydroxylic liquids. For n-propyl alcohol $\Delta H_E{}^*(\tau) \approx \Delta H^*(\eta) \approx 4{\cdot}3 \pm 0{\cdot}1$ kcal mole^{-1} but for (supercooled) glycerol near 230°K, $\Delta H_E{}^*(\tau) = 32$ kcal mole^{-1} : $\Delta H^*(\eta) = 21$ kcal mole^{-1}.

2.4.4 'Viscosity' and Molecular Rotation in Condensed Phases

A further insight into the role of viscosity in controlling molecular rotations in the liquid phase is provided by the use of Batschinski's relation. Batschinski's relation $\eta = A/(v-b)$ where A and b are constants and v is the specific volume of the liquid offers a very good representation of the temperature dependence of the viscosity (η). This relation is one

Table 2.4

Temperature dependence of dipolar relaxation times (in 10^{-12} sec) and of viscosity (in poise) in hydroxylic liquids

	Water		Methyl alcohol		Ethyl alcohol	
t°C	τ	$(T\tau/\eta) \times 10^7$	τ	$(T\tau/\eta) \times 10^6$	τ	$T\tau/\eta$
−10	32·3	3·19	152	4·02	270	3·19
0	22·4	3·43	116	3·87	222	3·44
10	16·6	3·61	92	3·76	178	3·48
20	12·6	3·69	73	3·56	138	3·39
30	9·5	3·62	58	3·38	107	3.28
40	7·5	3·61	(44)	(3.00)	83	3·13
50	6·2	3·65	(33)	(2.62)	64	2·95

† See S. Mallikarjun and N. E. Hill, *Trans. Faraday Soc.*, **61**, 1389 (1965).

of the starting points of the free-volume treatment of liquid properties. It has been established[54] that a similar relation is applicable to the dipole relaxation process: $\rho = 2\mathbf{k}T\tau = A'/(v-b')$. The conclusion suggested by Mallikarjun and Hill[55] from a study of nine rigid polar, non-specifically interacting molecular liquids was that the Batschinski relation and its τ-value analogue represented the temperature dependence of η and of τ more effectively than did the Arrhenius–Eyring–Bauer exponential relations. As the terms b and b' represent respectively the minimum specific volumes needed for viscous flow and for molecular rotation a comparison of them is of interest: Table 2.5. Real differences are found and these are larger the more nearly spherical the molecular shape. This comparison provides a natural link with those molecular systems where the molecules rotate in at least one phase in the solid state.

Before considering the evidence pertaining to molecular rotation in solids some further analysis of the viscosity factor as it can operate on molecular rotation in the liquid state is desirable. This has been provided by Nora Hill on the basis of Andrade's model of the viscosity process[56]. Applied to pure liquids the model envisages the rotational reorientation of one molecule to occur in unison with another—a molecular waltz which presupposes a sticky-collision of the partners. The volume parameter r^3 of the Debye molecular model is replaced by $\kappa^2\sigma$ where κ is the radius of gyration of the molecule and σ is a mean intermolecular distance (or effective molecular diameter): the relaxation time is given by

$$\tau = \frac{3(3-\sqrt{2})}{2\mathbf{k}T}\eta\kappa^2\sigma$$

Table 2.5

Comparison of 'zero free volume' factors for viscosity (b) and dipole rotation (b') in polar liquids

Liquid	b(cm^3 g^{-1})	b'(cm^3 g^{-1})	$(b-b')/b$ per cent
CCl_3CH_3	0·689	0·642	6·8
$CH_3{\cdot}CCl(NO_2){\cdot}CH_3$	1·061	0·978	7·8
$CHCl_3$	0·584	0·541	7·4
C_6H_5Cl	0·834	0·799	4·2
C_6H_5Br	0·623	0·611	1·9
$C_6H_5NO_2$	0·799	0·799	0·0

In this case again $(T\tau/\eta)$ is expected to be constant. On the assumption that $\kappa^2\sigma = r^3$, the new equation gives τ values which are approximately 0·2 those of Debye's equation. This is clearly in the right direction.

Whilst in a pure liquid there is only one value of η which can be called into play, for a solute B in a solvent A the frictional force relevant to the reorientation of B is not that related to the viscosity of the medium A, which depends upon (A–A) molecular interactions, but rather the restriction which arises in the (B–A) interaction. This is an essential feature in accounting for the variation of τ for a polar solute when it is studied in various solvents: frequently $\tau_1/\tau_2 \neq \eta_1/\eta_2$. This inequality is very marked when high viscosity media are considered: medicinal paraffin with a bulk viscosity some two hundred times that of benzene often increases the molecular relaxation time relative to that in benzene by less than a factor of five. Lamb[56a] has emphasized the often neglected feature that the viscosity itself is time-dependent and that, especially for the more viscous media, the relevant value for molecular relaxation can be quite different from the 'static' bulk value. The large anomalies implied by the use of the bulk viscosity are markedly reduced by the Hill treatment whose essential features are revealed in the expression for τ(B) in the medium A.

$$\tau(\mathrm{B}) = \frac{3}{\mathbf{k}T}\,\frac{I_{\mathrm{AB}}\,.\,I_{\mathrm{B}}}{I_{\mathrm{AB}}+I_{\mathrm{B}}}\,.\,\frac{m_{\mathrm{A}}+m_{\mathrm{B}}}{m_{\mathrm{A}}\,.\,m_{\mathrm{B}}}\,.\,\eta_{\mathrm{AB}}\,.\,\sigma_{\mathrm{AB}}$$

I_{AB} is the moment of inertia of molecule B about the centre of mass of A in the collisional state; the m's are molecular masses; σ_{AB} is the mean separation of A and adjacent B molecules; finally, η_{AB} is the 'mutual viscosity' defined by Andrade's considerations from the viscosity of the solution (η_{s}):

$$\eta_{\mathrm{s}} = f_{\mathrm{A}}{}^2\,.\,\eta_{\mathrm{A}}\frac{\sigma_{\mathrm{A}}}{\sigma_{\mathrm{s}}}+f_{\mathrm{B}}{}^2\,.\,\eta_{\mathrm{B}}\,.\,\frac{\sigma_{\mathrm{B}}}{\sigma_{\mathrm{s}}}+2f_{\mathrm{A}}f_{\mathrm{B}}\eta_{\mathrm{AB}}\frac{\sigma_{\mathrm{AB}}}{\sigma_{\mathrm{s}}}$$

The f's are the mole-fractions of the components and σ_{s} is a mean intermolecular distance for all molecules in the solution. Thus η_{AB} can be evaluated from measured η_{s} values—especially in the dilute concentration range: $f_{\mathrm{A}} \simeq 1{\cdot}0$; $f_{\mathrm{B}}{}^2 \ll f_{\mathrm{A}}f_{\mathrm{B}}$.

The original test of this relation is shown in Figure 2.4. In so far as the solute volumes were constant, the Debye equation predicts a proportionality between τ and η_{A}: to a similar approximation, the Hill relation provides linearity between τ and η_{AB}. The far better conformity to the

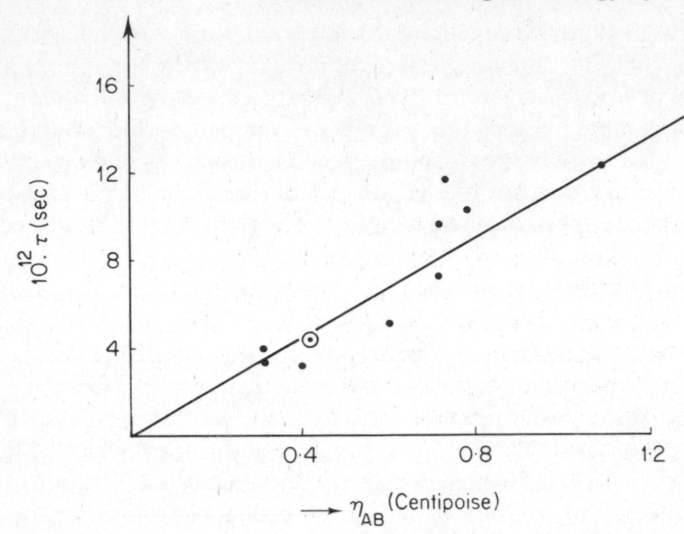

Figure 2.4 Relaxation time and mutual viscosity. Data for six different solutes and four solvents are represented. For the point marked by a circle both scales should be multiplied by ten

latter relation reflects the greater physical reality of the Hill representation of the molecular reorientation.

Further studies by Meakins and Smyth[57] have shown the quantitative limitation of the Hill equation quoted for $\tau(\mathrm{B})$. Comparison of the absolute values predicted with those observed gave ratios which for some larger solute molecules reached six or even nine. In precisely these cases where the solute molecule is appreciably larger than the solvent molecule, the $\tau(\mathrm{obs})$ conformed best with the simple Debye equation. It was when the solute was comparable with or smaller than the solvent molecule that the Hill evaluation showed to advantage.

Much further and more detailed appraisal of the molecular geometry and interactions are needed before a quantitative understanding of the molecular orientational process in solutions is possible. It would appear that the molecular rotational behaviour is dependent upon a number of intimate features including not only the spatial distribution of the molecules but the details of their local interactions. For the illumination that might be provided on these features it is important that carefully planned studies of the molecular relaxation times should be made with a view to establishing the molecular variables.

2.4.5 Rotator Phase Systems

In liquids the rotational degree of freedom is normally fully developed, i.e. there is complete, if hindered, molecular rotation about all inertial axes. In the crystalline state, the reverse is normally true; with the exception of limited molecular libration, rotation is absent. There are notable exceptions to this latter condition including long-chain molecules that rotate about one axis (i.e. their length) and pseudo-spherical molecules that rotate about all axes[58]. The (hindered) rotation of molecules in the crystal lattice is shown by such features as a solid-phase transition (with a rotational heat of melting), and increased specific heat, a decrease in heat of sublimation, an abnormally small heat of melting, the appearance of plasticity in the crystal, etc. X-ray diffraction shows the onset of molecular rotational mobility at the transition point.

Dielectric studies are especially suited to the study of this condition in polar media. Whilst the permittivity (ε_0) of a polar liquid normally falls sharply at the freezing point, those whose molecules lose no rotational mobility in the solid show only a small change (due to change of density) in passing through the freezing point (Figure 2.5). Whilst the ε_0 value shows the molecules to be reorientating completely in the solid state (until a lower transition temperature is reached) the measurements of the dielectric relaxation times confirm that the rotational process is not retarded in the crystalline rotator phase. The data for some typical examples summarized in Table 2.6 suffice to show that, surprisingly, the activation energy for the molecular reorientation is in some instances actually less in the solid than in the liquid phase: such is very definitely the case for succinonitrile where (in kcal mole^{-1}) $\Delta H_E^*(\tau)$ is $3{\cdot}2 \pm 0{\cdot}5$ in

Table 2.6

Data for some liquids and their rotator phase solids

	ε_0 at f.p.		10^{12} τ(sec) at f.p.		$\Delta H_E^*(\tau)$ kcal mole^{-1}		$\Delta H^*(\eta)$ kcal mole^{-1}
	liq.	solid	liq.	solid	liq.	solid	liq.
$(CH_3)_2CCl_2$	$15{\cdot}3_7$	$15{\cdot}2_7$	12·7	13·6	1·25	1·4	2·50
CCl_3CH_3	$9{\cdot}2_2$	$8{\cdot}8_9$	11·0	12·4	1·12	(1·1)	2·55
$(CH_3)_3CNO_2$	$24{\cdot}3_3$	23·3	7·9	10·2	0·81	0·5	3·36
$(CH_3)_2CClNO_2$	29·0	29·0	20·7	21·5	1·46	1·31	3·25

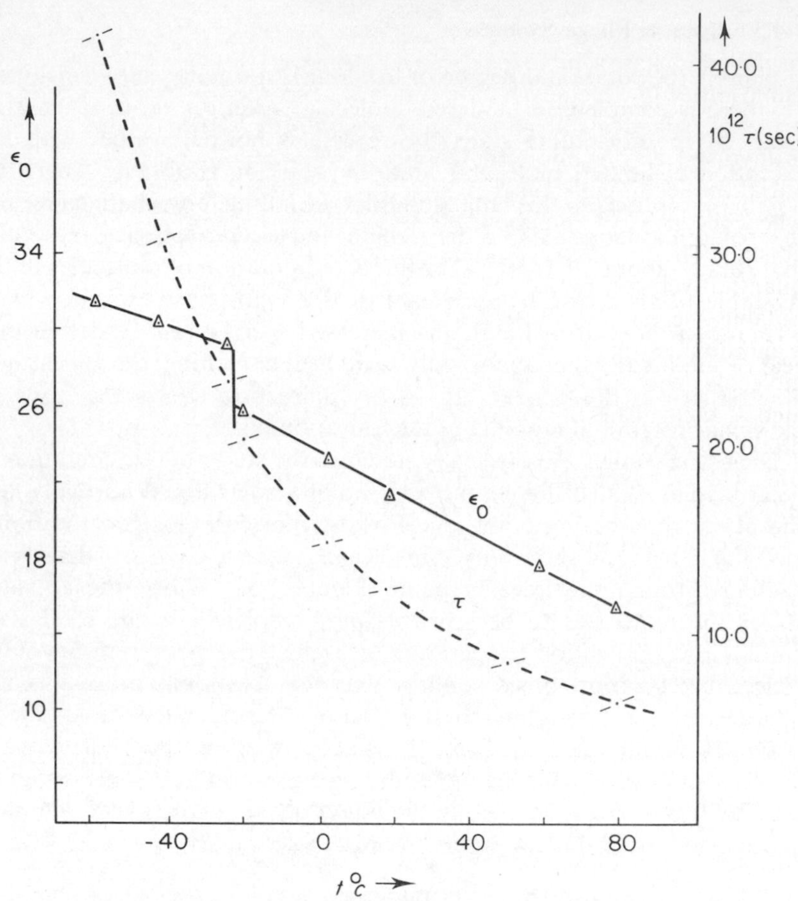

Figure 2.5 Temperature dependence of ε_0 and τ in the neighbourhood of the freezing-point for a typical rotator phase compound 2-chloro-2-nitropropane, $CH_3 . CCl . NO_2 . CH_3$.

the liquid and $2{\cdot}1 \pm 0{\cdot}5$ in the solid. Table 2.2 also shows another result characteristic of such cases as can be measured over temperature ranges in both liquid and solid: there is a pronounced difference, $\Delta H_E^*(\tau) < \Delta H^*(\eta)$. The divergence between these activation energies is far larger than the usual 10 or 20%: it is ascribed in these cases to the influence of the pseudo-spherical form of the molecules (Cl, NO_2 and CH_3 are of closely similar volumes) allowing them to rotate with the minimum translational displacement of their neighbours.

The same shape factor is relevant in giving the $\Delta H_E{}^*(\tau)$ smaller values in the solid state: these spherical molecules crystallize in a face-centred cubic lattice in which each molecule is surrounded by a very symmetrical array of (spherical) neighbours. A minimum variation of local energy is involved as one molecule rotates in this environment and a larger variation, leading to a larger activation energy, is present in the liquid where the array of neighbours is necessarily less symmetrical than in the crystal. These qualitative ideas do not appear to have been developed quantitatively for the spherical rotators although that has been done with notable success for rotation about one axis. Nevertheless an order of magnitude estimate of the interaction energies between the rotating molecules and of those between the non-rotating molecules when separated at their distances in the crystal lattice suggests that it is the dipole–dipole energies which lead to 'rotational freezing' at the lower temperature, when the thermal energy is not sufficient to overcome their attractive action[59].†

It is clear that the rotator phase provides much insight into molecular rotational processes in condensed phases. Thus it is particularly relevant that in the Hill comparison of the minimum volume terms b and b', most normal liquids show only small differences between them (Class I liquids): but many liquids (Class II) do show an appreciable difference, $b > b'$: i.e. a smaller minimum volume is needed for molecular rotation than is needed for the combined rotation–translation movement involved in viscous flow. The Class II liquids are essentially those which on freezing retain molecular rotation in the first crystalline phase. Some examples are quoted in Table 2.7[56]. Although there are well-known diffculties in defining the molecular free-volume quantitatively for a liquid, it is clear from these examples that this factor, which is $(v-b)$ or $(v-b')$, provides a criterion of much value in describing molecular behaviour in condensed phases.

Table 2.7

Percentage values of $(b-b')/b$ for Class I and Class II liquids

Class I:	furan, 1·3; tetrahydrofuran, 0·0; bromnaphthalene 0·0; quinoline 0·0
Class II:	*tert*-butylchloride, 28; 2,2-dichloropropane, 10·3; *tert*-butylbromide, 9·5; thiophene, 16·2

† C. Brot and I. Darmon have completed a full assessment in molecular terms of some typical rotator phases (to be published 1970).

Further insight into the molecular rotational mechanism is provided by the study of camphane derivatives which retain a rotator phase over a wider temperature range than do the substituted methanes[60]. Some typical data for *d*-camphor are summarized in Table 2.8. In this table n is the factor representing the distribution of relaxation times, i.e. the departure from the simple Debye form ($n = 1{\cdot}00$) of the absorptions which is represented in the Cole–Cole equation

$$\varepsilon^* = \varepsilon_\infty + \frac{(\varepsilon_0 - \varepsilon_\infty)}{1+(j\omega\tau)^n}$$

An uncertainty of the order of $\pm 0{\cdot}01$ in the n-values represents the practical limit of discrimination. Characteristic behaviour is illustrated in Table 2.8 where the n values show an increasing departure from unity as the temperature falls. Although this may be partially due to imperfections (strains and fissures) increasing in the specimen as the temperature falls, part of the effect is probably real and, as such, indicates the onset of local anisotropy in the crystal field. Furthermore, distinct departures from linearity in the Arrhenius–Bauer–Eyring activation energy plots are established: i.e. $\log \tau$ *vs* $(10^3/T)$ is significantly curved. A good representation of the data is given by the linearity of $\log \tau$ *vs* $(10^3/(T-T_\infty))$ where $T_\infty = 135°\text{K}$. This form of temperature dependence although frequently used in an empirical fashion, is recognized as arising in certain cooperative processes.

Some general assessment can be based on the data of Table 2.9. The ranges in the values for $\Delta H_E{}^*$ represent the total curvature in the $\log (T\tau)$ *vs* $(10^3/T)$ plots, and the uncertainty in $\Delta S_E{}^*$ is derived from the range in $\Delta H_E{}^*$. The barrier to rotation is, for three of the lattices, no larger than chlorobenzene would encounter in a benzene solution. The smaller the mean $\Delta H_E{}^*$, the larger the uncertainty in it. This and other indications

Table 2.8

Dielectric dispersion data for polycrystalline *d*-camphor

t°C	ε_0	ε_∞	ε''_{max}	n	$10^{12}\tau$ sec
80	9·82	2·74	3·54	1·00	11·0
50	10·5	2·82	3·84	$0{\cdot}99_5$	14·1
20	$11{\cdot}3_5$	$2{\cdot}9_2$	4·10	0·98	19·8
0	11·9	$3{\cdot}0_0$	4·17	0·96	28·0
−20	12·4	3·07	4·18	0·93	43·0

Table 2.9

Eyring equation $\left[\tau = \frac{h}{\mathbf{k}T}\exp\left(\frac{\Delta H_E^*}{RT} - \frac{\Delta S_E^*}{R}\right)\right]$ parameters for hindered molecular rotation in polycrystalline camphane derivatives

Compound	Camphene	Camphor	Bornyl chloride	Isoborneol
ΔH_E^* (kcal mole^{-1})	$2{\cdot}2 \pm 0{\cdot}7$	$1{\cdot}8 \pm 0{\cdot}7$	$2{\cdot}5 \pm 0{\cdot}3$	$5{\cdot}5 \pm 0{\cdot}1$
ΔS_E^* (cal deg^{-1} mole^{-1})	$-1{\cdot}9 \pm 2{\cdot}0$	$+2{\cdot}9 \pm 2{\cdot}8$	$-1{\cdot}9 \pm 1{\cdot}0$	$+6{\cdot}1 \pm 0{\cdot}4$

confirm that the 'true' activation energy is not being defined and that a far more detailed representation of the activation process is required—and is justifiable—than that provided by the Eyring equation.

Without overloading the observational data a realistic appraisal could be undertaken in terms of a simple Bragg–Williams cooperative process. This could be reduced to

$$\mu(\text{obs}) = f_r \,.\, \mu_0^2$$

$$\tau = \text{A} \exp\left(\frac{[1 - f_r]V}{RT}\right)$$

where f_r = fraction of molecules rotating in the lattice and V the maximum barrier height in the absence of rotation. The permittivities (ε_0) at each temperature allowed an estimate of f_r, assuming $\mu_0 = 2{\cdot}75$D from solution data: and such f_r's quantitatively reproduced the $\log \tau$ *vs* $(10^3/T)$ plot. Thus the available data were well fitted by the simple Bragg–Williams relations, but a critical examination showed the experimental factors were not sufficiently precise to justify their acceptance as quantitative confirmation of the model. This situation offers much scope for further experimental and theoretical assessment and could clearly provide a much needed simplification of the conditions of molecular reorientation in condensed phases†.

The transition from the polar molecule lattice of camphor to that of the isolated molecule in dilute solution was also followed by the experimental study of camphor–camphene solid solutions where miscibility is complete and the small dielectric absorption of the (rotator-phase) camphene can adequately be taken into account. On dilution of crystalline camphor

† See C. Brot and I. Darmon, footnote p. 179.

with increasing proportions of camphene, the n values move progressively to 1·00 and the curvature in log τ *vs* $(10^3/T)$ is similarly reduced.

Isoborneol in its temperature variation shows the reverse situation, i.e. as the temperature falls the increasing incidence of strong intermolecular forces become apparent. Above 50°C the dielectric absorption shows a distribution of relaxation times, the n-values at 90°, 70° and 50° being 0·92 : 0·88 : 0·80. Over this region $\Delta H_E{}^*$ is essentially constant at 5.5 kcal mole^{-1}. This value is much higher than that for the other camphane derivatives in Table 2.9, despite the much smaller molecular dipole moment of isoborneol (μ = 1·65D) compared with that of camphor (μ = 2·95D). The clear indication is that localized hydroxyl-group interactions (OH—OH), of the hydrogen-bond character, markedly increase the barrier to rotation. The increment it appears to contribute (*ca* 4 kcal mole^{-1}) fits this suggestion. Below 50°C the absorption becomes markedly broader and the indications are that it is now incorporating a number of different rate processes. This agrees with the temperature pattern of the permittivity[61] which itself shows a gradual transition to a non-rotator (or maybe a 'glass rotator') state between 50 and 15°C. This system, again, could with profit be far more thoroughly studied.

One further class of rotator-phase systems must be mentioned. The hydrogen halides have been the particular concern of Cole and his collaborators[62]. All but hydrogen fluoride show from their permittivities complete rotational mobility in the phase or phases near the melting point. For hydrogen fluoride the (μ^2/d^3) term expressing the order of magnitude of the dipole–dipole interaction energy is much larger than $\mathbf{k}T$ (m.p.) and there is, on this initial basis, a strong hydrogen-bonding situation in the hydrogen fluoride lattice. Some of the relevant factors for the other

Table 2.10

Arrhenius equation [$\tau = A \exp(\Delta H_A{}^*/RT)$] parameters for molecular rotation in some low temperature phases

Molecule and crystal phase	Temp. range (°K)	$-\log_{10} A$	$\Delta H_A{}^*$ (kcal mole^{-1})
HCl	63–100	13·00	2·6
HBr(III) 1st max	63–89	11·2	2·7
2nd max	63–89	12·1	1·6
DI(III)	62–75	13·3	2·2
DI(II)	78–100	15·6	3·1

molecules are given in Table 2.10. The hydrogen halide molecules are close to spherical in form and (especially HBr and HI) have small dipole moments. Adam has estimated[63] that in the HBr lattice perhaps only some 13% of the net interaction arises from the dipole–dipole term. Nevertheless he is able to give a satisfactory account of the situation in terms of a lattice diffusion model. The essential relations are two:

$$k = (1/\tau) = \nu_0 \,.\, \exp(-\Delta H^*/RT)$$

$$\Delta H^* = zu - a(\Delta u)$$

Here z represents the number of adjacent molecules or sites involved in the reorientation process: zu is the activation energy in the absence of vacancies and (Δu) is the reduction in it per vacancy of which the number is a. [The result is very similar to a simple Bragg–Williams relation.] The two absorptions encountered in the HBr(II) phase are accounted for by the alternative values $a = 0$ and $a = 1$.

2.4.6 Long-chain Organic Compounds

These compounds provide models for uniaxial rotation. Immediately below their freezing points many long-chain compounds of the formula $CH_3(CH_2)_nCH_2X$ form a crystalline phase in which the whole molecule is rotating about its long axis which is usually perpendicular to the planes in which lie the end-groups CH_3 and X (halogen, OH, CN, NH_2, etc.). The dielectric studies have been made almost exclusively on polycrystalline specimens. Meakins[64] and Dryden[65], Smyth[66] and Hoffman[67] have been prominent in these studies. Some of the systematic findings have been summarized graphically by Dryden and Meakins. The isostructural conditions in the homologues, i.e. for different numbers of carbon atoms in the chain of the molecule (n in the general formula), lead to a regular dependence of the absorption frequencies (Figure 2.6) and of the associated activation energies (Figure 2.7) upon the chain length. There then necessarily follows a well-defined (essentially linear) dependence for the loss maximum of log (frequency) upon the activation energy: this is equivalent to a linear dependence of entropy of activation upon enthalpy of activation. One particular interest of these relations is the increments they provide for these parameters per (CH_2) group. These are factors whose orders of magnitude an adequate model must reproduce. Also, it becomes possible to extract contributions to the

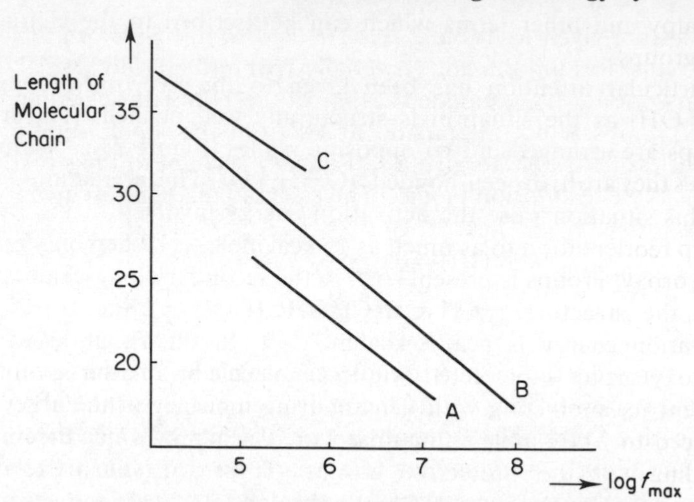

Figure 2.6 Variation of the frequency (f_{max}Hz) for ε''(max) with the number of (C—C) links in the chain for β-phase aliphatic compounds: A = ketones in crystalline hydrocarbons; B = crystalline esters; C = crystalline ethers

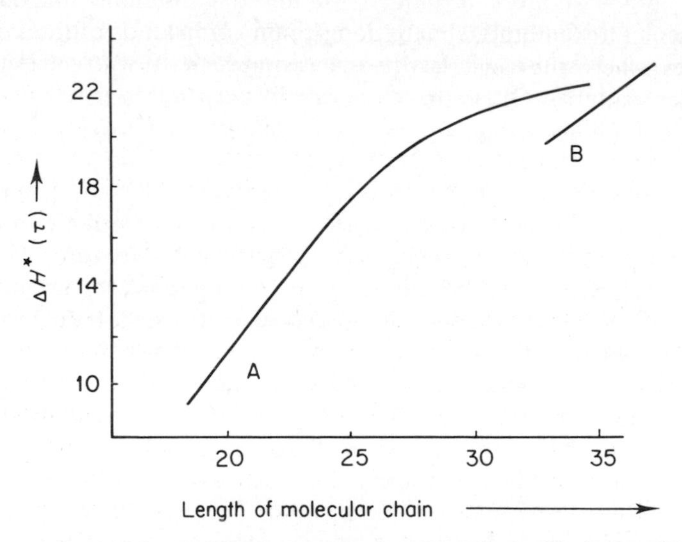

Figure 2.7 Variation of $\Delta H^*(\tau)$ in kcal mole^{-1} with the number of (C—C) links in the chain for β-phase aliphatic compounds: A = crystalline esters; B = crystalline ethers

enthalpy and other terms which can be ascribed to the characteristic end-groups.

Particular attention has been given to the hydroxylic compounds (X = OH) as the situation is structurally well defined. The hydroxyl groups are arranged in two opposing planes (Figure 2.8). Within these planes they are hydrogen-bonded: (O—H···O). The interactions favoured by this situation raise the activation energy involved in the hydroxyl-group reorientation to as much as 15 kcal mole^{-1}. When only one plane of hydroxyl groups is present—i.e. in the secondary long-chain alcohols, with the structure $—CH_2CH_2CHOHCH_2CH_2—$, the corresponding activation energy is near 5 kcal mole^{-1}. In the situation where the hydroxyl group is 'protected' within a molecule by adjacent groups which prevent its interacting with neighbouring molecules, the energy can be reduced to 2 kcal mole^{-1}, or raised to 10 kcal mole^{-1} if there is strong bonding with the 'protective' groups. These are typical data relating to the behaviour of an individual group and its local interactions: even more general interest pertains to the factors controlling the rotational motion of the whole chain. These will be treated later.

2.4.7 Molecular Rotation in Clathrates

The structures embraced by the description 'clathrate' are those crystalline phases where the rigid lattice incorporates cavities or channels in a regular pattern. Often the structure is only stable when a molecule

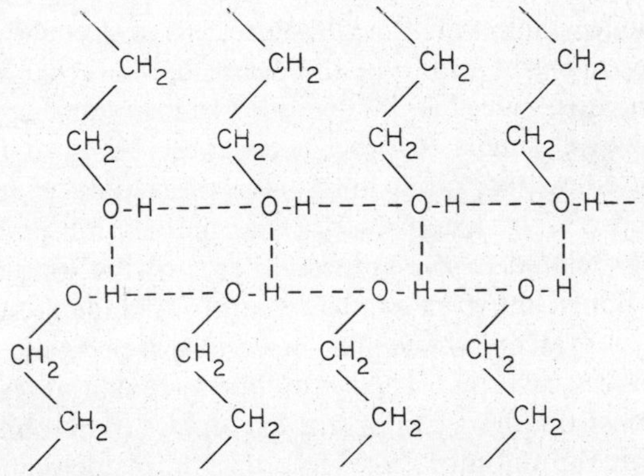

```
     \          \          \         \
      CH2        CH2        CH2       CH2
     /          /          /         /
   CH2        CH2        CH2       CH2
     \          \          \         \
     O-H------O-H-------O-H-------O-H--
     |          |          |         |
     |          |          |         |
   O-H------O-H-------O-H-------O-H
   /          /          /         /
 CH2        CH2        CH2       CH2
   \          \          \         \
    CH2        CH2        CH2       CH2
   /          /          /         /
```

Figure 2.8 Schematic representation of (O—H···O) bonding in long-chain primary alcohols

different from that of the main lattice is present within the cavity or channel. Such a molecule must not be so small as to make escape from the cavity and through the lattice likely: neither must it distort the lattice too greatly by its presence. Very frequently the molecule encaged has a considerable degree of rotational and also some translational mobility. As the lattice is essentially rigid, the clathrates offer the opportunity of studying the behaviour of individual molecules in well-defined local force fields.

The first of the two structurally most important types of clathrates are those in which the cavities are isolated and often approximately spherical in form so that they encage a single guest molecule. Of this character are the quinol clathrates in which the *p*-dihydroxybenzene molecule (quinol) forms the host lattice. Another such system is provided by the 'gas hydrates' which are essentially ice-clathrates, with the qualification that the modified (cubic unit cell) ice lattice does not occur in the absence of guest molecules filling a substantial proportion (*ca* 35%) of the cavities. No useful purpose is served by thinking of these structures as clathrate compounds or any other form of 'compound': not only is there no stoichiometric control of the composition, there is no specific interaction between the components which are engaged together only in a space-filling capacity and that itself can have noticeably variable limits. This is especially true of the clathrates based on Dianin's compound in which the separated cavities, being unusually large, can contain a variable number, one, two or possibly three molecules, of the guest species. In these circumstances it is feasible to consider studying the interactions of two molecules held together in close proximity within the larger cavity. The second structural type to be considered is that of the channel-form. Here the host—urea is the best-known example—crystallizes with continuous open channels in the lattice, these being at least partially filled by some elongated, often long-chain molecules, which are appropriately shaped. Such structures provide models for uniaxial rotators and are closely related to the rotator phases of the long-chain organic compounds. Whilst the greater proportion of detailed study has so far been given to the systems whose host-molecules are organic, the inorganic world has similar structures. The synthetic molecular sieves of the zeolite type are amongst the most important examples: the mobility of foreign molecules within the anionic lattice has not yet been adequately clarified[68].

Thanks to the simplification provided by the regular lattice field, the molecular rotation within clathrates is of special interest. Meakins and Dryden[69] early showed that relatively very small barriers can arise even in the rotation of long molecules within the distinctly polar urea lattice (see Table 2.11). The feature emphasized by these data is that the total interacting energy for the molecule and its medium is almost irrelevant in the molecular rotational motion in comparison with the variation of the energy with rotational angle. If the total interaction is small the variation with angular motion cannot, of course, be large: but even large interaction energies can accompany ready rotation when they are not angularly dependent. This aspect is repeatedly emphasized by the dielectric study of clathrates. The contrast between the activation energies for methyl alcohol and methyl cyanide in quinol shows how changes in intermolecular distances can influence molecular barriers. The van der Waals length of methyl alcohol is about 4·7 Å: that for methyl cyanide about 5·7 Å: the latter dimension can fit into the cavity in quinol (*c*-dimension of 5·46 Å) only by some distortion of the cavity to a prolate ellipsoidal form. Even then the rotational freedom is greatly reduced so that τ(methyl cyanide)/τ(methyl alcohol) is *ca* 10^7 in quinol. Furthermore the rotational movement is restricted to axes perpendicular to the *c*-axis and this gives the single quinol–methyl cyanide clathrate crystal a very high dielectric anisotropy ($\varepsilon_1 = 3{\cdot}4 : \varepsilon_2 = 34$).

Further systematic studies of the rotation in quinol clathrates have been made by K. Williams[70]. He encountered the difficulty that small molecules such as H_2S, SO_2 and HCl rotate so freely in their cages that only at or below liquid air temperatures was there clear evidence of the

Table 2.11

Dryden and Meakins data for the Arrhenius representation

$$\left[\frac{1}{\tau} = \mathrm{A} \exp\left(-\frac{\Delta H_{\mathrm{A}}^*}{RT}\right)\right]$$

of molecular rotation in clathrates

Host	Guest	A (sec^{-1})	ΔH_A^* (kcal $mole^{-1}$)
Quinol	CH_3OH	4×10^{12}	2·3
	CH_3CN	$1{\cdot}6 \times 10^{15}$	18
Urea	$(C_{12}H_{23})_2O$	1×10^{10}	1·0
	$(C_{15}H_{31})_2CO$	$1{\cdot}6 \times 10^{10}$	1·4
	$(C_8H_{17})_2CO$	$3{\cdot}8 \times 10^{10}$	1·8
	n-$C_{18}H_{37}Br$	$1{\cdot}6 \times 10^{11}$	2·4
	$BrCH_2(CH_2)_8CH_2Br$	$1{\cdot}3 \times 10^{11}$	2·3

dispersion region being reached even at 8·5 GHz (λ = 3·5 cm). This general finding can be established from permittivity measurements only, as, taken in the form of the permittivity increment ($\Delta\varepsilon_0$) above the value for the quinol host lattice, an effective dipole moment can be evaluated for the encaged molecule: in this way it can be established that such molecules are rotating as fully (in the geometric sense) as in the gaseous or dilute solution phases. Table 2.12 gives some typical values for quinol clathrates. Sulphur dioxide although one of the largest molecules in this group showed a permittivity increment ($\Delta\varepsilon_0$) inversely proportional to the absolute temperature. This behaviour establishes that for the same temperature range the molecule either experiences no barrier to rotation or, what is more probably correct in this instance, has one which is symmetric, i.e. is repeated on rotation through π radians. The small dielectric absorption measurable at 8·5 GHz suggested a critical frequency in the range 150 to 200 GHz, i.e. $\bar{\nu}$ *ca* 5 cm^{-1}. The temperature effect on this absorption was so slight as to suggest an activation energy below 1 kcal mole^{-1}. Such results can only be accepted as preliminary indications: Allen has established that the rotational motion can give rise to a degenerate resonance absorption*. The full delineation of such features could lead to a notable increase in our appreciation of the transformation of the quantized rotational mode of the gas phase to the hindered relaxation process of the condensed phases[71].

The HCN molecules in quinol gave lower frequency absorptions whose form and frequency maximum could be observed from 210°K to 130°K. In contour the absorptions were accurately of Debye-type over the upper half of the temperature range but showed somewhat increased breadth at lower temperatures, until the β-value for the Fuoss–Kirkwood function became 0·94 at 110°K. From 270°K to 220°K the permittivity increment showed the HCN molecule to be rotating in a potential field whose

Table 2.12

Apparent dipole moments (in D) evaluated for encaged molecules in quinol clathrates

	HCl	H_2S	HCN	HCOOH	SO_2
μ(clathrate)	1·1	1·1	3·3	1·8	1·4
μ(lit)	1·08	1·17	2·93	1·77	1·60

* See also P. Rowland Davies, Faraday Soc. Oxford Discussion, 1969.

Table 2.13

Dielectric estimates of rotational relaxation times (τ) and activation enthalpies (ΔH_E*) for some guest molecules in quinol clathrates

Guest	$10^{12}\tau$ (sec) estimated for $-60°$C	ΔH_E* kcal mole^{-1}
HCl	*ca* 1	<1·0
SO_2	*ca* 1	<1·0
HCN	20	1·26
HCOOH	125	3·34±0·04

lowest minima were equal to within a factor of RT. Below 220°K an inequality between two extreme orientations (↔, ↔) became noticeable: the energy difference, evaluated from Fröhlich's model, was near 0·30 kcal mole^{-1}. The nature of the transition, if transition there is, near 220°K is not clear; it could be the incidence of anisotropic contraction of the cage as the temperature falls or it could be the onset of an order–disorder transition in the orientation of the guest molecules. An evaluation of the directly observed τ-values ($2\pi\nu_{critical}$) using the Eyring equation, led to $\Delta H_E^* = 1{\cdot}26 \pm 0{\cdot}20$ kcal mole^{-1}; $\Delta S_E^* = 2{\cdot}4 \pm 1{\cdot}0$ e.u. It is clear that the energy difference ($\Delta E = 0{\cdot}3$ kcal mole^{-1}) and the enthalpy of activation encountered in the molecular rotational process arise from the notable electric asymmetry of the HCN molecule which can be regarded as a well-pointed electric vane whose rotation explores the potential field in the quinol cavity. The data for some further molecules are summarized in Table 2.13. The case of formic acid deserves comment: the activation energy for the reorientation is now appreciable: even so, it is less than the

Table 2.14

Dielectric estimates of rotational relaxations (τ) and activation enthalpies (ΔH_E*) in ice-clathrates

System	Molecule	$10^{12}\tau$ (sec)	ΔH_E* kcal mole^{-1}	ΔS_E* e.u.
$C_4H_8O{\cdot}17H_2O$	H_2O	$1{\cdot}1 \times 10^7$ at 200°K	8·3±0·5	7·0
$C_4H_8O{\cdot}17H_2O$	C_4H_8O	3·0 at 155°K	0·27±0·05	−2·3
$C_3H_6O{\cdot}17H_2O$	C_3H_6O	2·1 at 123°K	0·17±0·10	−2·4
$C_2H_4O{\cdot}7H_2O$	C_2H_4O	4·2 at 155°K	0·46±0·10	−4·2

maximum interaction anticipated (i.e. the H-bond energy, —O—H···O). This situation recurs in the ice-clathrates.

Measurements on the ice-clathrates are summarized in Table 2.14. Davidson[72] and collaborators had studied the host dipole relaxation process in ice-clathrates and compared the results with those for other forms of ice. Here we are concerned with the far higher frequency process of the guest molecules. The significant finding is that a polar molecule such as tetrahydrofuran

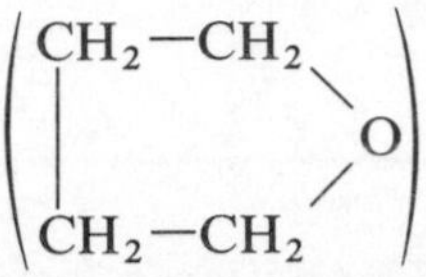

$$\left(\begin{array}{l} CH_2-CH_2 \\ \quad\quad\quad\quad\quad O \\ CH_2-CH_2 \end{array}\right)$$

shows no sign of hydrogen-bonding proclivities in the ice cavity. In this case Davidson has made a fairly detailed appraisal of the energy variation experienced by a point dipole of moment μ_D which rotates at the centre of the ice cavity. The approximately spherical symmetry of the latter and the number of hydroxyl groups distributed over its surface results in smaller energy variations than is implied by the observed ΔH_E^* value. Of course, the C_4H_8O molecular dipole is far from being located at the centre, but the pseudo-spherical symmetry of the cavity imposes an energy variation which is very small. This particular cavity has a free diameter (van der Waals) of 6·6 Å whilst the effective diameter of the tetrahydrofuran is only about 5·6 Å. Thus the exceptionally low barriers to rotation in the clathrates are another expression of the dominant influence of a 'free volume' factor in molecular motion.

The uniaxial rotation of long-chain compounds has already been described. The best analysis of this molecular process and a model to whose completeness many later studies will aim, is that of the long-chain compounds in urea lattice channels. The crystal structure of this clathrate is clearly defined. Urea molecules hydrogen-bonded into an hexagonal array (Figure 2.9) provide the ample channel into which the long-chain compounds fit. Meakins studied these systems dielectrically, Table 2.11, and one of the rotators evaluated from first principles by Lauritsen[73] was the hentriacontan-16-one long-chain ketone: $CH_3(CH_2)_{14}CO(CH_2)_{14}CH_3$. The evaluation is in two steps. Firstly the geometry of the interaction as the ketone rotates is fully established and the energy variation calculated. The energy values can be estimated either on group-interaction functions of the Lennard-Jones form,

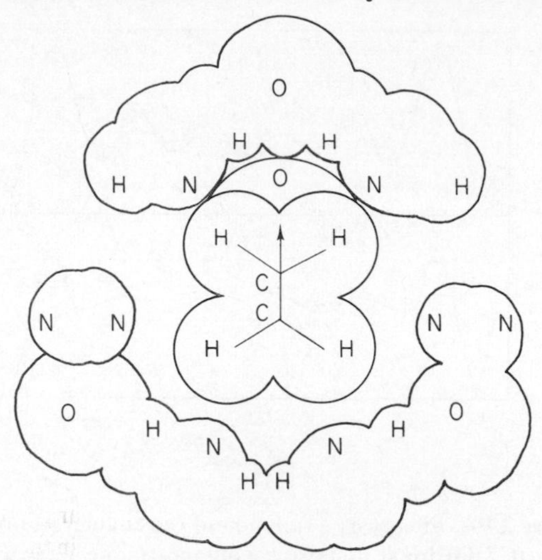

Figure 2.9 Hexagonal array of urea molecules with a section of the long-chain included ketone: van der Waals contours

$V(r) = c[\frac{1}{2}(r_0/r)^{12} - (r_0/r)^6]$; or from estimates of the 'volume overlap' of the groups and the attendant repulsion energies. The former proves to be the more satisfactory and offers the prospect of relating the parameters of the Lennard-Jones function to the specific details for the groups (dipole moments, polarizabilities, excitation energies and repulsion terms). Given the energy variation as the molecule rotates (the essential part being the interaction of the (C=O) group with the urea cage) a systematic statistical mechanical calculation can be made of the individual rotational jumps available to the molecule giving finally four Debye-type loss components for an energy contour with five separated minima in 2π radians. Each of these carries its own contribution to the observed resultant dipole relaxation. The agreement between the sum of the superposed features and the experimental curve (Figure 2.10) is as satisfactory as such an *a priori* calculation could currently achieve.

The Hoffman–Lauritzen model is of wide significance as it demonstrates the quantitative success of a comparatively simple evaluation and representation of molecular fields in the condensed state. A great deal more needs to be done in such quantitative analysis of well-defined molecular environments. The various forms of clathrates are particularly attractive

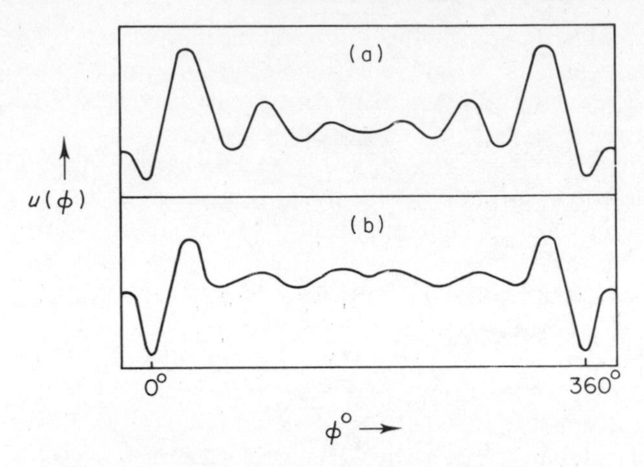

Figure 2.10 Energy as a function of the angular position of the (C=O) dipole in hexatriacontane-16-one in the urea clathrate: (a) calculated from 6–12 energy function; (b) deduced from dielectric data

in this respect; the rotator phase solids and the spherical-molecule liquids derived from them represent further approachable stages in the direction of the typical polar liquid. It is reasonable to hope that much will be achieved by such further calculations for molecular models of dielectric relaxation.

2.4.8 Molecular Reorientation in Polymers

In sharp contrast to clathrates with their well defined and regular placement of guest molecules, the amorphous polymer state provides an almost completely random molecular environment. In the absence of any crystallinity it is possible to distribute solute molecules through such a polymer matrix and Edwards[74] has studied, in particular, the situation when polar molecules are molecularly dispersed in amorphous polystyrene. The qualitative aspects are significant especially in relation to molecular rotation processes in disordered—i.e. glass or liquid—condensed phases.

A polystyrene matrix is essentially devoid of dielectric absorption and the observed factors (ε' and ε'') show, in the instances evaluated by Edwards, increments proportional to concentration up to about 10 per cent by weight of solute: these and other essential criteria establish the molecular distribution of the solute in the polystyrene. The data for camphor, which is a rigid polar molecule of almost spherical form, provided revealing features. The permittivity increments ($\Delta\varepsilon_0$) showed the solute, up to 7·5% by weight in polystyrene, to contribute to the specific polarization to the same degree as it would in liquid benzene. The dielectric absorption was confined to a single feature in the microwave region which became appreciably broader than the simple Debye contour as the temperature fell: β(Fuoss–Kirkwood) = 0·78 at 69°C: 0·44 at −20°C. In polystyrene at 20°C, τ(camphor) = $38 \pm 2 \times 10^{-12}$ sec: in one of the lowest viscosity solvents (n-heptane, η_{20} = 0·41 c.p.) at 20°C, τ(camphor) = $7{\cdot}0 \times 10^{-12}$ sec. The trivial influence of the enormous change in 'viscosity' from heptane to amorphous polystyrene at 20°C is largely due to the spherical form of the camphor molecule: once having achieved solubility and lodged itself between the polystyrene chains, there is the minimum constraint to hinder the rotation of the spherical molecule. The apparent (Eyring) activation energy expressing the temperature dependence of τ(camphor) in polystyrene from 20 to 70°C was $\Delta H_E^* = 0{\cdot}9 \pm 0{\cdot}3$ kcal mole^{-1}. Other relevant data for camphor are shown in Table 2.15.

The dependence of τ and ΔH_E^* for the molecular rotation upon shape and size of the solute in the polymer is very pronounced. The rotational motion of the rigid molecular frame ($C_{10}H_7O$—) of β-naphthol gave τ(20°C) = $1{\cdot}0 \times 10^{-7}$ sec and $\Delta H_E^* = 11{\cdot}2 \pm 0{\cdot}8$ kcal mole^{-1} although its volume is comparable with that of the camphor. A separate well-resolved absorption in the β-naphthol case was clearly assignable to the independent

$$\left(\begin{array}{c} \quad H \\ \;/ \\ -O \end{array} \right)$$

Table 2.15

	Crystalline camphor	Camphor in solid solution in camphene	Camphor in polystyrene	Camphor in cyclohexane solution
$10^{12}\tau$ (sec) at 20°C	19·8	15·4	38	7·0
ΔH_E^* kcal mole^{-1}	1·8	1·2	0·9	—

group rotation: $\tau(20°\text{C}) = 51 \times 10^{-12}$ sec, $\Delta H_E^* = 0{\cdot}5$ kcal mole^{-1}. Other molecules and their molecular reorientational absorptions in polystyrene are given in Table 2.16†. These absorptions account for the total molecular rotational process which, in anthrone, appeared in two separate steps, the one at higher temperature with the high activation energy moving rapidly towards a merger with the other: at any temperature above that of the merger, the single absorption will have the character of a rigid polar molecule in a high viscosity medium. In such studies, the polar solute is clearly being used as a probe to follow changes in the molecular behaviour of the polymer matrix[74]. The 'activation energies' must be regarded with caution in that, whilst they adequately represent the temperature coefficient of the τ's over at least 20° (e.g. in tetracyclone), it is clear that a wider temperature range would impose variations on ΔH_E^* owing to the temperature variation of the polystyrene properties. Furthermore, these energies measure the amount needed to displace the polystyrene chains sufficiently to allow the solute molecule to rotate, and accordingly they are related to the molecular rotational volume for the solute—being approximately proportional to the excess of that volume over a minimum of about 100 Å^3. Such a correlation can be expected to hold, at most, only for rigid molecules of one particular form. The uppermost value ($\Delta H_E^* \sim 130$ kcal mole^{-1}) may well represent the limit set by the activation energy for the viscous flow of the (solute plasticized) polystyrene.

Molecular reorientation and essentially rotational-type motions have been extensively studied in polar polymers. It is not intended to do more than refer to such studies here[75].

Table 2.16

Molecular rotational relaxations observed in polystyrene

Molecule	$10^6 . \tau$(mean)	β	t°C	ΔH_E^* apparent kcal mole^{-1}
Anthrone (i)	1·0	0·22	30	7·4
(ii)	142	0·27	90	59
Cholest-4-en-3-one	428	0·30	80	52
Tetraphenylcyclo pentadienone	1170	0·22	90	130

† See also Mansel Davies and J. W. L. Swain, *Chem. Phys. Lett.*, **2**, 42 (1968).

2.4.9 Dipole Rotations in Water and in Ice

The importance of water as a medium necessitates a summary of its dielectric dispersion as this clearly reveals the major features of molecular rotational capabilities. Despite the established spectrum of its molecular interactions in the liquid state[76], the dielectric behaviour of water is phenomenologically very simple. It shows a single relaxation process of almost ideally the simple Debye type (the distribution factor between 0° and 60° is 0·98 to 1·00) and the activation energy $\Delta H_E{}^* = 5{\cdot}0$ kcal mole^{-1}, a value which scarcely differs for D_2O: see Table 2.17. Saxton[77] emphasized the surprising agreement between the free energy of activation for the dielectric dispersion and that for the viscosity—as evaluated on Eyring's treatments:

$$\tau = \frac{h}{\mathbf{k}T}\exp(\Delta G_\mu/RT); \qquad \eta = \frac{hN}{V}\exp(\Delta G_\eta/RT)$$

$\Delta G_\mu = 1{\cdot}11\,\Delta G_\eta$ which also ensures $\Delta H_\mu{}^* \approx \Delta H_\eta{}^*$. A coincidental feature is provided by the use of Debye's equation $\tau = (3v/\mathbf{k}T)\eta$, as this gives a molecular volume (v) closely similar to that of gas-collisional estimates,

$$v = \frac{4\pi}{3}(1{\cdot}35\ \text{Å})^3 = 10{\cdot}3\ \text{Å}^3$$

The permittivity of water is unusually high[79] for the liquid at t°C:

$$\varepsilon_0(H_2O) = 87{\cdot}740 \quad 0{\cdot}4008t + 9{\cdot}398(10^{-4})t^2 + 1{\cdot}410(10^{-6})t^3$$

The Onsager treatment of the specific polarization of a polar medium (which is a significant improvement on the Mossotti–Clausius–Debye version) leads to the expression

$$\frac{(\varepsilon_0 - \varepsilon_\infty)(2\varepsilon_0 + \varepsilon_\infty)}{\varepsilon_0(\varepsilon_\infty + 2)^2} = \frac{4\pi N}{9V\mathbf{k}T}\mu^2$$

Table 2.17

Dielectric relaxation times for liquid water[78]

t°C	0	10	20	30	40	50	60
10^{12} (sec)	17·7	12·6	9·2$_4$	7·2$_2$	5·7$_8$	4·7$_2$	3·8$_7$
	4·9	5·1	5·9	4·9	4·8	4·4	—

where V is the molar volume, N the Avogadro number, μ the effective molecular dipole moment, ε_0 and ε_∞ are the low and high frequency permittivities, i.e. before the commencement and after the completion of the dipole dispersion. The interpretation of the dielectric properties of water in molecular terms depends on the value accepted for ε_∞. The experimental uncertainties in the older microwave measurements (1952) are reflected in Table 2.17 where individual values are of uncertainty *ca* $\pm 0{\cdot}06$: the mean value 4·55 was favoured†. This value has now been completely confirmed by maser source (30 cm^{-1}) measurements[80] of the refractive index which give for that frequency $\varepsilon(H_2O) = 4{\cdot}55 \pm 0{\cdot}08$. Such new absorption as occurs beyond 20 cm^{-1} is not of a relaxation type, but rather of a resonance character which means that its contribution to ε_0 is appropriately considered to be an 'atomic polarization' term as it arises from a vibrational or librational motion of or within the molecule. This defines $\varepsilon_\infty = 4{\cdot}55$ and that value gives in Onsager's equation $\mu(H_2O) = 1{\cdot}80 \pm 0{\cdot}03$D between 0° and 50°C. This value shows no real difference from $\mu(H_2O$ vapour$) = 1{\cdot}83$D. This further surprising conformity of liquid water to the simplest expectations perhaps reflects two conditions:

(i) The water molecules are all bonded to such a degree that dipole reorientation requires the rupture of at least one hydrogen bond giving the significantly large (for a liquid) $\Delta H_E^* = 5{\cdot}0$ kcal $mole^{-1}$. This serves to define an activated state for the process rather narrowly and gives the single relaxation time condition.

(ii) The total amplitude of the dispersion $(\varepsilon_0 - \varepsilon_\infty)$ conforms to the observed gas-phase dipole moment because, whilst the process consists essentially of an (O—H) group rotation

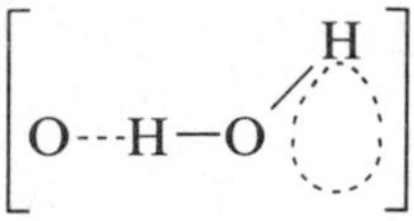

$$\left[\ O\cdots H—O\overset{H}{\diagup}\ \right]$$

a succession of such group rotations suffices to completely reorientate the water dipole. It is, of course, assumed in this that the net molecular dipole moment is not seriously influenced by the hydrogen-bonding interactions of the condensed phase: this is not unlikely.

The essential aspect of these results is that the molecular dipole rotation which amounts to the total molecular reorientation in water takes place

† Professor J. B. Hasted has critically reviewed the data: Aberystwyth Meeting Dielectrics Discussion Group (1970).

as a single rate process not itself reflecting the variety of geometric configurations possible in the hydrogen-bonded system. Whilst proton-jumping is an important part of the ionic conduction mechanism there is no sign of its contribution to the dielectric properties of water as such.

The influence of a simple electrolyte[81] on the dipole polarization of water is summarized for sodium chloride in Table 2.18. The dispersion remains characterized by a single relaxation time and ε_∞ does not appear to differ significantly from the value in water. The overall effect of the ions is to give the medium the properties found in water at a higher temperature—e.g. the 0·5M aqueous NaCl has ε_0 comparable to water at 30°C. This has led to a supposed 'structure-breaking' role for the ions, but this is almost certainly an inappropriate description. Such is the electrostatic field in the neighbourhood of the ions that the forces on the water dipoles are much in excess of those in the hydrogen-bond. Accordingly, a number of molecules are 'frozen' around each ion giving a reduced dipole contribution or ε_0 value. The $\Delta H^*(\tau)$ for the 3·0M NaCl solution is some 0·9 kcal mole^{-1} less than for water itself. This does reflect the different structural situation of at least that category of hydroxyl groups via which the dipole reorientation takes place.

In ordinary ice (at least eight other crystalline forms are known at higher than atmospheric pressures[82] the permittivity continues to increase beyond the water values as the temperature falls: Table 2.19. Even so,

Table 2.18

Dielectric parameters for aqueous sodium chloride

τ's in units of 10^{-12} sec

Conc	0·0M		0·50M		1·00M		1·50M		2·00M		3·00M	
t°C	ε_0	τ	ε_0	τ	ε_0	τ	ε_0	τ	ε_0	τ	ε_0	τ
0	88	18·7	77	17·1	69	16·4	62	15·3	56	14·4	46	13·0
10	84	13·6	74	12·2	66	11·8	60	11·3	54	10·9	43	10·3
20	80	10·1	71	9·2	63	9·0	57	8·7	51	8·3	41	7·6
30	77	7·5	68	7·2	60	7·1	55	6·9	49	6·7	39	6·2
40	73	5·9	65	5·7	58	5·6	52	5·5	47	5·4	37	5·2

the observed dipole moment is sufficiently large for the specific polarization,

$$\left(\frac{\varepsilon_0-1}{\varepsilon_0+2}\right) \sim 1{\cdot}0,$$

to require only the complete reorientation of one molecule in about 10^7. It is significant that $\Delta H_E{}^*$ for $(H_2O)_{ice}$ = 13·0 kcal mole^{-1} : for $(D_2O)_{ice}$ = 13·2 kcal mole^{-1}. These figures clearly eliminate the rôle of proton jumping or quantum-mechanical leakage as a contributing feature in the process. The rotation of individual hydroxyl groups is again the essential element in the molecular reorientation. Bjerrum[83] indicated how this could occur in the crystalline phase by cooperative mechanisms originating at defects in the ideal crystal lattice structure. One such pattern of L and D defect movements is represented figuratively: Figure 2.11. (a) → (b) → (c) etc. The estimated activation energy for such a mechanism agrees with that observed to an acceptable margin of uncertainty. Powles[84] *inter alia* has treated the model quantitatively and successfully matched the observed ε_0 values. It should be noted that in the absence of defects in a crystal, its permittivity (at low field strengths) would be the ε_∞ and not the ε_0 value†.

Particularly interesting (and unusual) dielectric and electrolytic conduction properties appear for ice containing small concentrations of hydrogen fluoride[85]. Being isoelectronic with water, close fitment into the lattice is possible, and this takes place under the form $(H_3\overset{+}{O})\,(\overline{F})$.

2.4.10 Dipole Rotations in Adsorbed Phases

A number of aspects make the study of molecular rotation in adsorbed phases of particular interest. Unless the adsorbed species influences the dielectric properties of its adsorbent, rather sensitive dielectric methods

Table 2.19

Dielectric parameters for ordinary ice

t°C	−0.1	−20.9	−44·7	−65·8
ε_0	91·5	97·4	104	133
ε_∞	3·10	3·10	3·10	3·1
$10^5\ \tau$ (sec)	2·2	16·4	252	4500

†See E. Whalley in *Physics of Ice*, Plenum Press, New York, 1969.

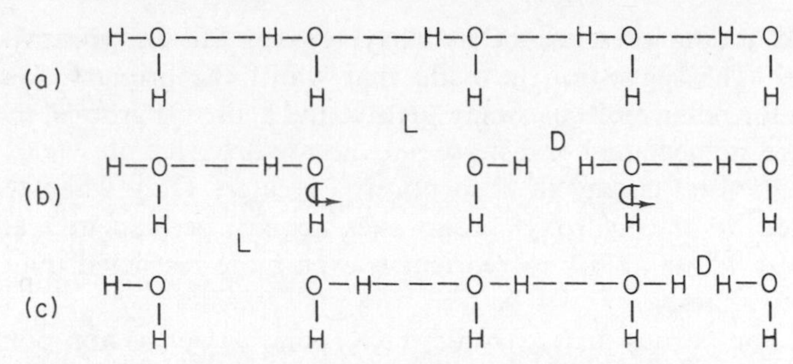

Figure 2.11 Schematic representation of the formation and displacement of Bjerrum defects in the ice lattice

are needed to detect the properties of the adsorbed phase. Difficulties also arise in the quantitative interpretation of the data, often commencing with the uncertainty in the density to be ascribed to the adsorbed phase. Nevertheless in an increasing number of instances (e.g. Al_2O_3, SiO_2, etc) where adsorption of polar molecules attains large volumes, significant deductions can be made. These studies have been reviewed[86].

The observational data are often used to evaluate an effective specific polarization for the component phases (p_i) on the basis of additivity in terms of volume fractions (v_i):

$$p(\text{obs}) = \sum v_i p_i$$

Alternatively, the permittivity increment ($\Delta\varepsilon_0$) plotted against v_2 for the adsorbate can be used to evaluate an 'effective permittivity', $\Delta\varepsilon_0/\Delta v_2$. Polar molecules ($H_2O$, C_2H_5Cl, CH_3COCH_3) on silica gel show two regions of linearity in a plot of $\Delta\varepsilon_0$ against Δv_2. The linearity is maintained even over regions where considerable changes in the heat of adsorption occurs and it follows that the permittivity is not responding to variations in molecular interactions which are frequently considerable. The $\Delta\varepsilon_0/\Delta v_2$ slopes are relatively insensitive to temperature so that differences from the liquid or gaseous states in the reorientational behaviour of the adsorbed phase are considerable. Kurbatov has suggested the adsorbed molecules may reorient by a libratory oscillation between defined orientational sights. Molecular motion so restricted would give the observed reduction of the 'effective permittivity' below the liquid values and could perhaps explain its temperature constancy.

As v_2 increases the second linear region in $\Delta\varepsilon_0/\Delta v_2$ is of lower slope than the first. On silica (and silicate) adsorbents this second region is ascribed to the presence of hydroxyl groups. It disappears in their absence. The suggestion is made that whilst the primary adsorption process for polar molecules may involve the hydroxyl groups, the molecules are immediately transferred to non-hydroxylic sites (e.g. $Si—\bar{O}$) where the total energy of adsorption is greater. Only when these are saturated do the hydroxyl group sites acquire permanently adsorbed molecules whose ability to reorient is even more restricted than on the alternative sites.

On a non-porous surface (rutile, TiO_2) polar as well as non-polar gases again show two regions in the $\Delta\varepsilon_0$ *vs* Δv_2 plots. In these instances the second region has the higher effective permittivity (i.e. $\Delta\varepsilon_0/\Delta v_2$ slope), its value approaching that for the liquid state, and, similarly to that phase, it shows a decrease in the value with rising temperature. The interpretation offered is that the first region of low permittivity increment corresponds to the formation of a unimolecular layer, the second region corresponding to increasing multilayer formation. Unfortunately, dispersion studies do not appear to be available: they could provide much further insight into the behaviour of the adsorbed molecules.

Water adsorbed on many polar porous oxides (SiO_2, Al_2O_3, TiO_2) has frequently been studied dielectrically, the studies including dielectric absorption measurements over wide frequency ranges[87]. A number of absorptions are found corresponding to different reorientational processes for the same adsorbed molecules, or, more probably, the differing rotational relaxations of differently adsorbed water molecules. In silica gel, two such processes have activation energies of 14 and 20 kcal mole^{-1}. They are both present over an appreciable concentration range (e.g. 7 to 24% H_2O by weight), for the whole of which range capillary condensation probably occurs*.

REFERENCES

1. J. H. van Vleck, *Theory of Electric and Magnetic Susceptibilities*, Oxford University Press, London, 1932.
2. R. H. Cole, *Progress in Dielectrics*, Vol. 3, Heywood, London 1961, provides one of the best summaries: see also Böttcher's book, p. 323 et seq; Nora Hill, W. E. Vaughan, A. H. Price and Mansel Davies, *Dielectric Properties and Molecular Behaviour*, Van Nostrand Reinhold, London, 1969. Other summaries: R. H. Cole, *Ann. Rev. Phys. Chem.*, **11**, 149 (1960); C. P. Smyth, *Ann. Rev. Phys. Chem.*, **17**, 433 (1966).

* For studies of zeolites with absorbed phases see: B. Morris, *J. Phys. Chem. Solids*, **30**, 73, 89, 103 (1969).

3. J. G. Kirkwood, *J. Chem. Phys.*, **4**, 592 (1936).
4. J. H. van Vleck, *J. Chem. Phys.*, **5**, 991 (1937); A. D. Buckingham and J. A. Pople, *Trans. Faraday Soc.*, **51**, 1179 (1955); R. W. Zwanzig, *J. Chem. Phys.*, **25**, 211 (1956); **27**, 821 (1957); T. L. Hill, *J. Chem. Phys.*, **28**, 61 (1958).
5. W. F. Brown, 'Dielectrics,' in *Encylopaedia of Physics*, Vol. XVII, Springer, Berlin, 1956; C. P. Smyth, *Dielectric Behaviour and Structure*, McGraw-Hill, New York, 1955, Chap. 3.
6. See Cole, reference 2; H. Fröhlich, *Theory of Dielectrics*, Oxford University Press, London, 1958; J. S. Dryden and R. J. Meakins, *Rev. Pure Appl. Chem.*, **7**, 15 (1957) gives a review of some observational data.
7. M. Davies, *J. Chim. Phys.*, **63**, 67 (1966).
8. M. Davies and A. Edwards, *Trans. Faraday Soc.*, **63**, 2063 (1967).
9. R. W. Wood, *Physical Optics*, Macmillan, New York, 1911 and later editions, Chap. XIV; R. G. Breene, *The Shift and Shape of Spectral Lines*, Pergamon Press, London, 1961.
10. H. A. Kramers and W. Heisenberg, *Z. Physik*, **31**, 681 (1925); S. A. Korff and G. Breit, *Rev. Modern Physics*, **4**, 471 (1932).
11. G. Birnbaum and A. A. Maryott, *Phys. Rev.*, **92**, 270 (1953); *J. Chem. Phys.*, **31**, 617 (1959); **32**, 686 (1960); **36**, 2032 (1962); **41**, 154 (1964).
12. G. Birnbaum, A. A. Maryott and P. F. Wacker, *J. Chem. Phys.*, **22**, 1782 (1954).
13. H. A. Gebbie and N. W. B. Stone, *Proc. Phys. Soc.* (*London*), **82**, 543 (1963).
14. A. A. Maryott and S. J. Kryder, *J. Chem. Phys.*, **41**, 1580 (1964).
15. G. Herzberg, *Infra-red and Raman Spectra of Polyatomic Molecules*, D. Van Nostrand, New York, 1945.
16. T. M. Sugden and C. N. Kenney, *Microwave Spectroscopy of Gases*, D. Van Nostrand, London, 1965.
17. D. M. Dennison and J. S. Kohler, *Phys. Rev.*, **57**, 1106 (1940); D. M. Dennison and D. G. Burk, *Phys. Rev.*, **84**, 408 (1951); *J. Mol. Spectr.*, **3**, 299 (1959); D. M. Dennison and E. V. Ivash, *J. Chem. Phys.*, **21**, 1804 (1953); *Phys. Rev.*, **89**, 895A, (1953).
18. E. B. Wilson, Jr., *Chem. Rev.*, **27**, 17 (1940); E. B. Wilson, Jr., C. C. Lin and D. R. Lide, *J. Chem. Phys.*, **23**, 136 (1955).
19. Reference 16, p. 319 provides a bibliography.
20. C. H. Townes and A. L. Schawlow, *Microwave Spectroscopy*, McGraw-Hill, New York, 1955; W. Gordy, *Rev. Mod. Phys.*, **20**, 765 (1948); W. Gordy, W. V. Smith and R. Trambarulo, *Microwave Spectroscopy*, Wiley, New York, 1953; D. J. E. Ingram, *Spectroscopy at Radio and Microwave Frequencies*, Butterworth, London, 1955.
21. K. H. Illinger, *Progress in Dielectrics*, Vol. 4, Heywood, London, 1962.
22. R. Karplus, *Phys. Rev.*, **73**, 1120 (1948).
23. P. W. Anderson, *Phys. Rev.*, **76**, 647 (1949).
24. J. P. Gordon, H. J. Zeiger and C. H. Townes, *Phys. Rev.*, **95**, 282 (1954); R. Karplus, *Phys. Rev.*, **74**, 223 (1948).
25. J. H. van Vleck and V. F. Weisskopf, *Rev. Mod. Physics*, **17**, 227 (1945); J. H. van Vleck and H. Morgenau, *Phys. Rev.*, **76**, 1211 (1949); H. Fröhlich, reference 6.

26. A. von Hippel, *Dielectrics and Waves*, Wiley, New York, p. 178.
27. B. Bleaney and J. H. N. Loubser, *Proc. Phys. Soc.* (*London*), **63A**, 483 (1950); G. Birnbaum, *Phys. Rev.*, **77**, 145 (1950); G. Birnbaum and A. A. Maryott, *Phys. Rev.*, **92**, 270 (1953).
28. F. W. Heineken and A. Battaglia, *Physica*, **24**, 589 (1958); F. W. Heineken and W. H. de Wijn, *Arch. Sci.* (*Geneve*), **11**, 102 (1958).
29. G. Birnbaum and A. A. Maryott, *J. Chem. Phys.*, **24**, 1022 (1956); **27**, 360 (1957); **29**, 1422 (1958); A. A. Maryott, A. Estin and G. Birnbaum, *J. Chem. Phys.*, **32**, 1501 (1960).
30. K. S. Cole and R. H. Cole, *J. Chem. Phys.*, **9**, 341 (1941).
31. G. Birnbaum and A. A. Maryott, *J. Phys. Chem.*, **64**, 1778 (1960): see also reference 29.
32. J. E. Boggs, C. M. Crain and J. E. Whiteford, *J. Phys. Chem.*, **61**, 482 (1957); J. E. Boggs, C. M. Crain and C. M. Thompson, *J. Phys. Chem.*, **61**, 1625 (1957); J. E. Boggs and H. C. Agnew, *J. Phys. Chem.*, **63**, 1127 (1959).
33. R. M. Fuoss and J. G. Kirkwood, *J. Am. Chem. Soc.*, **63**, 385 (1941).
34. D. W. Davidson and R. H. Cole, *J. Chem. Phys.*, **19**, 1484 (1951); D. W. Davidson, *Can. J. Chem.*, **39**, 571 (1961).
35. N. E. Hill, *Proc. Phys. Soc.* (*London*), **240A**, 101 (1957); see also R. H. Cole, reference 2.
36. S. H. Glarum, *J. Chem. Phys.*, **33**, 1371 (1960); R. H. Cole, *J. Chem. Phys.*, **42**, 637 (1965).
37. F. Perrin, *J. Phys. Radium*, **5**, 497 (1934); A. Budo, *Phys. Z.*, **39**, 706 (1938); **40**, 603 (1939).
38. M. Davies and C. Clemett, *J. Phys. Chem. Solids*, **18**, 80 (1961).
39. E. Bauer, *Cahiers Phys.*, **20**, 1, 1944.
40. J. D. Hoffman and H. G. Pfeiffer, *J. Chem. Phys.*, **22**, 132 (1954); J. D. Hoffman, *J. Chem. Phys.*, **23**, 1331 (1955); J. D. Hoffman and B. M. Axilrod, *J. Res. Nat. Bur. Stand.*, **58**, 61 (1957); J. I. Lauritzen, Jr., *J. Chem. Phys.*, **28**, 118 (1958).
41. Y. Rocard, *J. Phys.*, **4**, 247 (1933); J. G. Powles, *Trans. Faraday Soc.*, **44**, 802 (1945).
42. (i) C. Brot, B. Lassier, G. W. Chantry and H. A. Gebbie, *Spectrochim. Acta.*, **24A**, 295 (1968); (ii) G. W. Chantry, H. A. Gebbie, B. Lassier and G. Wyllie, *Nature*, **214**, 163 (1967).
43. B. Bleaney, J. H. N. Loubser and R. P. Penrose, *Proc. Phys. Soc.* (*London*), **59**, 185 (1947).
44. D. H. Whiffen, *Trans. Faraday Soc.*, **46**, 124 (1950).
45. L. Hartshorn, J. V. L. Parry and L. Essen, *Proc. Phys. Soc.* (*London*), **68**, 422 (1958).
45a. J. Ph. Poley, *J. Appl. Sci. Res.*, **B4**, 337 (1955).
46. C. J. F. Böttcher, *Physica*, **6**, 59 (1939); See also Böttcher's book, p. 324.
47. C. H. Cartwright and J. Errera, *Proc. Roy. Soc.* (*London*), **154A**, 138 (1936).
48. H. A. Gebbie, N. W. B. Stone, F. D. Findlay and E. C. Pyatt, *Nature*, **205**, 377 (1965); G. W. Chantry and H. A. Gebbie, *Nature*, **208**, 378 (1965);

J. E. Chamberlain, E. B. C. Werner, H. A. Gebbie and W. Slough, *Trans. Faraday Soc.*, **63**, 2605 (1967).
49. Y. Leroy and E. Constant, *Compt. Rend.*, **262**, 1391 (1966); **264B**, 533 (1967); Y. Leroy, *thesis*, University of Lille, 1967; E. Decamps, A. Hadri and J. M. Munier, *Spectrochim. Acta*, **20**, 373 (1964).
50. S. G. Kroon and J. van der Elsken, *Chem. Phys. Lett.*, **1**, 285 (1967).
50a. J. E. Chamberlain, M. Davies, H. A. Gebbie and G. W. F. Pardoe, *Trans. Faraday Soc.*, **64**, 847 (1968).
51. N. E. Hill, *Proc. Phys. Soc.* (*London*), **82**, 723 (1963).
52. M. Sharp (Aberystwyth), *Ph.D. Thesis*, University of Wales, 1968.
53. A. J. Petro and C. P. Smyth, *J. Am. Chem. Soc.*, **79**, 6142 (1957).
54. N. E. Hill, *Trans. Faraday Soc.*, **55**, 2000 (1959).
55. S. Mallikarjun and N. E. Hill, *Trans. Faraday Soc.*, **61**, 1389 (1965).
56. N. E. Hill, *Proc. Phys. Soc.* (*London*), **67B**, 149 (1954).
56a. J. Lamb, 'Molecular Relaxation Processes,' *Chem. Soc.* (*London*) *Spec. Publ.*, **20**, 1966, p. 13. A. J. Barlow and J. Lamb, *Discussions Faraday Soc.*, **43**, 223 (1967).
57. R. J. Meakins, *Trans. Faraday Soc.*, **54**, 1160 (1958); D. A. Pitt and C. P. Smyth, *J. Am. Chem. Soc.*, **81**, 783 (1959).
58. J. Timmermans, *Les Constantes Physiques des Composés Organiques Crystallisés*, Masson et Cie, Paris, 1953.
59. C. Clemett and M. Davies, *Trans. Faraday Soc.*, **58**, 1705 (1962).
60. C. Clemett and M. Davis, *Trans. Faraday Soc.*, **58**, 1718 (1962).
61. W. A. Yager and S. O. Morgan, *J. Am. Chem. Soc.*, **57**, 2071 (1935).
62. R. H. Cole and S. Havriliak, *Discussions Faraday Soc.*, **23**, 31 (1957).
63. G. Adam, *J. Chem. Phys.*, **43**, 662 (1965).
64. R. J. Meakins, *Progress in Dielectrics*, Vol. 3, Heywood, London, 1961.
65. J. S. Dryden and R. J. Meakins, reference 6.
66. C. P. Smyth, *J. Phys. Radium*, **63**, 59 (1966); reference 5.
67. J. D. Hoffman, 'Molecular Relaxation Processes,' *Chem Soc.* (*London*) *Spec. Publ.*, **20**, 1966.
68. See, however, B. Morris, *Ph.D. thesis*, University of Wales, 1967.
69. J. S. Dryden and R. J. Meakins, *Nature*, **169**, 324 (1952); J. S. Dryden, *Trans. Faraday Soc.*, **49**, 1333 (1953); R. J. Meakins, *Trans. Faraday Soc.*, **51**, 953 (1955).
70. M. Davies and K. Williams, *Trans. Faraday Soc.*, **64**, 529 (1968).
71. S. J. Allen, *J. Chem. Phys.*, **44**, 394 (1966); J. C. Burgiel, H. Meyer and P. L. Richards, *J. Chem. Phys.*, **43**, 4291 (1965).
72. D. W. Davidson, see summary and references in 'Molecular Relaxation Processes,' reference 67.
73. J. I. Lauritzen, Jr., *J. Chem. Phys.*, **28**, 118 (1958).
74. M. Davies and A. Edwards, *Trans. Faraday Soc.*, **63**, 2163 (1967).
75. Reviews are available: R. M. Fuoss, in *Chemistry of Large Molecules* (Ed. R. E. Burk and O. Grummett), Interscience, New York, 1943; A. J. Curtis, *Progress in Dielectrics*, Vol. 2, Heywood, London, 1960; W. H. Stockmayer, I.U.P.A.C. Symposium on Macromolecular Chemistry, Tokyo, September 1966.

76. Summaries and references can be found in: J. L. Kavanau, *Water and Solute-Water Interactions*, Holden-Day, San Francisco, 1964; G. C. Pimentel and A. L. McClellan, *The Hydrogen Bond*, Freeman, San Francisco, 1960; Dielectric properties are summarized in: J. B. Hasted, *Progress in Dielectrics*, Vol. 3, Heywood, London, 1961; D. Eisenberg and W. Kauzmann, *The Structure and Properties of Water*, Oxford University Press, 1969.
77. J. A. Saxton, *Physical and Meteorological Societies' Report*, 1947: pp. 292, 306, 316; J. A. Lane and J. A. Saxton, *Proc. Roy. Soc.* (*London*), **213A**, 400, 472 (1952); **214A**, 531 (1952).
78. E. H. Grant, T. J. Buchanan and H. F. Cook, *J. Chem. Phys.*, **26**, 156 (1957).
79. A. A. Maryott and C. G. Malmberg, *J. Res. Nat. Bur. Stand.*, **56**, 1 (1956).
80. J. E. Chamberlain, G. W. Chantry, H. A. Gebbie, N. W. B. Stone, T. B. Taylor and G. Wyllie, *Nature*, **210**, 790 (1966).
81. J. B. Hasted, D. M. Ritson and C. H. Collie, *J. Chem. Phys.*, **16**, 1 (1945); J. A. Saxton and J. A. Lane, *Proc. Roy. Soc.* (*London*), **214A**, 531 (1952); R. Pottel, *Ber. Bunsen Ges. Phys. Chem.*, **69**, 363 (1965); **71**, 135 (1967); **73**, 437 (1969).
82. R. P. Auty and R. H. Cole, *J. Chem. Phys.*, **20**, 1309 (1952).
83. N. Bjerrum, *Mat. Fys. Medd. Dansk. Vid. Selsk.*, **27**, 3 (1951).
84. J. G. Powles, *J. Chem. Phys.*, **20**, 1302 (1952); see also J. B. Hasted, *Progress in Dielectrics*, Vol. 3, Heywood, London, 1961.
85. H. Granicher, C. Jaccard, P. Scherrer and A. Steinemann, *Discussions Faraday Soc.*, **23**, 50 (1957).
86. R. L. McIntosh, *Dielectric Behaviour of Physically Adsorbed Gases*, E. Arnold, London, 1966.
87. J. Le Bot and S. Le Montagner, *Compt. rend*, **233**, 862 (1951). K. Kamiyoshi and J. Ripoche, *J. Phys.*, **19**, 943 (1958).

3

Storage and Dissipation of Energy in Viscoelastic Materials

Adi Eisenberg

3.1. INTRODUCTION

Viscoelastic materials, and all high polymers are normally viscoelastic, are subject to most of the same energy dissipation mechanisms that are encountered in materials composed of small molecules; in addition, however, a range of mechanisms are encountered which are unique to this group. Since the universal mechanisms have been discussed in the preceding chapters, only those which are directly connected with the phenomenon of viscoelasticity will be described here.

This chapter is not meant to be a comprehensive review of the field; rather, it is designed for those readers who, while possessing a broad background in physics and chemistry, are not familiar with the phenomenon of viscoelasticity and the accompanying relaxation mechanisms. For this reason, an introductory treatment will be attempted, with a discussion of only several representative mechanisms, and, insofar as possible, books or review articles will be chosen as references rather than the original literature.

First of all, a brief introduction into the nature of viscoelasticity will be presented, primarily from the phenomenological point of view. Then, the nature of elasticity of long-chain molecules will be discussed, since this represents a unique mechanism of energy storage. A brief experimental discussion of viscoelasticity will follow, and then, some specific relaxation mechanisms will be described, from both the macroscopic and molecular points of view. Finally, the temperature dependence of relaxation processes will be introduced, as well as the glass transition in polymers.

3.2 THE PHENOMENON OF VISCOELASTICITY[1–5]

In looking at the simplest idealizations of materials, we can distinguish simple elastic (Hookean) and simple viscous (Newtonian) bodies. An elastic solid is characterized by the linear relation between stress and strain, as expressed by Hooke's law

$$f = Es \tag{3.1}$$

where f is the stress or force per unit area (F/A) and s the strain, or the change in length divided by the original length ($\Delta L/L_0$) in a simple extension experiment. When deformed as a result of the application of an external force, an ideally elastic solid maintains its deformation without change as long as the force is maintained. By contrast, a Newtonian liquid, upon imposition of a stress, f, is deformed at a constant rate, i.e. it experiences a continuous change in the strain denoted by $\mathrm{d}s/\mathrm{d}t$. The rate of deformation is proportional to the force, and the proportionality constant is the viscosity of the liquid, η, i.e.

$$f = \eta \frac{\mathrm{d}s}{\mathrm{d}t} \tag{3.2}$$

The simplest viscoelastic material can be thought of as a simple combination of an elastic element (a spring) and a viscous element (a dashpot, which can be thought of as a cylinder filled with a viscous liquid and fitted with a loosely fitting piston) of the type shown in Figure 3.1. If these elements are attached in series, the system is known as a Maxwell

Figure 3.1 The spring–dashpot model for a viscoelastic material

body. While models of this type do not provide any physical insight into mechanism of relaxation or energy dissipation, they are nonetheless very instructive in demonstrating the phenomenological aspects of visco-elasticity. For this reason, the equations describing one type of relaxation behaviour of a Maxwell body will be presented and solved. In the following sections the molecular picture will be developed, particularly with regard to mechanisms of energy storage and dissipation.

Confining our attention to the spring, for the time being, we can rewrite equation (3.1) for a shear deformation rather than a tensile deformation as

$$s_s = f_s \cdot (1/G) \tag{3.3}$$

where G is the shear modulus, and the subscript s refers to the spring. Since we will be dealing with time-dependent phenomena, it is convenient to differentiate equation (3.3) with respect to time, i.e.

$$\mathrm{d}s_s/\mathrm{d}t = (1/G)(\mathrm{d}f_s/\mathrm{d}t) \tag{3.4}$$

For the viscous element, equation (3.2) is already in its time-dependent form. If the model in Figure 3.1 is deformed, in some time-dependent manner, perhaps by stretching as a result of the application of a force, then the total rate of deformation, $\mathrm{d}s/\mathrm{d}t$ equals the sum of the rates of deformation or rates of strain of the individual components, i.e.

$$\frac{\mathrm{d}s}{\mathrm{d}t} = \frac{\mathrm{d}s_s}{\mathrm{d}t} + \frac{\mathrm{d}s_d}{\mathrm{d}t} = \frac{1}{G}\frac{\mathrm{d}f}{\mathrm{d}t} + \frac{1}{\eta}f \tag{3.5}$$

the subscript d denoting the dashpot. The subscripts on the strain have been dropped since the strain is equal on all members of the model because they are linked in series.

Equation (3.5) is a general expression from which the behaviour of a Maxwell body can be calculated for a wide range of conditions. Of great interest is the solution for the stress relaxation experiment in which at $t = 0$ the sample is suddenly deformed and maintained at a constant deformation with time. As might be expected, the force needed to keep the body at this constant deformation will decrease with time. Initially, the spring takes up all the strain, the dashpot acting as a rigid body; as a function of time, however, the spring contracts at the expense of the dash-pot, the force decreasing in proportion to the contraction of the spring. In this process, incidentally, the elastic energy stored in the model is dissipated as heat; this is the principal mechanism of energy dissipation with which this article will deal, although specific additional mechanisms, including side-chain motions, will also be discussed.

In solving equation (3.5) for the case of stress relaxation, it is important to keep in mind that once the stretching has been accomplished, $\mathrm{d}s/\mathrm{d}t$ is zero, while f varies. The equation thus becomes

$$\frac{\mathrm{d}f}{\mathrm{d}t} = -\frac{G}{\eta}f \tag{3.6}$$

the solution for which is

$$f = f_0 \exp\left[-(\mathrm{G}/\eta)t\right] \tag{3.7}$$

using the boundary conditions that $f = f_0$ at $t = 0$. If we divide both sides of the equation by s, which in this case does not vary with time, then the equation can be recast in terms of the modulus, rather than the stress. Looking at equation (3.7), it is evident that η/G had the dimension of time and, due to the form of the equation, this value is known as the relaxation time, τ, of the system.

As might be expected, real materials do not, by and large, exhibit a relaxation behaviour which is Maxwellian, although by an appropriate combination of springs and dashpots the behaviour of a wide range of real materials can be approximated[2]. A special class of rubbers, i.e. those which relax by bond interchange, are approximately Maxwellian[2]; this will be discussed more extensively in Section 3. . .

3.3 RUBBER ELASTICITY[6,7]

Having devoted the first part of the chapter to the phenomenology of viscoelasticity, it is advisable at this point to enquire into the method of energy storage by polymeric molecules, particularly with regard to mechanisms which are not encountered in materials of a low molecular weight. Specifically, this section will be devoted to the storage of mechanical energy, i.e. to rubber elasticity. Only a most elementary version of the theory will be presented, with the sole aim of acquainting the reader with the basic ideas rather than the latest research results. One might say that the present section will be devoted to the nature of the spring in the model representation of viscoelasticity. First, however, it is necessary to enquire into the shape of a free polymer molecule, and the response of the free molecule toward an externally imposed change in the end-to-end distance. These considerations, coupled with several simplifying assumptions, will show that elasticity of rubbery polymers is primarily an entropy phenomenon involving conformations of polymer molecules with unstrained angles or bonds, in contrast to the elasticity of crystalline solids or glasses in which mechanical energy is stored by changing internuclear distances from their equilibrium positions or by the deformation of bond angles.

3.3.1 The Free Polymer Molecule

In looking at the shapes of free polymer molecules, it is perhaps most illuminating to begin with a consideration of the low molecular weight hydrocarbons. In the ethane molecule, the distance between the first and last carbon atom is fixed, and the same is true of the propane molecule. In butane, however, the problem is somewhat more complicated. Even if we assume that a C—C bond has only three low-energy positions relative to the preceding C—C bond (one *trans* and two *gauche* positions), then the butane molecule has three different configurations, two of which yield the same end-to-end distance while the third (in which the last bond is *trans* relative to the first) yields a different one. One could already define an average end-to-end distance in which the *gauche* configuration would have twice the weight of the *trans* configuration provided the energies were the same.

In introducing the quantitative treatment of very long chains, it is more convenient to treat a hypothetical chain in which each atom along the backbone represents a universal joint and in which the atoms occupy no volume. The model can then be made to conform to physical reality by imposing certain obvious restrictions, for instance, a tetrahedral value for the C—C—C angle, and so on. But this is only a minor modification. In dealing with a very large number of bonds, it is convenient first to look for the root mean square (r.m.s.) projection of a single bond of the idealized chain onto an arbitrary axis. The average projection, of course, will be zero, since each direction is equally probable, the r.m.s. projection, however, will not. It should be recalled, though, that this r.m.s. projection onto an arbitrary axis does have a sign, so that in estimating the total end-to-end distance of a chain we cannot simply multiply the r.m.s. projection by the total number of bonds, but by the excess of bonds projecting in one direction over the other. The problem has thus become analogous to the determination of the excess of heads over tails (or vice versa) in a certain number of tosses of a coin.

In computing the r.m.s. projection of a single bond onto the x axis, l_x, we utilize the following relationship:

$$\overline{l_x^2} = \int_{-1}^{+1} l_x^2 p(l_x) \mathrm{d}(l_x) \tag{3.8}$$

Referring to Figure 3.2, we imagine that the bond in question is of length l and projects along the direction of the unit vector (heavy line) at an angle ψ from the x axis and ϕ from the y axis. l_x is then the projection of this one bond onto the x axis ($= l \cos \psi$), and $p(l_x)\,\mathrm{d}(l_x)$ is the probability of the projection of the end of the bond lying between l_x and $l_x + \mathrm{d}l_x$. It is seen intuitively that $p(l_x)\,\mathrm{d}(l_x)$ is proportional to the size of the solid angle of bond directions over which l_x lies in the particular range, which is indicated by the shaded area on the surface of a hemisphere a quarter of which is illustrated in Figure 3.2, and is given by

$$p(l_x)\,\mathrm{d}(l_x) = \frac{\int_0^{2\pi} \sin \psi \,\mathrm{d}\psi \,\mathrm{d}\phi}{4\pi} \tag{3.9a}$$

The integration in the numerator is performed over ϕ and the result divided by 4π to yield solid angle rather than surface area. Upon

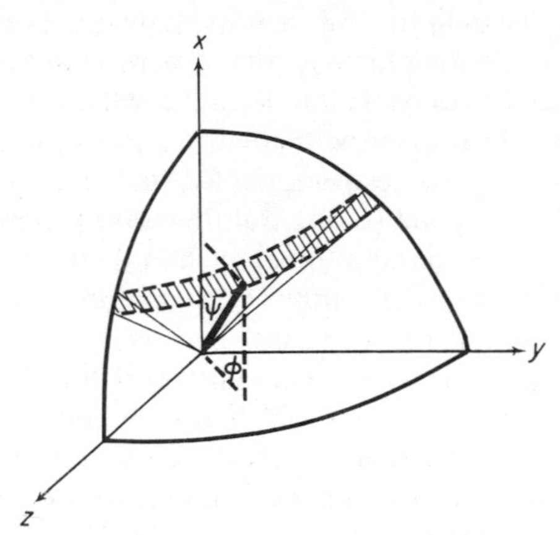

Figure 3.2 Illustration of the probability $p(l_x)\,\mathrm{d}(l_x)$ of finding the projection of a unit vector (heavy line) at an angle ψ with the x axis (equation 3.9)

integration and simplification we obtain:

$$p(l_x)\,\mathrm{d}(l_x) = \tfrac{1}{2}\sin\psi\,\mathrm{d}\psi \tag{3.9b}$$

Inserting this expression into equation (3.8) and recalling that

$$l_x{}^2 = l^2\cos^2\psi \tag{3.10}$$

we obtain for the r.m.s. projection of a single bond

$$\sqrt{\overline{l_x^2}} = \sqrt{l^2/3} \tag{3.11}$$

We must now calculate the excess of bonds pointing in one direction (n_+) over the other (n_-) for a certain total number of bonds n; by multiplying this excess ($n_+ - n_-$) by $\sqrt{\overline{l_x{}^2}}$ we will obtain the end-to-end distance of a chain. The problem of calculating the excess of heads over tails (or vice versa) in a certain number of tosses of a coin is a standard exercise given in textbooks on statistics. It is a Gaussian distribution, which, when multiplied by the r.m.s. projection, yields:

$$W(x)\,\mathrm{d}x = \frac{\beta}{\pi^{1/2}}\,\mathrm{e}^{-\beta^2 x^2}\,\mathrm{d}x \tag{3.12}$$

where $W(x)\,\mathrm{d}x$ is the probability of encountering the projection of the second end of a chain of n links onto the x axis (the first end being at the origin) between x and $x+\mathrm{d}x$, and β is given by equation (3.13).

$$\beta = \left(\frac{1}{2nl_x^2}\right)^{1/2} \tag{3.13}$$

Equation (3.12) can be easily extended into three dimensions by considering that for a large enough number of links the position of the projection of the second chain-end on any axis is independent of the position of the projection on any other. Since the equations for $W(y)\,\mathrm{d}y$ and $W(z)\,\mathrm{d}z$ are analogous to $W(x)\,\mathrm{d}x$, the probability of finding the second chain-end in a volume element between x and $x+\mathrm{d}x$, y and $y+\mathrm{d}y$, and z and $z+\mathrm{d}z$ is simply the product of the individual probabilities, i.e.:

$$\begin{aligned} W(x, y, z)\,\mathrm{d}x\,\mathrm{d}y\,\mathrm{d}z &= W(x)\,\mathrm{d}x\,W(y)\,\mathrm{d}y\,W(z)\,\mathrm{d}z \\ &= \left(\frac{\beta}{\pi^{1/2}}\right)^3 \mathrm{e}^{-\beta(x^2+y^2+z^2)}\,\mathrm{d}x\,\mathrm{d}y\,\mathrm{d}z \\ &= \left(\frac{\beta}{\pi^{1/2}}\right)^3 \mathrm{e}^{-\beta^2 r^2}\,\mathrm{d}x\,\mathrm{d}y\,\mathrm{d}z \end{aligned} \tag{3.14a}$$

since $r^2 = x^2+y^2+z^2$.

A plot of equation (3.14a) is reproduced in Figure 3.3, indicating that the probability per unit volume of finding the second chain-end is greatest at the origin of the coordinate system, i.e. exactly where the first chain-end is located. Since in this idealized chain under discussion here all configurations are of equal energy, this must mean that the total number of chain configurations consistent with a specific (vectorial) end-to-end distance is greatest for an end-to-end distance of zero, and decreases as the end-to-end distance increases. This fact will be of great importance in the subsequent discussion of rubber elasticity. It should be noted, however, that the average end-to-end distance independent of direction is not zero,

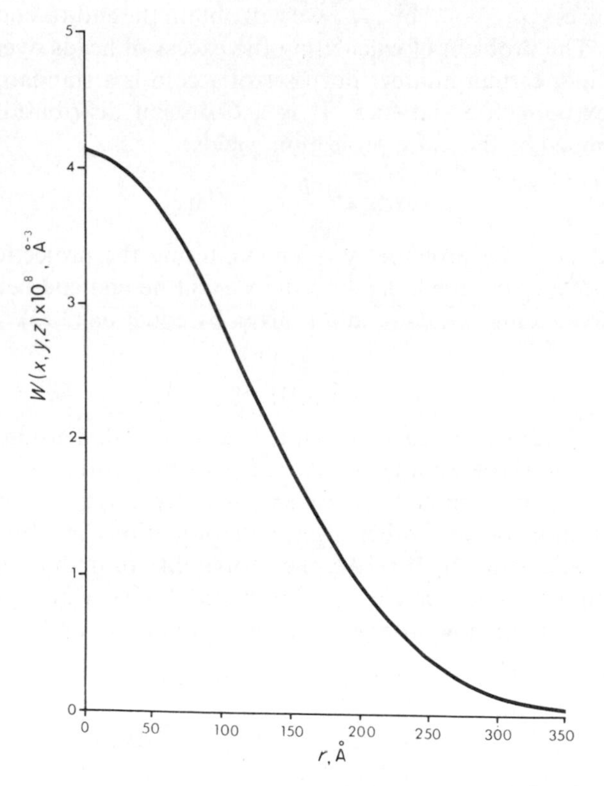

Figure 3.3 Probability per unit volume ($Å^{-3}$) of finding second chain-end at a distance r from the first for a freely jointed chain of 10^4 links of 2 Å each (equation 3.14a)

since the probability of encountering the second chain-end at a given distance is obtained by multiplying the probability per unit volume at that distance by the volume of a spherical shell the radius of which is the distance from the origin, i.e.:

$$W(r)\,\mathrm{d}r = W(x, y, z) \cdot 4\pi r^2\,\mathrm{d}r \tag{3.14b}$$

which yields a maximum at $r = 1/\beta$, the most probable value of r. It is also possible to obtain the average value of the radius (not only the most probable value) by performing the following operation

$$\overline{r^2} = \frac{\int_0^\infty r^2 W(r)\,\mathrm{d}r}{\int_0^\infty W(r)\,\mathrm{d}r} \tag{3.15}$$

which yields

$$\overline{r^2} = 3/2\beta^2 = nl^2 \tag{3.16}$$

3.3.2 Elasticity of a Single Chain

Before proceeding with the discussion of the elasticity of bulk rubber, it is advisable to apply ideas developed in the preceding section in estimating the behaviour of a single polymer chain upon change of its end-to-end distance. Since the probability per unit volume of finding the second chain-end is at a maximum in the vicinity of the first and decreases with distance from the first, the total number of configurations consistent with an end-to-end distance of zero must be a maximum, decreasing (in a manner shown in Figure 3.3) as the distance between the ends increases. Since all configurations of the idealized chain are of equal energy, changes in the Helmholtz free energy associated with changes in the end-to-end distance will be determined solely by the entropy term, and this, in turn, is proportional to the log of the number of configurations available to the chain under particular conditions of constraint. Based on previous considerations, we can say that the number of configurations is proportional to the probability of finding the second chain-end at a particular distance from the first, thus, the entropy term will be governed by the log of the probability density. From equation (3.14), we can write (since $s = -\mathbf{k}\ln W$)

$$s = \mathrm{c} - \mathbf{k}\beta^2 r^2 \tag{3.17}$$

where s is the entropy of the chain, c is an arbitrary constant which includes the size of the differential volume element, etc., but which is of no importance since only relative values of the entropy are involved and **k** is Boltzmann's constant. It should be stressed that there is no unique way in which the entropy of a molecule can be defined, i.e. that any definition is applicable only to the particular problem under consideration.

From equation (A.5) (Appendix), we see that $\mathrm{d}A = \mathrm{d}E - T\,\mathrm{d}S$. This equation can be applied to a single chain in which no internal energy changes occur upon extension. Thus, we get

$$\mathrm{d}A = -T\,\mathrm{d}s = \mathrm{d}W \quad \text{(Appendix)} \tag{3.18}$$

which enables us to calculate the work required to move the second chain-end from r to $r + \mathrm{d}r$ while keeping the first one fixed. This work is (equations 3.17 and 3.18)

$$\frac{\mathrm{d}W}{\mathrm{d}r} = \frac{\mathrm{d}A}{\mathrm{d}r} = -T\frac{\mathrm{d}s}{\mathrm{d}r} = 2\mathbf{k}T\beta^2 r \tag{3.19}$$

Obviously, work is done in increasing the end-to-end distance by $\mathrm{d}r$, the work being:

$$\mathrm{d}W = f\,\mathrm{d}r \tag{3.20}$$

where f is the force against which the work is done. This force is $\mathrm{d}W/\mathrm{d}r$, or

$$f = 2\mathbf{k}T\beta^2 r \tag{3.21}$$

Thus, we see that a coiled polymer molecule is quite analogous to a spring of zero length in that a displacement of its ends requires work to be done against a force acting along a line joining the ends; the force, incidentally, is proportional to the distance between the ends.

3.3.3 Bulk Elasticity

In order to apply the ideas developed in the preceding section to elasticity of bulk samples, the concept of a network must be introduced first. In practice, rubbers are vulcanized to minimize irreversible flow. On a molecular level, vulcanization involves a chemical reaction which bonds two (or more) chains to each other, usually through an intermediate fragment; the chain bond is commonly referred to as a cross-link. Even in the absence of chemical vulcanization, polymer chains of sufficient length are physically entangled, the entanglements acting as crosslinks. It is useful at this point to define a chain not in the manner in which the word was used before, but as the segment between crosslinks, the ends of the chain being the points at which it is attached to other segments. We can thus visualize the rubber as a collection of partly stretched springs, tied at the ends, the whole system being in a state of equilibrium. Upon deformation of the sample, for instance by stretching, some of the springs experience a net increase in elongation, others a net decrease. As will be shown below, however, on the average the chains are elongated and work is required to accomplish this net elongation. A correlation of the work with the distance will yield an expression for the modulus of elasticity of a rubber as a function of some fundamental properties of the polymer, such as the molecular weight, density and temperature.

It is convenient to consider a cube of rubber dimensions $L_0 \times L_0 \times L_0$ and to stretch this cube so that its length is αL_0 while its other dimensions are $\alpha^{-1/2} L_0$, the volume remaining the same. It is clear that a sphere, placed at some point within the unstrained cube, would become an ellipsoid upon stretching. Also, if one were to connect vectorially the two ends of each chain, and then gather all vectors (keeping their orientation in space but moving only their origins) to a common point, one would obtain a bundle of sticks or arrows (of a distribution of lengths) which would look like a fuzzy sphere, the average radius being r_0. Upon stretching, this sphere would also become an ellipsoid, the components of the vectors in the direction of stretching having become elongated, while those perpendicular to it have become shortened. If we make the assumption that the original radius r_0 of this sphere of averaged vectors is proportional to L_0, and that upon extension the major axis of the ellipsoid is αL_0 while the minor axes are $\alpha^{-1/2} L_0$, we can calculate the root mean square value of the radius of the ellipse $\sqrt{\overline{r'^2}}$ where r' is the final radius of the ellipse (which, of course, depends on direction). It is important to note that it is the overall average that is important here because while work will have to be done to stretch some of the chains, the decrease in lengths of the others will release energy and only the net work is needed to compute the modulus.

The calculation of the $\overline{r'^2}$ is performed in the following manner:

The equation for a sphere in cartesian coordinates is

$$r_0^2 = x^2 + y^2 + z^2 \tag{3.22}$$

while for the ellipsoid it is

$$r'^2 = x'^2 + y'^2 + z'^2 \tag{3.23}$$

In spherical coordinates (Figure 3.2), x, y and z become

$$x = r_0 \cos \psi \tag{3.24}$$

$$y = r_0 \sin \psi \cos \phi \tag{3.25}$$

$$z = r_0 \sin \psi \sin \phi \tag{3.26}$$

Upon transformation from sphere to ellipsoid, we obtain

$$x' = \alpha x = \alpha r_0 \cos \psi \tag{3.27}$$

$$y' = \alpha^{-1/2} y = \alpha^{-1/2} r_0 \sin \psi \cos \phi \tag{3.28}$$

$$x' = \alpha^{-1/2} z = \alpha^{-1/2} r_0 \sin \psi \sin \phi \tag{3.29}$$

The mean square value of the final radius is then obtained from

$$\overline{r'^2} = \frac{\int\int r'^2 \text{ (element of solid angle)}}{\int\int \text{(element of solid angle)}} \tag{3.30}$$

where the element of a solid angle is

$$\mathrm{d}s = \sin \psi \, \mathrm{d}\phi \, \mathrm{d}\psi \tag{3.31}$$

Equation (3.29) thus becomes

$$\overline{r'^2} = \frac{r_0^2 \int_0^{2\pi} \int_0^{\pi} (\alpha^2 \cos^2 \psi + \alpha^{-1} \sin^2 \psi \sin \psi \, \mathrm{d}\psi \, \mathrm{d}\phi}{\int_0^{2\pi} \int_0^{\pi} \sin \psi \, \mathrm{d}\psi \, \mathrm{d}\phi} \tag{3.32}$$

In the usual manner, the integration over ϕ is performed from 0 to 2π, while that over ψ from 0 to π. The result is

$$\overline{r'^2} = (r_0^2/3)(\alpha^2 + 2\alpha^{-1}) \tag{3.33}$$

Equation (3.33) shows that in the process of stretching a cube of rubber of dimension $L_0 \times L_0 \times L_0$ to a rectangular prism the length of which is αL_0, the average value of the vectorial end-to-end distance (from crosslink to crosslink) changes from r_0 to r', the value of which is given in terms of r_0 and α by equation (3.33). The entropy of stretching can be computed by subtracting the entropy before stretching from that after stretching, which, for a single chain, can be obtained from equation (3.17).

The entropy change of a chain Δs is thus

$$\Delta s = [c - \mathbf{k}\beta^2 r'^2] - [c - \mathbf{k}\beta^2 r_0{}^2] \tag{3.34}$$

For N chains, the above value has to be multiplied by N. Recalling that $\beta^2 r_0{}^2 = 3/2$ (equation 3.16) (where the subscript zero indicated the average unstretched end-to-end distance in the rubber), we obtain

$$\Delta S = -\frac{N\mathbf{k}}{2}\left[\alpha^2 + \frac{2}{\alpha} - 3\right]. \tag{3.35}$$

where S is used for the macroscopic sample in contrast to s for a single chain.

In the Appendix it is shown that $\mathrm{d}A = \mathrm{d}W$. For a rubbery sample (neglecting PV work) the reversible work done on the system is, just as in the case of a single chain (equation 3.20),

$$\mathrm{d}W = +f\,\mathrm{d}L = \mathrm{d}A \tag{3.36}$$

(rather than $= P\,\mathrm{d}V$ as it would be in the case of an ideal gas). For an isothermal process

$$\mathrm{d}A = \mathrm{d}E - T\,\mathrm{d}s \tag{3.37}$$

and since for the chains under discussion here, a change in the end-to-end distance does not involve any stretching of bonds (or bending of bonds out of the tetrahedral angle for the idealized hydrocarbon chain) the $\mathrm{d}E$ term falls out as it did for a single chain in equation (3.18). Thus, $\mathrm{d}A$ becomes equal to $-T\,\mathrm{d}S$, and we obtain (utilizing equations 3.36 and 3.37)

$$f = -T\left(\frac{\partial S}{\partial L}\right)_T \tag{3.38}$$

We wish to obtain equation (3.38) in terms of α rather than L; recalling that

$$\alpha = L/L_0 \tag{3.39}$$

we obtain

$$\mathrm{d}L = L_0\,\mathrm{d}\alpha \tag{3.40}$$

$(\partial S/\partial L)$ thus becomes $(1/L_0)(\partial S/\partial \alpha)$ and equation (3.38) can be rewritten as

$$f = -\frac{T}{L_0}\left(\frac{\partial S}{\partial \alpha}\right)_T \tag{3.41}$$

Differentiation of equation (3.35) with respect to α and substitution into equation (3.41) yields

$$f = \frac{N\mathbf{k}T}{L_0}\left(\alpha - \frac{1}{\alpha^2}\right) \tag{3.42}$$

To obtain an expression for the stress (force per unit area) rather than the force, we divide both sides of equation (3.40) by the original cross-section of the sample (which contains N chains) and obtain (after setting $L_0 A_0 = V$, the volume)

$$\text{Stress} = \frac{f}{A_0} = \frac{N\mathbf{k}T}{V}\left(\alpha - \frac{1}{\alpha^2}\right) \tag{3.43}$$

where the subscript zero has been dropped from the V since the sample is assumed to be incompressible. It can easily be verified that

$$\frac{N}{V} = \rho \frac{N_{av}}{M_c} \tag{3.44}$$

where ρ is the density of the sample, $N_{av} = 6{\cdot}02 \times 10^{23}$ and M_c the average molecular weight of the chains between crosslinks. Recalling that $N_{av} \cdot \mathbf{k} = R$, the gas constant, equation (3.41) becomes

$$\text{Stress} = \rho \frac{RT}{M_c}\left(\alpha - \frac{1}{\alpha^2}\right) \tag{3.45}$$

In the limit of small extensions, α can be set equal to $1 + \Delta L/L_0$ and the term $1/\alpha^2$ expanded as a power series. Neglecting terms containing powers of ΔL we obtain

$$\alpha - \frac{1}{\alpha^2} = 3\frac{\Delta L}{L_0} \tag{3.46}$$

$\Delta L/L_0$ is the strain for simple extensions and the stress divided by the strain is, by definition, the Young's modulus. We thus obtain

$$E = 3\rho \frac{RT}{M_c} \tag{3.47}$$

an expression for the Young's modulus of an ideal rubber. This equation fits natural rubber remarkably well in the region of low values of α. It should be pointed out that for incompressible rubbers $E = 3G$, where G is the shear modulus.

It is evident that the storage of energy in a rubbery material is, at least to a first approximation, associated with a decrease in entropy upon changes in the mean end-to-end distance of a polymer chain. No internal energy changes need be involved, such as would be encountered in the stretching of bonds or the bending of bond angles, and the modulus is very simply related to the density, temperature and chain-length between chemical crosslinks in the network. In the absence of these crosslinks, the material would still act as a rubber, but only for relatively short times. The physical entanglements would, with time, disentangle, leading to a permanent or irrecoverable deformation in, for instance, stress relaxation experiments. The material would thus be viscoelastic, the disentanglement of the chains and their concurrent slippage relative to each other providing a mechanism of energy dissipation. The subsequent sections will deal with that mechanism, from both a phenomenological and a theoretical point of view, as well as with a number of other mechanisms such as side-chain motion or chemical bond interchange.

3.4 EXPERIMENTAL VISCOELASTICITY

Having addressed ourselves to the problem of rubber elasticity, the next item to be discussed is the following: How does the polymer dissipate the energy stored in the manner discussed in Section 3.3. In this section, therefore, several methods utilized in the study of viscoelasticity will be presented, along with representative results.

3.4.1 Stress Relaxation

Perhaps the simplest technique that can be utilized in the study of viscoelasticity is stress relaxation (see Section 3.2). Experimentally, the sample is deformed at $t = 0$ and kept at constant deformation as a function of time. If the experiment is performed at various temperatures, and plotted as log of the modulus *vs* log time, results such as those shown in Figure 3.4 are obtained. These results are for polymeric selenium[8] (of molecular weight of $1{\cdot}9 \times 10^5$), a material possessing only a backbone and no side-chains, but they are characteristic of an enormous range of polymers,

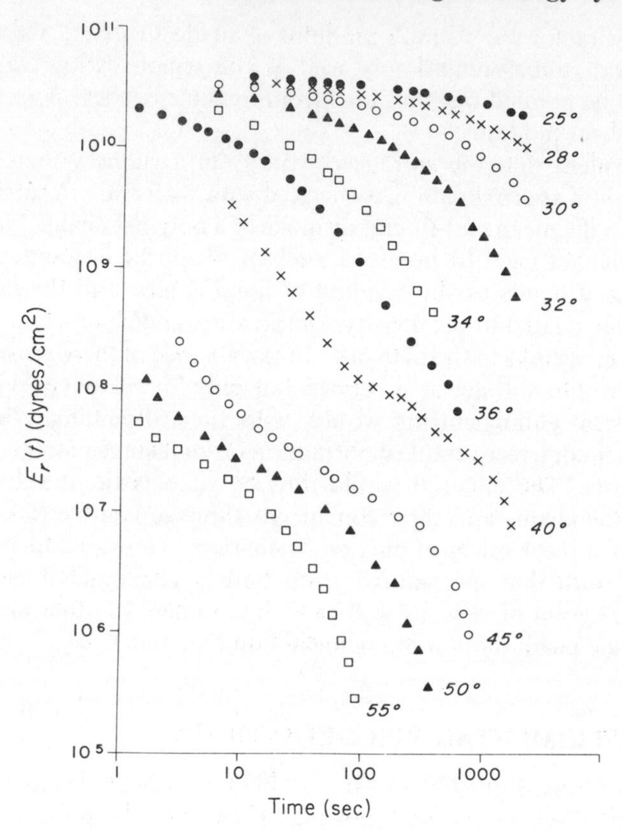

Figure 3.4 Stress relaxation curves for selenium

even those possessing highly complex pendant groups. This type of behaviour, therefore, is not very sensitive to the detailed structure of the chain: it is due, rather, to the diffusional motions of the chain or of parts thereof. Since only one relaxation mechanism is involved, the effect of temperature will be simply to accelerate (or decelerate) the process by some factor depending on the changes in temperature. In other words, the effect of changing the temperature is equivalent to multiplying time by some factor; or, on a logarithmic time scale, of shifting the curve horizontally by some amount. Referring to Figure 3.4, if we wished to

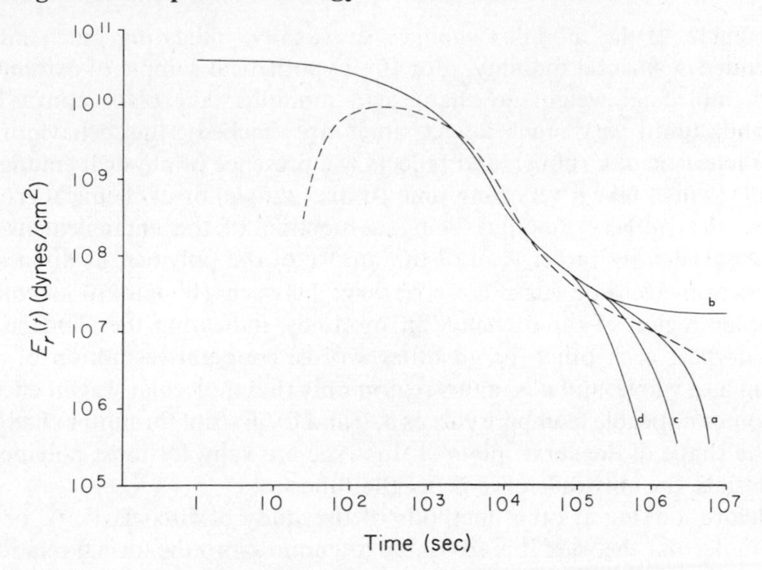

Figure 3.5 Stress relaxation master curves for selenium. Curve a: maximum experimentally obtainable molecular weight; curves c,d: lower molecular weights; curve b: hypothetical sample of extremely high molecular weight; dashed line: distribution of relaxation times for curve a

know how the sample would behave at 30°C at times beyond those accessible experimentally, all we would have to do is to hold the 30°C curve constant and shift the others horizontally relative to that one until overlap is achieved. The results of such a procedure, shown in Figure 3.5 (curve a), indicate how the sample would behave at 30°C over a time scale of 1 to 10^7 seconds. Any other temperature could have been chosen and the curve would look identical, except for shifts along the log time axis. Line b in Figure 3.5 indicates the anticipated results for a sample of a very much higher molecular weight, while lines c and d show the results for samples of lower molecular weights.

A study of Figure 3.5 reveals several interesting features. For short times (up to 10^2 seconds) the modulus has a value of *ca* 10^{11} dynes/cm^2, characteristic of glasses, and varies only very slightly with time; this indicates that in that time region segmental motion is highly restricted, the small change in modulus being due to motion of very small segments of the chain, perhaps isolated Se units. Between 10^2 and 10^5 seconds (for curve a) the modulus changes drastically, indicating much more extended segmental mobility. For the hypothetical sample of extremely high molecular weight no changes in modulus take place above 10^6 seconds (until very much longer times are reached); this behaviour is characteristic of a rubber, and reflects the presence of physical entanglements (which take a very long time to disentangle) or of chemical crosslinks, the rubbery modulus being a function of the entanglement or crosslink density rather than of the nature of the polymer, as discussed in Section 3.3. For curve a we observe between 10^5 and 10^7 seconds another region of rapid change in modulus, indicating the slippage of chains past each other, or, in other words, cooperative motion of the chain as a whole, and it is in this region only that molecular weight effects becomes noticable (compare curves a, c and d). Except for minor changes in the shape of the curve, plots of this type are valid for most polymeric materials, the only difference being the time scale.

Before looking at other methods of the study of viscoelasticity, or at the molecular theories, it is advisable to enquire into the formal relationship between the model presented in Figure 3.1 and a result of the type represented by Figure 3.5. It is clear from equation (3.7) that a plot of $\log f$ or $\log E$ *vs* linear time would be linear, the intercept being f_0 (or E_0) and the slope being $(1/2{\cdot}303)\,(G/\eta)$. A plot of the terminal part of curve a of Figure 3.5 as $\log E$ *vs* linear time is shown in Figure 3.6, indicating that at least the terminal portion of the plot (above $1{\cdot}5 \times 10^6$ seconds and below 10^7 dynes/cm^2) is represented by a Maxwell body. In order to elucidate the earlier portion of the curve (below $1{\cdot}5 \times 10^6$ seconds), we can treat the plot of Figure 3.6 in a manner quite analogous to the treatment of radioactive decay of several isotopes[2] on a plot of the log of the disintegration rate *vs* time: the terminal linear portion, representing the longest-lived isotope is subtracted from the experimental curve; the linear portion of that subtracted value yielding the next-longest-lived isotope, which, when subtracted from the remaining curve, yields the third longest-lived isotope, etc. In exactly the same manner an experimental curve can be decomposed into a number of Maxwell bodies or perhaps even more complex assemblies of models. It should be stressed, however, that no physical insight into the relaxation mechanisms is gained by that procedure. Alternatively, it is possible by utilizing various mathematical approximations to obtain a continuous spectrum or distribution of relaxation times[2,3]. The latter is also shown in Figure 3.5 as the dashed line for curve a. Line e is the first approximation to the distribution of relaxation times

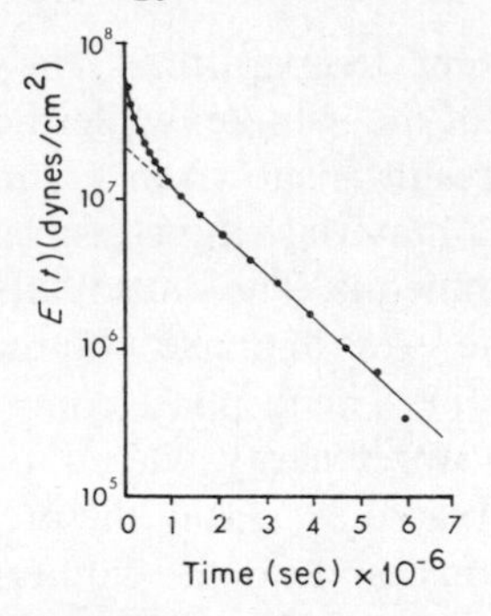

Figure 3.6 Terminal portion of curve a of Figure 3.5 on semilog plot (equation 3.7)

computed from the relation

$$H_1(\tau) = -E_r(t)\left[\frac{\mathrm{d}\log E_r(t)}{\mathrm{d}\log t}\right]_{t=\tau} \tag{3.48}$$

It should be added that the relaxation time obtained in the linear plot shown in Figure 3.6 can be correlated with the molecular weight. Above the point at which entanglements appear in the polymer, the logarithm of the longest relaxation time is proportional to 3·4 times the log of the molecular weight for any series of polymers. Incidentally, the bulk viscosity varies in the same way with the molecular weight[9].

3.4.2 Forced-Vibration Studies

We see that the dissipation of energy in viscoelastic systems is very strongly time and temperature dependent; although the curves presented here dealt specifically with selenium, a wide range of organic polymers would yield very similar curves in stress relaxation. That technique, as mentioned before, is not very sensitive to detailed features of the chain, but does give a very good idea of the long-time behaviour. If we wish to acquaint ourselves with relaxation mechanisms due to the finer details of the structure, we resort to some type of dynamic study, for example a forced vibration in which the sample is, for instance, sheared and allowed to relax alternately at some definite frequency as a function of temperature or as a function of frequency at a constant temperature. The strain may be imposed sinusoidally at one end of the sample, while the stress can be measured at the other. A typical result is shown in Figure 3.7, in which it can be seen that for viscoelastic materials the stress lags behind the strain, i.e. the response is not instantaneous. The sinusoidal stress can be decomposed into two components: the stress in phase with the strain and the stress 90° out of phase with the strain. The in-phase component is a function of the ability of the material to store energy, while the out-of-phase component indicates the energy dissipated as heat in the process.

A wide range of functions can be defined in terms of the strain, in-phase stress, out-of-phase stress and phase angle. For the purposes of this discussion, the most important are the complex modulus G^*, defined as the ratio of the peak stress to peak strain; the storage modulus G', defined as the ratio of the peak in-phase stress to the peak strain, and the loss modulus G'', defined as peak out-of-phase stress to the peak strain. Due to the phase angle between G' and G'', it is clear that

$$G^{*2} = G'^2 + G''^2 \tag{3.49}$$

Finally, the phase angle between stress and strain is defined as δ, such that

$$\tan \delta = G''/G' \tag{3.50}$$

Equations (3.49) and (3.50) are illustrated graphically in Figure 3.8, and the results for a representative material, polymethylmethacrylate are shown in Figure 3.9, summarizing the results of several investigations of that material actually obtained using free vibration studies (see next section) at approximately 1 cycle per second[10–13]. While the G' curve is strongly reminiscent of the type of result which might have been obtained

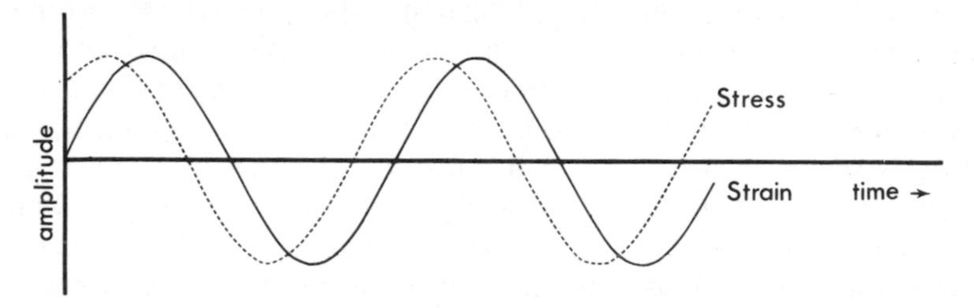

Figure 3.7 Plot illustrating the phase difference between stress and strain in forced vibration experiments

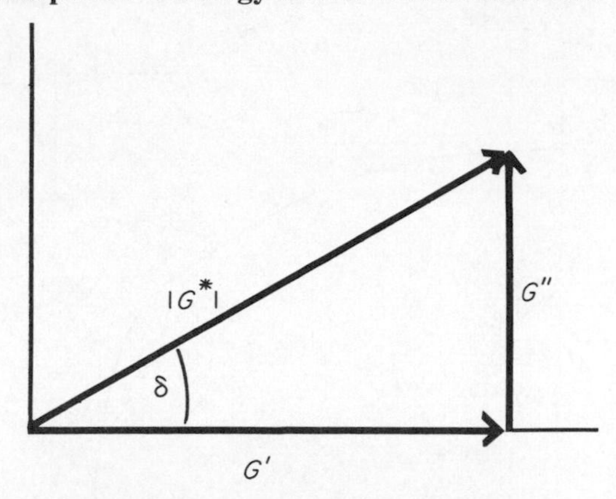

Figure 3.8 Graph illustrating relationship between G', G'', $|G^*|$, and tan δ (equations 3.47 and 3.48)

in stress relaxation, the tan δ curve provides a wealth of additional information which will be discussed in greater detail in the next section. Suffice it to say that each peak in that curve represents distinct dissipation mechanism, and can frequently be correlated with a specific molecular motion of a distinct chemical group.

It should be pointed out that at higher frequencies the peaks will be moved to higher temperatures, the shift yielding the activation energy of a particular process. This is discussed more fully in Section 3.6.2 and illustrated in Figure 3.13.

3.4.3 Free-Vibration Studies

A somewhat simpler method for the study of dynamic-mechanical properties of polymers is the torsional pendulum, in which the sample, perhaps with a torsion wire in series, is made part of a freely oscillating system, the frequency and amplitude of oscillation depending on the modulus and the inertia of the system as a whole. For viscoelastic materials, it is observed that once the pendulum has been set in motion, the amplitudes of successive waves decrease exponentially, the logarithmic decrease in amplitude being proportional to the energy dissipated in one cycle, and thus to the internal friction of the material. It should be mentioned that the detailed mathematical solutions for vibrating systems

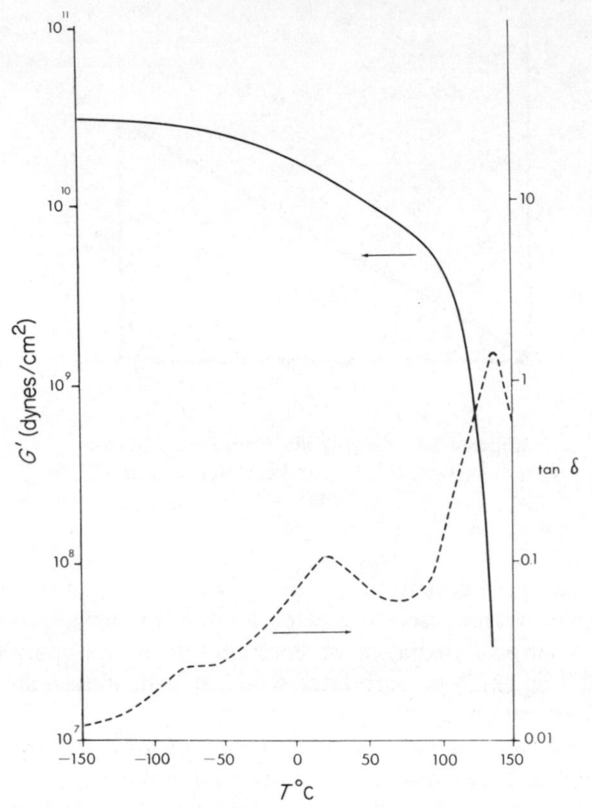

Figure 3.9 G' and tan δ for Polymethylmethacrylate[10–13]

containing viscoelastic bodies can be obtained from solutions of the differential equations describing the motion of a damped harmonic oscillator (for instance a mass on a spring suspended inside a viscous medium) with or without external excitation. The torsional pendulum, in contrast to driven systems, cannot be made to operate over wide frequency ranges, and can therefore not be used conveniently in the study of activation energies of molecular motions.

3.5 MOLECULAR INTERPRETATIONS

As yet, no theories of viscoelasticity have appeared which give a quantitative correlation of the relaxation behaviour with molecular structure. Our understanding of the molecular basis of viscoelasticity has progressed far enough, however, so that three examples of at least a semiquantitative nature can be given here. The first will deal with treatments of what can conveniently be referred to as the featureless chain; the second will describe the identification of several peaks in the tan δ curve in terms of specific molecular motions, and, finally, the third will deal with chemical relaxation mechanisms (bond interchange).

3.5.1 Theories of the Featureless Chain[14–20]

In attempts to predict the viscoelastic behaviour of polymer chains, Rouse[15] and Bueche[16] started with the following simplifying assumption: the end-to-end distance of a free polymer molecule is known to fluctuate, the fluctuation being Gaussian. Instead of treating the polymer chain on a link-to-link basis, which was mathematically impossible, they broke down the polymer chain of molecular weight M_i into N_i submolecules, each of which, in turn, still exhibits a Gaussian fluctuation of the end-to-end distance and thus acts as a Hookean spring. The original polymer molecule can thus be regarded as a collection of beads (which possess the mass of the submolecule) connected by hookean springs, a segment of which is shown schematically in Figure 3.10. The spring constant K can be computed from equation (3.19); setting

$$b^2 = 3/2\beta^2 \tag{3.51}$$

we obtain, by dividing f/r

$$K = 3\mathbf{k}T/b^2 \tag{3.52}$$

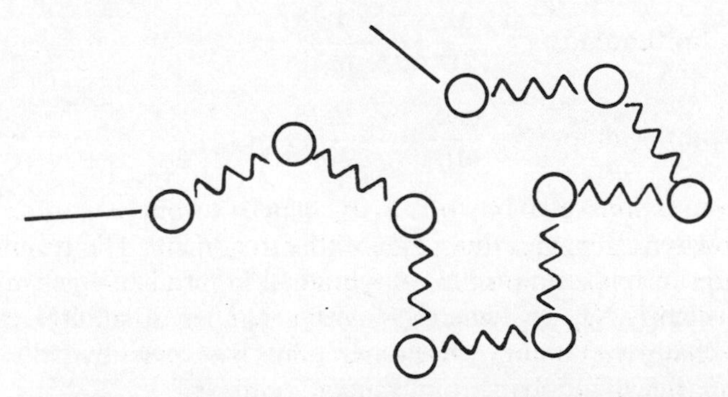

Figure 3.10 Illustration of model used in the development of the Rouse–Bueche–Zimm theories of viscoelasticity[15–17]

where b^2 is simply the average end-to-end distance of the submolecule (see equation 3.16).

Perhaps the most important parameter characteristic of viscoelastic behaviour that can be calculated is the recovery time τ (which is twice the corresponding relaxation time)[14]. This is done by subjecting the polymer molecule (bead and spring model) in a viscous medium to a shear gradient, allowing the springs to stretch, and setting up equations for the forces on each bead. Confining ourselves to the x-direction, the hookean restoring force on the beads is:

$$\text{bead no. } 0 \quad F_{0x} = (-3\mathbf{k}T/b^2)(x_0 - x_1) \tag{3.53}$$

$$\text{bead no. } i \quad F_{ix} = (-3\mathbf{k}T/b^2)(-x_{i-1} + 2x_i - x_{i+1}) \tag{3.54}$$

$$\text{bead no. } N \quad F_{Nx} = (-3\mathbf{k}T/b^2)(-x_{N-1} + x_N) \tag{3.55}$$

Since the beads are moving in a viscous medium, the assumption is made that each bead experiences a force due to the viscous drag which is proportional to its velocity in the medium. If ρ is the friction factor, then the force on the molecule is

$$F_i = \rho(\mathrm{d}x_i/\mathrm{d}t) \tag{3.56}$$

At cessation of flow, the shear field is stopped and the molecules begin to recover. At that point, it is assumed that beads are not subject to any acceleration, so that the elastic and viscous forces balance. The velocity of the beads can thus be given as

$$\text{for bead no. } 0 \quad \frac{\mathrm{d}x_0}{\mathrm{d}t} = -\frac{3\mathbf{k}T}{\rho b^2}(x_0 - x_1) \tag{3.57}$$

$$\text{for bead no. } i \quad \frac{\mathrm{d}x_i}{\mathrm{d}t} = -\frac{3\mathbf{k}T}{\rho b^2}(-x_{i-1} + 2x_i - x_{i+1}) \tag{3.58}$$

$$\text{for bead no. } N \quad \frac{\mathrm{d}x_N}{\mathrm{d}t} = -\frac{3\mathbf{k}T}{\rho b^2}(-x_{N-1} + x_N) \tag{3.59}$$

These equations can be solved by matrix techniques; the solution itself, however, is beyond the scope of this treatment. The results for the retardation or relaxation times are obtained in terms of eigenvalues of a matrix which is $N_i \times N_i$, where N_i is the number of submolecules into which a chain containing i monomer units has been divded. The p'th relaxation time of a polymer molecule of i units is

$$\tau_{p,i} = (b^2/24\rho\mathbf{k}T)\csc^2[p\pi/2(N_i + 1)] \tag{3.60}$$

We see that a discrete spectrum of relaxation times is obtained, i.e. the material behaves as if it consisted of a series of Maxwell models.

At this point b, ρ and N_i still remain; they must be eliminated since neither can be determined experimentally. Their elimination, by methods given in the literature[14] yields

$$\tau_{p,i} = 6\eta M_i^2/\pi^2 cRTM_w p^2 \tag{3.61}$$

where η is the bulk viscosity, c the concentration (or density) of the polymer, M_i the molecular weight of the fraction under consideration and M_w the weight average molecular weight. Once $\tau_{p,i}$ has been obtained, the viscoelastic functions of interest can be calculated.

The Rouse–Bueche theory has been extended by several investigators. Among these Zimm[17] included hydrodynamic interactions, Tobolsky and Aklonis[18] proposed a double Rouse function for bulk polymers and Shen, Hall and De Wames[19] extended the theory even further by considering factors neglected in the original derivations. Very recently Blatz[20] discussed the dissipation of energy in polymeric material directly on the basis of the above-mentioned theories.

These theories have been particularly useful in predicting G' and G'' of dilute polymer solution as a function of frequency (ω). The results of the Rouse treatment and of the Zimm treatment differ somewhat in terms of the limiting slopes of the G' and G'' *vs* ω. The Rouse theory is particularly applicable to good solvents since it does not consider hydrodynamic shielding, while the Zimm theory, which does, applies best for poor solvents where the molecule occupies a smaller volume. The agreement between theory and experiment in some cases has been truly striking. Unfortunately, however, these theories still cannot predict the detailed behaviour of a polymer chain from its structure alone. If we wish to find relaxation mechanisms connected to motions of, for instance, side-chains, we must resort to dynamic testing, a sample result of which was given in Figure 3.9.

3.5.2 Molecular Motions from Dynamic Studies—Fine Structure[3,21]

The success of dynamic mechanical studies is based on the fact that if the applied frequency is close to the frequency of some molecular motion, energy will be dissipated and a maximum in tan δ will be observed. The identification of the peaks, however, is not simple. In the case of polymethylmethacrylate, among others, the situation is fortunate in that a very wide variation of structural parameters is possible and has indeed been carried out.

Referring back to Figure 3.9 the highest temperature peak (α) occurs above 100°C for frequencies of approximately 1 cps. Since this temperature region is quite close to the dilatometric glass transition temperature (see Section 3.6.1), the peak has been identified as due to the motion of relatively short segments of the chain backbone. It might be mentioned that this situation is paralleled in most other polymers for which data are available.

The second peak (β), occurring at *ca* 20°C, has been identified by recourse to dielectric measurements. In dielectric work, the relative heights of the α and β peaks are reversed[22] suggesting that the β peak is due to the motion of groups which possesses a significant dipole moment, i.e. the $-COOCH_3$ side-chain. The peak at *ca* -100°C is only poorly identified, but it is known to be due to the presence of water, since it is absent in carefully dried specimens. N.m.r. spectroscopy incidentally has also been very useful in helping to identify dynamic mechanical peaks in polymers[23,24]. Additional peaks have been observed at still lower temperatures, and these, in turn, have been assigned to motions of the methyl group on the backbone and the side-chain (see Section 3.6.2).

It should be clear, by now, that mechanical relaxation spectroscopy, as this field has been called recently, rests, at this time, much more on an empirical than a theoretical foundation, reminiscent perhaps of optical spectroscopy prior to modern quantum mechanics. It is a rapidly growing field, however, and many new results can be expected in the near future, but a major theoretical breakthrough is needed before quantitative predictions of line positions, heights and shapes can be expected. It is an extremely valuable tool, however, for the elucidation of low-temperature relaxation mechanisms.

3.5.3 Chemical Bond Interchange

One additional relaxation mechanism should be mentioned in this section, the bond interchange mechanism. That mechanism is encountered in polymers possessing labile linkages such as the polysulfide rubbers[2] or sulfur[25], and also in otherwise stable polymers in the presence of bond interchange catalysts.

If the chemical relaxation occurs far above the glass transition temperature in a rubbery polymer, the phenomenology and interpretation is exceedingly simple. In the transition region, i.e. that region in which the modulus falls from a value of *ca* 10^{10} dynes/cm^2 to the rubbery plateau value, no effect of the chemical interchange is observed. In the rubbery region, however, a relaxation is found which, when the modulus is plotted as log E *vs* t, yields a straight line (Maxwellian behaviour) from which the relaxation time can be calculated. A plot of this relaxation time against $1/T$ shows that the relaxation mechanism is subject to an Arrhenius temperature dependence, the activation energy being that required for the rupture or reorganization of the labile bond.

If the bond interchange occurs at temperatures close to the glass transition, then the situation is far more complicated in that the contribution of the slippage mechanism has to be separated from that of the bond interchange mechanism, since both contribute to the flow. This has been accomplished recently for two systems[25].

3.6 FREQUENCY AND TEMPERATURE DEPENDENCE OF RELAXATION PROCESSES

The best way to gain an idea of the temperature dependence of molecular relaxation processes in polymeric materials is through a study of the temperature dependence of the bulk viscosity or some related parameter in viscoelasticity, and the temperature dependence of the maxima for the various loss peaks. Before proceeding with the discussion of the above topics, a brief digression into the area of the glass transition seems advisable, since that phenomenon has a strong relation to the temperature dependence of the primary relaxation phenomenon in polymers.

3.6.1 The Glass Transition[26]

Experimentally, the glass transition is frequently observed in the study of the volume of a polymer as a function of temperature. A plot such as the one shown in Figure 3.11 is usually obtained, and the intersection of the extrapolated straight-line segments is referred to as the glass transition. While it has some of the hallmarks of a second-order transition, it is not truly that because, as it is observed experimentally, it is kinetic in nature. We see this, for instance, from the variation of the glass transition with cooling rate. In a very rapid experiment the intersection lies at higher temperatures than in a very slow experiment. The glass transition values normally reported refer usually to a rate of *ca* 1°C/minute.

A number of other phenomena which accompany the glass transition are of interest here. In the vicinity of that region the material changes from

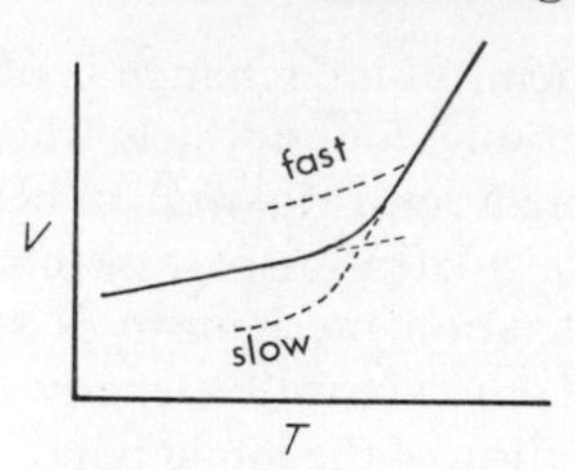

Figure 3.11 Volume–temperature relation for glass-forming material in the vicinity of the glass transition for normal cooling (—) and for fast and slow cooling rates (----)

a hard glassy solid to a plastic. The viscosity at T_g is very close to 10^{13} poise, and, as was mentioned in Section 3.5.2, tan δ exhibits a pronounced maximum.

Qualitatively, the glass transition can be understood best by considering the time requirements of molecular motions at various temperatures on the one hand and the rate of heat transfer on the other. The molecular motions responsible for heat transfer, while varying with temperature, are not very strongly dependent on it. By contrast, Brownian motions of whole segments of the backbone are very much time dependent. At a very high temperature, the material is a liquid, the rate of segmental motion being very high. As the temperature decreases, in a cooling experiment at constant cooling rate for instance, the rate of heat transfer does not change drastically, while that of molecular motions does slow down at some exponential rate. As long as the molecular motions which are responsible for increased packing efficiency (decrease in volume) as a function of decreasing temperature are faster than the rate of cooling, the material exhibits all the properties of an equilibrium liquid, i.e. in the dependence of volume only on temperature. However, as soon as the rate of molecular motion becomes of the same order of magnitude as the rate of cooling, the sample does not have enough time to avail itself of the decrease in temperature to pack more efficiently, and a non-equilibrium volume is assumed which is a function not only of the temperature but also of the rate of cooling. It should be noted that close to the glass transition temperature, volume after-effects become pronounced; i.e. a sample quenched from a high temperature to the vicinity of T_g may undergo a volume contraction long after thermal equilibrium is reached[27]. At still lower temperatures the sample assumes all the characteristics of a solid (except for long-range order) and the expansion coefficient is of the same order as that for molecular crystalline materials.

While the preceding provides a rough qualitative explanation of the observed phenomena, a quantitative understanding of the glass transition has still not been achieved. Various explanations based on a constant free volume at T_g, or a constant excess configurational entropy or other parameters have been offered, all of which seem to fit the experimental data, at least for homopolymers, quite well. In this section, only the free volume approach will be presented; while the configurational entropy approach will be discussed briefly.

3.6.2 The WLF equation[28]

Although originally the WLF equation was developed from a different point of view, it is useful for our purposes to start the derivation from the finding of Doolittle that the viscosity is a function only of the free volume, i.e.

$$\ln \eta = \ln \mathrm{A} + \mathrm{B} V_0 / V_f \tag{3.62}$$

where η is the viscosity, A and B are constants and V_0 and V_f are the occupied and free volumes, respectively. For the ratio of two viscosities

$$\ln \frac{\eta_1}{\eta_2} = \left(B \frac{V_0}{V_{f,1}} - \frac{V_0}{V_{f,2}} \right) \tag{3.63}$$

where the subscripts 1 and 2 refer to two different conditions, for instance, temperatures. The fractional free volume, f, can be expressed as

$$f = \frac{V_f}{V_0 + V_f} \tag{3.64}$$

and, since V_0 is very much larger than V_f,

$$\ln \frac{\eta_1}{\eta_2} \approx \mathrm{B} \left(\frac{1}{f_1} - \frac{1}{f_2} \right) \tag{3.65}$$

We can take the glass transition as a standard temperature (point 1) and develop the equation by expressing the fractional free volume in terms of the fractional free volume at T_g and the free volume expansion coefficient α_f as

$$f = f_g + \alpha_f(T - T_g) \tag{3.66}$$

Setting $\ln(\eta_{T_g}/\eta_T) = \ln \alpha_T$, we get

$$\ln \alpha_T = B\left(\frac{1}{f_g} - \frac{1}{f_g + \alpha_f(T - T_g)}\right) \tag{3.67}$$

which, upon rearrangement and conversion to decimal logs, becomes

$$\log \alpha_T = \frac{B}{2{\cdot}303 f_g}\left(\frac{T - T_g}{f_g/\alpha_f + T - T_g}\right) \tag{3.68}$$

i.e.

$$= C_1\left(\frac{T - T_g}{C_2 - T - T_g}\right) \tag{3.69}$$

For a very wide range of materials it turns out that $B \approx 1$, $C_1 = 17{\cdot}4$ and $C_2 = 51{\cdot}6$ which means that

$$f_g = 0{\cdot}025 \tag{3.70}$$

$$\alpha_f = 4{\cdot}8 \times 10^{-4}\ \mathrm{deg}^{-1} \tag{3.71}$$

i.e. the fractional free volume at T_g is of the order of 2·5% while the free volume expansion coefficient (experimentally equivalent to the differences of the liquid and glassy expansion coefficients) is $4{\cdot}8 \times 10^{-4}\ \mathrm{deg}^{-1}$.

The above results indicate that the glass transition occurs at an approximately constant free volume of *ca* 2·5%, but, perhaps even more importantly, that the temperature dependence of the viscosity, and by the same token of the energy dissipation mechanisms utilized in the flow process, obey approximately the relation given in equation (3.67) and reproduced graphically for selenium in Figure 3.12[8]. It should be pointed out that the shift utilized in the construction of a master curve for viscoelastic materials (Figures 3.4 and 3.5) also obey the WLF equation. As a matter of fact, the experimental points shown in Figure 3.12 are the shift factors for selenium.

In a somewhat later development, Gibbs, and DiMarzio[29] suggested that while the experimentally observed glass transition may well be kinetic, upon infinitely slow cooling a thermodynamic second-order transition will be reached which will have most of the characteristics of the glass

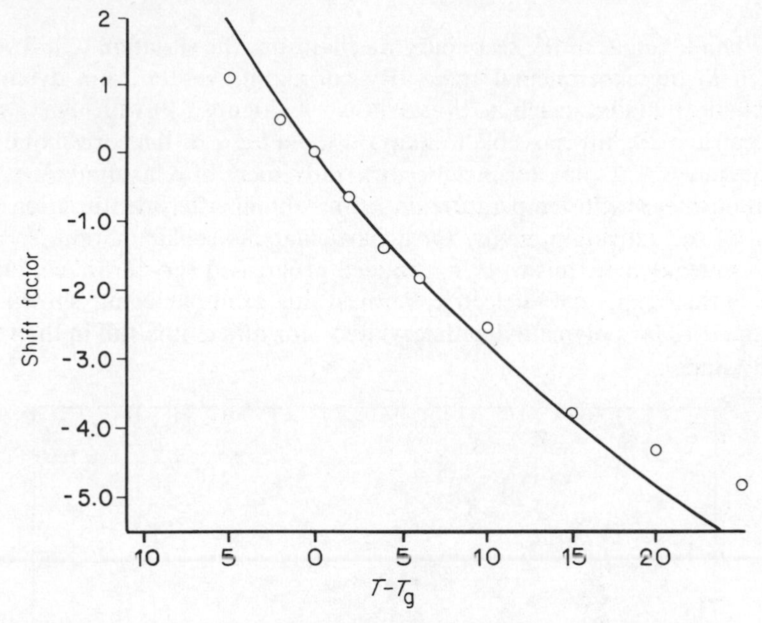

Figure 3.12 The WLF Equation (equation 3.67) with Universal constants[28]. The points are experimental results for selenium[8]

transition except that it would lie at lower temperatures. They arrived at this conclusion by computing the configurational entropy of a polymer chain utilizing a lattice model and two independent parameters, the hole energy (energy needed to separate neighbouring non-bonded segments) and the flex energy (differences in the troughs of potential energy curves for segmental rotation). They showed that the configurational entropy reaches zero at some temperature T_2 above zero °K, and that this temperature bears the hallmarks of a thermodynamic glass transition. Subsequently, Gibbs and Adam[30] extended that theory taking into account the size of the cooperatively rearranging region in glass-forming materials and showing that the size of this region is determined by configurational factors and can thus be expressed again in terms of the configurational entropy. The result they obtained is very close to that of the WLF theory, with C_2 being very slightly temperature dependent. It can thus be seen that our understanding of the temperature dependence of the primary energy dissipation mechanism rests on excellent foundation.

When it comes to the secondary mechanisms, the situation is still very much in the experimental stage. By correlating results from dynamic mechanical studies (such as those shown in Figure 3.9) with n.m.r. and dielectric work, it is possible to map out on a log τ or the corresponding frequency *vs* $1/T$ plot, for instance, the movement of relaxation times (τ) or frequencies with temperature and thus obtain, after identification, an idea of the activation energy for a particular molecular motion. While for some polymers this work has indeed progressed very far (in conjunction with n.m.r. and dielectric studies), an example being shown in Figure 3.13 for polymethylmethacrylate[31], for others it is still in the very early stages.

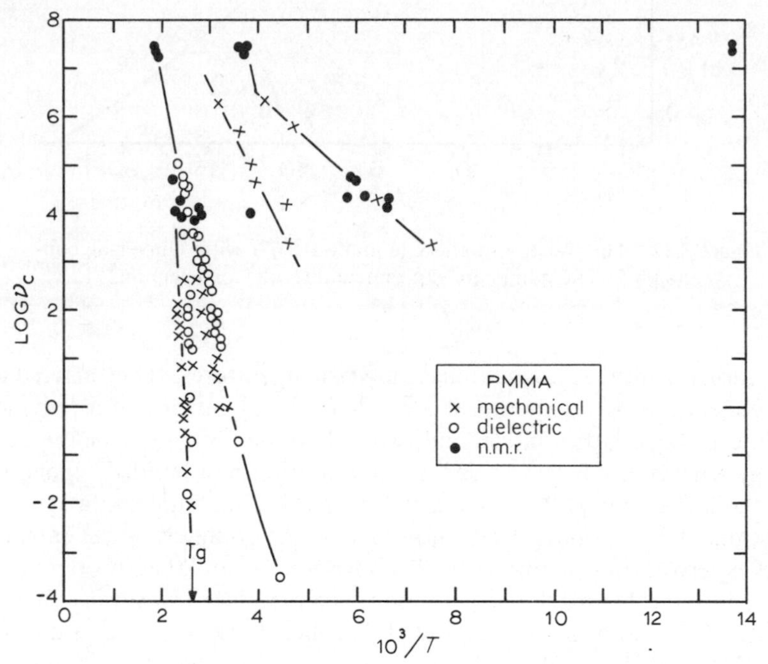

Figure 3.13 Temperature dependence of correlation frequency *vs* $1000/T$ for polymethylmethacrylate[31] ● n.m.r., ○ Dielectric, × Mechanical. The first line (highest temperature) corresponds to motion of the chain backbone segments, the second is due to motions of the ester group, the third to adsorbed water and the fourth to methyl groups attached to the main chain. The methyl groups attached to the side chain are active only at the highest frequencies even at very low temperatures[31].

3.7 SUMMARY

In summary, this chapter has attempted to describe the storage and dissipation of energy in viscoelastic materials, specifically polymers. Due to their high molecular weight, these materials possess, in addition to the mechanisms encountered in low molecular weight compounds, a range of additional mechanisms both for the storage and the dissipation of energy.

In the rubbery region, neighbouring segments of polymer chains are capable of practically unrestricted rotation and the chain can assume a very wide range of end-to-end distances, the probability of a particular length decreasing as the length increases. By virtue of this variation in probability, an entropy decrease is effected upon stretching of a chain which makes the chain act like a one-dimensional entropic spring, and which is thus responsible for the long-range elasticity of bulk rubber, the modulus being a function of the temperature, density and molecular weight between crosslinks rather than the chemical nature of the chain. This is the primary energy storage mechanism of rubbery materials.

In the absence of crosslinks, rubbers are viscoelastic over macroscopic time scales, the primary mechanism of viscous flow being the slippage of chains relative to each other. Several different techniques for the study of viscoelasticity are available, among which stress relaxation is most useful for studies of the macroscopic behaviour associated with motions of backbone segments or of the whole chain, while dynamic mechanical studies, whether in the form of forced or free vibrations, provide in addition information about motions of small groups such as side-chains or parts of side-chains. Information about the latter is also obtained from dielectric and n.m.r. studies.

Several theories relating to backbone motion do exist, but it is, as yet, impossible to predict the detailed behaviour of a material from structural considerations alone. Specific molecular mechanisms, particularly those relating to side-chain motion, can be identified, however, and so can relaxations due to chemical interchange along the backbone.

The temperature dependence of the main energy dissipation mechanism, which is due to relaxation (associated with the glass transition), is definitely non-Arrhenius, but is given by the WLF equation or the relation of Gibbs and Adam. Motions of side-chains may well be of the Arrhenius type, although in most cases data are not precise enough.

APPENDIX

The first law of thermodynamics can be expressed as

$$dE = dQ + dW \tag{A.1}$$

and the second law as

$$dQ = T\,dS \tag{A.2}$$

where the symbols have the usual meanings (i.e. E = internal energy, Q = heat, W = work, S = entropy). Combining these equations, we get

$$dE = T\,dS + dW \tag{A.3}$$

The Helmholtz free energy, A, is defined as

$$A = E - TS \tag{A.4}$$

which for isothermal processes, becomes

$$dA = dE - T\,dS \tag{A.5}$$

Comparing equations (A.3) and (A.5), we see that

$$dA = dW \tag{A.6}$$

REFERENCES

1. T. Alfrey, *Mechanical Behaviour of High Polymers*, Interscience, New York, 1948.
2. A. V. Tobolsky, *Properties and Structure of Polymers*, John Wiley, New York, 1960.
3. J. D. Ferry, *Viscoelastic Properties of Polymers*, John Wiley, New York, 2nd ed., 1969.
4. F. Bueche, *Physical Properties of Polymers*, Interscience, New York, 1962.
5. L. E. Nielsen, *Mechanical Properties of Polymers*, Reinhold, New York, 1962.
6. L. R. G. Treloar, *The Physics of Rubber Elasticity*, Oxford University Press, London, 1958.
7. P. J. Flory, *Principles of Polymer Chemistry*, Cornell University Press, Ithaca, N.Y., 1953.
8. A. Eisenberg and L. A. Teter, *J. Am. Chem. Soc.*, **87**, 2108 (1965).
9. T. G. Fox and V. R. Allan, *J. Chem. Phys.*, **41**, 344 (1964).
10. S. Iwayanagi and T. Hideshima, *J. Phys. Soc. Japan*, **8**, 365 (1953).
11. W. G. Gall and N. G. McCrum, *J. Polymer Sci.*, **50**, 489 (1961).
12. S. G. Turley and H. Keskkula, *J. Polymer Sci., C*, **14**, 69 (1966).
13. K. Schmieder and K. Wolf, *Kolloid-Z.*, **65**, 127 (1952).
14. W. L. Peticolas, *Rubber Chem. Technol.*, **36**, 1422 (1963).

15. P. E. Rouse, Jr., *J. Chem., Phys.*, **21**, 1272 (1953).
16. F. Bueche, *J. Chem. Phys.*, **22**, 603 (1954).
17. B. Zimm, *J. Chem. Phys.*, **24**, 269 (1956).
18. A. V. Tobolsky and J. J. Aklonis, *J. Phys. Chem.*, **68**, 1970 (1964).
19. M. C. Shen, W. F. Hall and R. E. De Wames, *J. Macromol Sci. Revs.*, C-2, 183 (1968).
20. P. Blatz, *Rubber Chem. Technol.*, (Rev's) in Press.
21. I. L. Hopkins and C. R. Kurkjian, *Physical Acoustics*, Academic Press, 1965, Vol. 2, p. 91.
22. K. Deutsch, E. A. W. Hoff and W. Reddish, *J. Polymer Sci.*, **13**, 565 (1954).
23. W. P. Slichter, *Advan. Polymer Sci.*, **1**, 35 (1958).
24. J. A. Sauer and A. E. Woodward, *Rev. Mod. Phys.*, **32**, 88 (1960).
25. A. Eisenberg, S. Saito and L. A. Teter, *J. Polymer Sci., C*, **14**, 323 (1966).
26. M. C. Shen and A. Eisenberg, *Progr. Solid State Chem.*, **3**, 394 (1963).
27. A. J. Kovacs, *Advan. Polymer Sci.*, **3**, 394 (1963).
28. M. L. Williams, R. F. Landel and J. D. Ferry, *J. Am. Chem. Soc.*, **77**, 3701 (1955).
29. J. H. Gibbs and E. A. DiMarzio, *J. Chem. Phys.*, **28**, 373 (1958).
30. G. Adam and J. H. Gibbs, *J. Chem. Phys.*, **43**, 139 (1965).
31. D. W. McCall, (1969). Dynamics of polymers. In Canter and Rush (Ed.), *Molecular Dynamics and Structure of Solids*, N.B.S. Special Publication No. 301.

4

Nuclear Magnetic Relaxation in Fluids

A. T. Bullock

4.1 INTRODUCTION

Studies of the rate of exchange of energy between an assembly of magnetic nuclei and the thermal degrees of freedom of their environment, conventionally referred to as 'the lattice', can provide much information about the molecular dynamics of the system. Theoretical descriptions relating the measured rates of energy exchange to statistical fluctuations of the lattice parameters have reached a high level of sophistication. The purpose of this chapter is to give a simple account of those spin–lattice relaxation mechanisms which are dependent on the molecular rotational degrees of freedom of the lattice. Rigour has been sacrificed when it has seemed more important to emphasize the physical nature of the models used.

4.1.1 Classical Equations of Motion

For a nucleus of magnetic moment $\boldsymbol{\mu}$ the equation of motion on the application of a steady magnetic field $\mathbf{H}$ is

$$\mathrm{d}\boldsymbol{\mu}/\mathrm{d}t = \gamma(\boldsymbol{\mu} \times \mathbf{H}) \tag{4.1}$$

where γ is the magnetogyric ratio of the nucleus. For a macroscopic assembly of such nuclei the magnetization $\mathbf{M}$ will have a time-dependence such that

$$\mathrm{d}\mathbf{M}/\mathrm{d}t = \gamma[\mathbf{M} \times \mathbf{H}] \tag{4.2}$$

The magnetization vector thus precesses about $\mathbf{H}$ with an angular frequency ω_0 given by

$$\omega_0 = \gamma H_0 \tag{4.3}$$

where H_0 is the magnitude of **H**.

The time-dependence of the components of **M** are thus

$$\frac{dM_x}{dt} = \omega_0 M_y \tag{4.4a}$$

$$\frac{dM_y}{dt} = -\omega_0 M_x \tag{4.4b}$$

$$\frac{dM_z}{dt} = 0 \tag{4.4c}$$

The tip of the magnetization vector moves clockwise round a cone with an angular velocity ω_0 as shown in Figure 4.1. In the usual convention, the z-axis is chosen to be parallel to **H**.

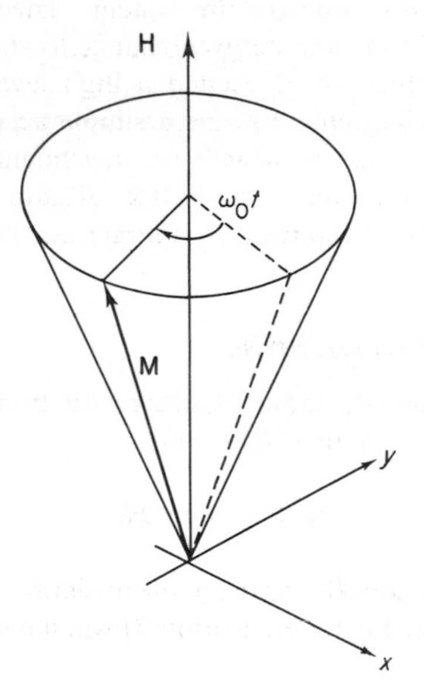

Figure 4.1 Precession of the magnetization vector about a steady magnetic field

The establishment of a nett magnetization M_0 parallel to the magnetic field requires the introduction of two relaxation times, T_1 and T_2. Introducing these into equations (4.4) gives

$$\frac{\mathrm{d}M_x}{\mathrm{d}t} = \omega_0 M_y - \frac{M_x}{T_2} \tag{4.5a}$$

$$\frac{\mathrm{d}M_y}{\mathrm{d}t} = -\omega_0 M_x - \frac{M_y}{T_2} \tag{4.5b}$$

$$\frac{\mathrm{d}M_z}{\mathrm{d}t} = -\frac{(M_z - M_0)}{T_1} \tag{4.5c}$$

Clearly M_z approaches an equilibrium value M_0 exponentially with a characteristic time T_1 whilst M_x and M_y decay to zero with a characteristic time T_2. These two relaxation times are the spin–lattice and spin–spin relaxation times respectively. In general T_1 and T_2 are not expected to be equal since changes in M_z involve an exchange of energy between the spin system and the thermal degrees of freedom of the surrounding medium whilst changes in the transverse components of the magnetization do not.

4.1.2 Quantum-Mechanical Description of Resonance

A nucleus described by a spin angular momentum operator **I**, in units of $\hbar$, possesses a magnetic moment $\boldsymbol{\mu}$ given by

$$\boldsymbol{\mu} = \gamma\hbar\mathbf{I} \tag{4.6}$$

The Hamiltonian for the potential energy of the nucleus in a magnetic field **H** is

$$\mathscr{H}_{(0)} = -\boldsymbol{\mu}\,.\,\mathbf{H} \tag{4.7}$$

Defining the z-axis as the direction of the field gives

$$\mathscr{H}_{(0)} = -\gamma\hbar H_0 I_z \tag{4.8}$$

Remembering that I_z has eigenvalues m, where m takes the values I, $I-1, \ldots -I$, it is seen that the allowed energy levels are given by

$$E_m = -\gamma\hbar H_0 m \tag{4.9}$$

Transitions between the levels may be induced by applying a time-dependent perturbation satisfying the usual conditions, the first of these being

$$\hbar\omega_0 = \Delta E \tag{4.10}$$

where ω_0 is the angular frequency of the perturbation and ΔE is the energy difference between the Zeeman levels concerned in the transition. The seond requirement is that the perturbation must have a non-vanishing matrix element between the two states. These conditions are met by applying a small alternating magnetic field perpendicular to the static field H_0. If this alternating field has an amplitude H_1 along the x-axis and an angular frequency ω, then the perturbation may be written

$$\mathscr{H}_{(t)} = -\gamma\hbar H_1 I_x \cos\omega t \tag{4.11}$$

The matrix elements of I_x between states m and m' are given by $\langle m|I_x|m'\rangle$ and are non-zero when $|m-m'| = 1$. Hence, from equations (4.9) and (4.10) the resonance condition is obtained.

$$\omega_0 = \gamma H_0 \tag{4.12}$$

To introduce the spin–lattice relaxation time into this quantum-mechanical description of nuclear magnetism we consider the case of an assembly of identical nuclei with $I = \frac{1}{2}$. In the absence of a static field the spin levels, described by the functions $|\alpha\rangle$ and $|\beta\rangle$ for $m = +\frac{1}{2}$ and $-\frac{1}{2}$ respectively, are degenerate and are therefore equally populated. To attain the expected Boltzmann distribution of populations when a field H_0 is applied there must be a nett transfer of nuclei in the upper state to the lower. The situation is readily described by means of Figure 4.2. The rate

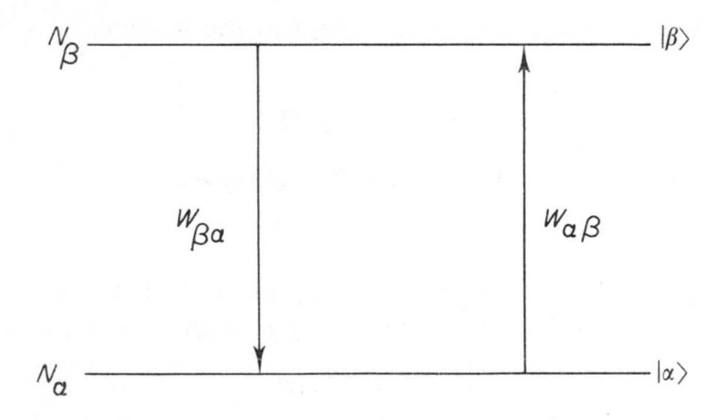

Figure 4.2 Lattice-induced transitions for a two-state system ($I = \frac{1}{2}$)

of change of N_α, the population of the lower state, is given by the first-order expression

$$\frac{dN_\alpha}{dt} = N_\beta W_{\beta\alpha} - N_\alpha W_{\alpha\beta} \tag{4.13}$$

where the W's are the lattice-induced transition probabilities. Two further variables are introduced, the total number of spins in the assembly and the population difference between the two states, thus

$$\begin{aligned} N &= N_\alpha + N_\beta \\ n &= N_\alpha - N_\beta \end{aligned} \tag{4.14}$$

From these two relationships it is readily shown that

$$N_\alpha = \tfrac{1}{2}(N+n); \qquad N_\beta = \tfrac{1}{2}(N-n) \tag{4.15}$$

Substituting these equations into (4.13) gives

$$\frac{dn}{dt} = -n(W_{\beta\alpha} + W_{\alpha\beta}) + N(W_{\beta\alpha} - W_{\alpha\beta}) \tag{4.16}$$

When the assembly of spins is in thermal equilibrium with the lattice the population difference, n_0, is equal to $N[(W_{\beta\alpha} - W_{\alpha\beta})/(W_{\beta\alpha} + W_{\beta\alpha})]$ and equation (4.16) may be rewritten

$$\frac{dn}{dt} = -\frac{(n-n_0)}{T_1} \tag{4.17}$$

in which $1/T_1$ is equal to $(W_{\beta\alpha} + W_{\alpha\beta})$. Clearly, assuming that the change of populations is governed by the first-order rate law of equation (4.13), an arbitrary population difference approaches the equilibrium value n_0 in an exponential manner with a time constant T_1. It is not possible to define a unique spin–lattice relaxation time for multilevel systems except under certain conditions. However, it is an experimental fact that such systems frequently reach thermal equilibrium with the lattice with an exponential time-dependence.

4.1.3 Relaxation Mechanisms

The Hamiltonian for a nucleus having a magnetic moment and situated in a magnetic field may be separated into two parts:

$$\mathscr{H} = \mathscr{H}_{(0)} + \mathscr{H}_{(t)} \tag{4.18}$$

where $\mathscr{H}_{(0)}$ is the Zeeman term which is independent of time and is given by equation (4.7). $\mathscr{H}_{(t)}$ represents the time-dependent magnetic or, in the case of nuclei $I \geqslant 1$ and possessing electric quadrupole moments, electric perturbation which is responsible for spin–lattice relaxation. We are currently concerned only with perturbations whose time dependence arises from molecular rotation or internal rotations of a molecule. These are:

(i) Dipole–dipole couplings with other magnetic nuclei in the same molecule modulated by molecular rotation.
(ii) Anisotropic chemical shift tensor. This is referred to a system of axes in the frame of the molecule and thus molecular rotation will modulate the Zeeman coupling of the nucleus to the static magnetic field H_0.
(iii) Nuclei possessing electric quadrupole moments will often experience a considerable electric field gradient due to the electron distribution in the molecule. Again, molecular tumbling will give the perturbation a dependence upon time.
(iv) Spin–rotational interaction. This arises from the coupling of the nucleus to the molecular magnetic field. This field is a consequence of the rotating charge distribution of the molecule. The coupling is modulated by changes in the angular velocity of the molecule brought about by molecular collisions.

The relaxation mechanisms (i), (ii) and (iii) have as a common factor perturbations which owe their time dependence to the fluctuations of the molecular coordinates within a fixed frame. On the other hand, the spin–rotational interaction is dependent upon fluctuations in the angular velocity of the molecule. There is no simple relationship between angular velocity and position in a fluid unless a specific model, such as Brownian rotational diffusion, is proposed. On these grounds it might be expected that measurements of contributions to T_1 from, say, the dipole–dipole mechanism and the spin–rotational mechanism in a fluid where both are operative will provide a test of models of molecular motion or, at the very least, extra information about the statistical nature of the motions of the molecules. It is for these reasons that the dependence of the spin–lattice relaxation time upon perturbations (i) and (iv) will be discussed in a semi-quantitative way whilst for the remaining two mechanisms results only will be given.

4.2 SPIN–LATTICE RELAXATION IN LIQUIDS

4.2.1 Dipole–dipole Relaxation

The Hamiltonian representing the interaction between two magnetic nuclei i and j may be written as

$$\mathscr{H}_{\mathrm{D}} = \gamma_i\gamma_j\hbar^2\left[\frac{\mathbf{I}_i \cdot \mathbf{I}_j}{r_{ij}^3} - \frac{3(\mathbf{I}_i \cdot \mathbf{r}_{ij})(\mathbf{I}_j \cdot \mathbf{r}_{ij})}{r_{ij}^5}\right] \tag{4.19}$$

in which $\gamma_{i,j}$ are the magnetogyric ratios of the nuclei, $\mathbf{I}_{i,j}$ are the nuclear spin angular momentum operators and r_{ij} is the magnitude of the internuclear vector $\mathbf{r}_{ij}$. It is usual to rewrite this Hamiltonian as

$$\mathscr{H}_{\mathrm{D}} = \frac{\gamma_i\gamma_j\hbar^2}{r_{ij}^3}[\mathrm{A}+\mathrm{B}+\mathrm{C}+\mathrm{D}+\mathrm{E}+\mathrm{F}] \tag{4.20}$$

where

$$\mathrm{A} = I_{zi}I_{zj}(1-3\cos^2\theta_{ij}) \tag{4.21a}$$

$$\mathrm{B} = -\tfrac{1}{4}[I_i{}^+I_j{}^- + I_i{}^-I_j{}^+](1-3\cos^2\theta_{ij}) \tag{4.21b}$$

$$\mathrm{C} = -\tfrac{3}{2}[I_i{}^+I_{zj} + I_{zi}I_j{}^+]\sin\theta_{ij}\cos\theta_{ij}\exp(-i\phi_{ij}) \tag{4.21c}$$

$$\mathrm{D} = -\tfrac{3}{2}[I_i{}^-I_{zj} + I_{zi}I_j{}^-]\sin\theta_{ij}\cos\theta_{ij}\exp(i\phi_{ij}) \tag{4.21d}$$

$$\mathrm{E} = -\tfrac{3}{4}I_i{}^+I_j{}^+\sin^2\theta_{ij}\exp(-2i\phi_{ij}) \tag{4.21e}$$

$$\mathrm{F} = -\tfrac{3}{4}I_i{}^-I_j{}^-\sin^2\theta_{ij}\exp(2i\theta_{ij}) \tag{4.21f}$$

In equations (4.21), I^+ and I^- are the shift operators, I_z has its previous meaning, and θ_{ij} and ϕ_{ij} are the polar and azimuthal angles made by the internuclear vector $\mathbf{r}$ in a system of Cartesian coordinates which has the direction of H_0 as the positive z direction.

If the nuclei concerned belong to the same molecule, then a time-dependence in θ_{ij}, ϕ_{ij} or r_{ij} or all three will result in modulation of $\mathscr{H}_{\mathrm{D}}$ and hence will provide a relaxation mechanism. In practice the small and rapid variations in r_{ij} are ineffective and it is fluctuations of θ_{ij} and ϕ_{ij} with time which lead to thermal equilibration of the spin system with the lattice. On the other hand, if the nuclei belong to different molecules, then the translational Brownian motion of the molecules ensures a modulation of r_{ij} which is effective in causing spin–lattice relaxation. However, this intermolecular process is not relevant to the present concern with rotational degrees of freedom.

The dipole–dipole Hamiltonian may be separated into time-dependent and time-independent parts

$$\mathscr{H}_{D}(t) = \sum_{n} a_{(n)} f_{(n)}(t) \tag{4.22}$$

The $a_{(n)}$ are functions of the spin operators alone and the time dependence is wholly carried by the $f_{(n)}(t)$ which are functions of the lattice variables r_{ij}, θ_{ij} and ϕ_{ij}.

In general, the probability of transitions induced by random lattice fluctuations is dependent upon the square of the matrix elements joining the states and the spectral density function $J(\omega)$. For two magnetically equivalent nuclei T_1 is given by[1]

$$(T_1)^{-1} = \tfrac{3}{2}\gamma^4\hbar^2 I(I+1)r^{-6}[J_1(\omega_0)+4J_2(2\omega_0)] \tag{4.23}$$

The spectral densities $J_n(\omega)$ are defined by the Fourier transforms

$$J_n(\omega) = \int_{-\infty}^{\infty} \exp(-\mathrm{i}\omega t)\,.\,g_n(t)\,.\,\mathrm{d}t \tag{4.24}$$

in which ω has the dimensions of angular frequency and $g_n(t)$ are the reduced correlation functions of the position functions in equations (4.21). These will be discussed later for various models of the molecular motion. For the present use we give a formal definition

$$g_{(n)}(t) = \overline{[f_{(n)}(t) f_{(n)}(t')^*]}/\overline{[f_{(n)}(t) f_{(n)}(t)^*]} \tag{4.25}$$

The correlation functions give the probability that the position function $f_{(n)}$ at time t does not change significantly in a period of time $(t'-t)$. In order to evaluate the spectral densities $J_1(\omega_0)$ and $J_2(2\omega_0)$ it is usually assumed that the correlation decays exponentially in both $g_1(t)$ and $g_2(t)$ and with the same time constant

$$g_1(t) = g_2(t) = \exp(-t/\tau_0) \tag{4.26}$$

where τ_c is called the correlation time. Evaluation of the Fourier transformation of equation (2.24) for $\omega = \omega_0, 2\omega_0$ followed by substitution in equation (4.23) gives

$$(T_1)^{-1} = \frac{2}{5}\gamma^4\hbar^2 I(I+1)r^{-6}\left[\frac{\tau_c}{1+(\omega_0\tau_c)^2}+\frac{4\tau_c}{1+(2\omega_0\tau_c)^2}\right] \tag{4.27}$$

Also, it may be shown that

$$(T_2)^{-1} = \frac{1}{10}\gamma^4\hbar^2 I(I+1)r^{-6}\left[6\tau_c + \frac{10\tau_c}{1+(\omega_0\tau_c)^2} + \frac{4\tau_c}{1+(2\omega_0\tau_c)^2}\right] \quad (4.28)$$

These equations are exact for $I = \frac{1}{2}$ and approximately correct for other values of I. It will be noted that in the expression for T_1 Fourier components at twice the resonance frequency contribute to the relaxation as well as those at the resonance frequency. The reason for this is that the total spin angular momentum, $I = (I_1+I_2)$ is conserved during tumbling. A singlet spin state is possible ($I = 0$) as is a triplet ($I = 1$) and it is the latter state which gives the nuclear magnetic resonance spectrum. Lattice fluctuations at $2\omega_0$ may induce transitions directly between the uppermost and lowest of the three triplet levels.

The assumption that the correlation decays exponentially with time is not entirely arbitrary. Such behaviour has been shown to be rigorously observed for the following models[2]:

(a) isotropic spin–spin interactions modulated by chemical exchange;
(b) thermally activated motion or rotation of a molecule between two configurations;
(c) hindered rotation involving more than two positions;
(d) rotational Brownian motion in the limit of high viscous rotational damping.

In addition Slichter has shown[3] that any two-valued randomly jumping function will have this property. Clearly this form of the correlation function is a very useful one and will be generally used in this chapter.

Figure 4.3 shows the behaviour of the spectral density function with frequency for long and short correlation times. $J(\omega)$ decreases very rapidly at a frequency of about τ_c^{-1}. For a given resonance frequency ω_0 the spin–lattice relaxation time will decrease as τ_c decreases from a high value towards ω_0^{-1}, will go through a minimum when $\omega_0\tau_c \approx 1$ and will finally increase. The variation of T_1 and T_2 with τ_c is shown in Figure 4.4. Nuclear magnetic resonance is usually observed at a frequency of about 10^6–10^8 Hz. For small molecules in liquids, τ_c is expected to be 10^{-10} s or shorter. Hence, it is often a good approximation to assume $\omega_0\tau_c \ll 1$ and equations (4.27) and (4.28) reduce to

$$(T_1)^{-1} = (T_2)^{-1} = 2\gamma^4\hbar^2 I(I+1)r^{-6}\tau_c \quad (4.29)$$

The correlations time has not yet been properly defined in terms of the rotational behaviour of the molecule. This is treated later. For the

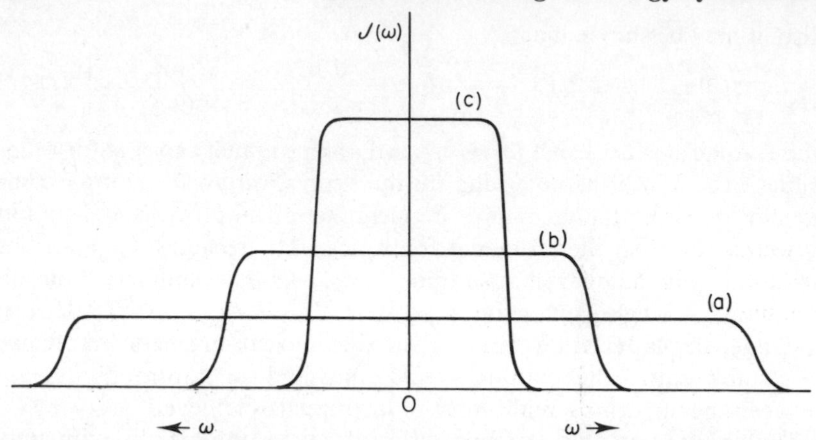

Figure 4.3 Spectral densities as a function of ω: (a) τ_c, the correlation time, short; (b) τ_c, intermediate; (c) τ_c, long.

present it is sufficient to say that τ_c characterizes the time-scale of molecular motion and is frequently defined as the mean time for a molecule to rotate through approximately one radian.

4.2.2 Anisotropic Chemical Shift

This relaxation mechanism is unique in being dependent upon the applied magnetic field H_0. The Zeeman coupling for a fictitious isolated nucleus given by equation (4.7) must be modified when considering nuclei in molecules. This is achieved by considering the shielding effect of the electrons in the molecule as a perturbation, namely $\gamma\hbar\mathbf{H}\,.\,\boldsymbol{\sigma}\,.\,\mathbf{I}$. The second rank tensor $\boldsymbol{\sigma}$ is the well-known chemical shift tensor having definite components in the molecular frame. In general, it is separable into an isotropic part $\sigma\mathbf{1}$ and an anisotropic part $\boldsymbol{\sigma}'$ ($\mathbf{1}$ is the unit tensor). Clearly, molecular tumbling coupled with this anisotropic shielding is equivalent to a fluctuating local field at the nucleus. Abragam has treated this relaxation mechanism[4]. Using a modified notation we write the components of $\boldsymbol{\sigma}'$ as

$$\sigma'_z; \qquad \sigma'_x = -\tfrac{1}{2}(1-n)\sigma'_z; \qquad \sigma'_y = -\tfrac{1}{2}(1+n)\sigma'_z$$

Abragam finds

$$(T_1)^{-1} = \frac{6}{40}\gamma^2 H_0{}^2 \sigma_z'^2 (1+n^2/3)\frac{2\tau_c}{1+(\omega_0\tau_c)^2} \tag{4.30}$$

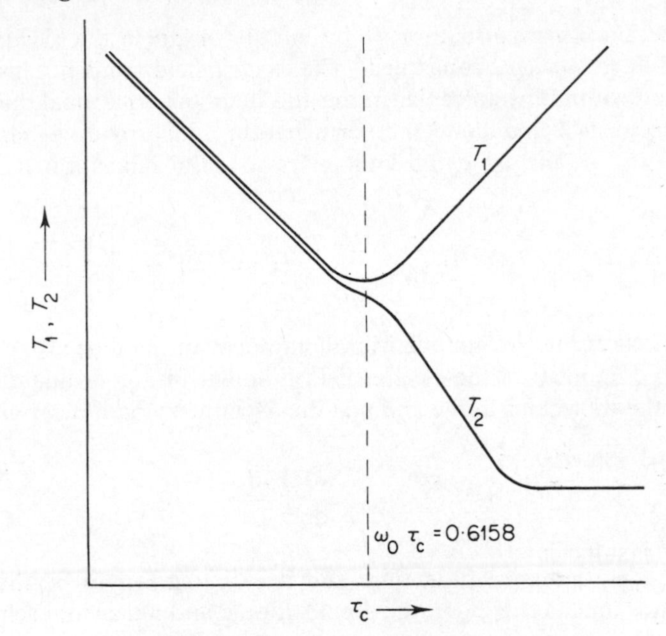

Figure 4.4 The variation of T_1 and T_2 with τ_c

McConnell and Holm[5] have also considered this relaxation mechanism. Where the chemical shift tensor has axial symmetry they find that

$$(T_1)^{-1} = \frac{2}{15}\gamma^2 H_0{}^2(\sigma_{\parallel} - \sigma_{\perp})^2 \frac{\tau_c}{1+(\omega_0\tau_c)^2} \tag{4.31}$$

where $\sigma_{\parallel}$ and $\sigma_{\perp}$ are components of the tensor parallel and perpendicular to the axis of symmetry. This is naturally a special case of equation (4.30).

Apart from the field-dependence of this relaxation process, it differs from the dipole–dipole mechanism in that for $\omega_0\tau_c \ll 1$, $T_1/T_2 = \frac{7}{6}$. This is not an important mechanism for ^{1}H spin relaxation but can be important for other nuclei, e.g. ^{19}F and ^{13}C, where shift anisotropies are large.

4.2.3 Electric Quadrupole Relaxation

Nuclei for which $I \geqslant 1$ possess electric quadrupole moments. Provided that the nucleus is in an environment of less than cubic symmetry then it will experience a perturbation which has its origin in the electric field gradient at the nucleus concerned. The electric field gradient tensor will be defined within the molecular frame and therefore rotational motion of the molecule will modulate the perturbation and provide a relaxation mechanism. Again using the limit $\omega_0\tau_c \ll 1$, the relaxation times are given by

$$(T_1)^{-1} = (T_2)^{-1} = \frac{3}{40}\frac{2I+3}{I^2(2I-1)}(1+\tilde{\eta}^2/3)\left(\frac{eQ}{\hbar}\cdot\frac{d^2V}{dz'^2}\right)^2\tau_c \qquad (4.32)$$

in which eQ is the electric quadrupole moment of the nucleus, d^2V/dz'^2 is the z-component of the electric field gradient in a coordinate system fixed in the molecular frame and $\tilde{\eta}$ is the asymmetry parameter given by

$$\tilde{\eta} = \frac{d^2V/dx'^2 - d^2V/dy'^2}{d^2V/dz'^2}$$

Finally, τ_c is the correlation time and may be shown to be the same correlation time occurring in the dipole–dipole and anisotropic chemical shift mechanisms.

The requirement of an asymmetrical environment for effective relaxation by this mechanism is rather too severe. The instantaneous electric field gradient in a molecule of cubic symmetry will take on a finite value during a molecular collision and this intermolecular perturbation will contribute to the total spin relaxation rate of a quadrupolar nucleus. It should be noted, however, that for nuclei ($I \geqslant 1$) in asymmetric environments, the intramolecular contribution given by equation (4.32) is much larger than the intermolecular effect. This feature endows this relaxation mechanism with a considerable advantage over the dipole–dipole mechanism since it implies that if a particular molecule has the properties required for efficient relaxation by quadrupole interaction with the lattice then intermolecular processes may be neglected and the rotational behaviour of the molecules in the pure liquid may be studied. For relaxation by dipolar interaction the inter- and intramolecular contributions are of the same order of magnitude and it is therefore difficult to obtain reliable values for τ_c in the pure liquid.

In general, for a quadrupolar nucleus in an asymmetrical environment, this mechanism frequently determines the spin–lattice relaxation rate even in the presence of other mechanisms.

4.2.4 Spin–Rotational Interaction

The three relaxation mechanisms described above have as a common feature correlation times which characterize the rate of change with time of the angular coordinates of the internuclear vectors. This situation is clearly analogous to that obtaining in the dielectric relaxation of dipolar molecules. The mechanism now to be discussed is unique in that the relevant correlation time pertains to the angular velocity rather than the angular position of the molecule. The following treatment and discussion is that given by Green and Powles[6], with minor alterations in notation.

The rotational motion of a molecule will produce a magnetic field by virtue of the rotation of the electronic and nuclear charges. If the angular velocity of the molecule is denoted by $\mathbf{\Omega}$ then an angular momentum index $\mathbf{J}$ may be defined thus:

$$I_0\mathbf{\Omega} = \mathbf{J}\hbar \tag{4.33}$$

in which I_0 is the moment of inertia of the molecule. For free rotation, $\mathbf{J}$ is the usual angular momentum operator with eigenvalues $0, 1, \ldots$. This is pertinent to dilute gases and will be treated later. For the moment, however, we examine the case of liquids in which, with very few exceptions, J is not a good quantum number since the rotational motion is interrupted so frequently by collisions that the rotational energy levels are 'smeared out'. The exceptions to this generalization are a few highly symmetrical light molecules. It is convenient to continue using $\mathbf{J}$ with the reservation that the motion may not be quantized.

$\mathbf{H}_{sr}$, the field produced at the nucleus by this mechanism, is given by

$$\mathbf{H}_{sr} \approx \frac{\mu_0}{R^3}\mathbf{J} \tag{4.34}$$

μ_0 being the nuclear magneton and R a measure of the molecular radius. It is of interest to compare the amplitude of this fluctuating field with that produced by direct dipolar coupling since the transition probabilities for spin–lattice relaxation by these mechanisms are directly proportional to the squares of these fields. In equation (4.34) μ_0/R^3 is of the order of the direct dipolar field. However, $(\bar{J}^2)^{1/2} \approx 30$ for a molecule at room temperature having a moment of inertia of 2×10^{-38} g cm^2. Clearly, the field produced at a nucleus by spin–rotational interaction may be one or two orders of magnitude greater than that produced by dipolar interaction. In spite of this, dipole–dipole interaction is frequently the dominant relaxation mechanism in liquids with the exception of those cases in which chemical exchange is occurring or which have nuclei possessing electric quadrupole moments. The reason is to be found in the relative value of the correlation times for the two mechanisms, that for the dipolar case being typically 10^{-11} seconds whilst that for the spin–rotational mechanism may be of the order 10^{-13} seconds.

The spin–rotational interaction gives rise to an additional term in the nuclear Hamiltonian:

$$\mathscr{H}_{sr} = -\boldsymbol{\mu} \,.\, \mathbf{H}_{sr} = -\hbar\gamma\mathbf{I} \,.\, \mathbf{H}_{sr} \tag{4.35}$$

The alternative, and equivalent, form is

$$\mathscr{H}_{sr} = -\hbar\mathbf{I} \,.\, \mathbf{C} \,.\, \mathbf{J} \tag{4.36}$$

where **C** is the spin–rotational interaction tensor*. A general expression for the spin–lattice relaxation time of a nucleus experiencing a randomly fluctuating field $H(t)$ with a correlation time τ is[7]

$$(T_1)^{-1} = \gamma^2[\overline{H_x{}^2} + \overline{H_y{}^2}]\tau/1 + \omega_0{}^2\tau^2 \tag{4.37}$$

where $\overline{H^2_{x,y}}$ are the mean square components of $H(t)$ along the x and y axes. For an isotropic fluctuating field $\overline{H_x{}^2} = \overline{H_y{}^2} = \overline{H_z{}^2} = \frac{1}{3}\overline{H^2(t)}$ and equation (4.37) reduced to

$$(T_1)^{-1} = \frac{2\gamma^2\overline{H^2(t)}\tau}{3(1+\omega_0{}^2\tau^2)} \tag{4.38}$$

It may be shown that[6]

$$\overline{H^2(t)} = \tfrac{1}{3}\gamma^{-2}\overline{J(J+1)}\sum_{v} C_{vv}{}^2 \tag{4.39}$$

in which $v = x, y, z$ and the bar represents an average over a statistical ensemble of identical systems.

For the special case of a spherical molecule having an axially symmetric spin–rotational tensor Green and Powles[6] derive the result

$$(T_1)_{sr}^{-1} = \frac{2\mathbf{k}T\hbar^{-2}I_0(2C_\perp{}^2 + C_\parallel{}^2)\tau_{sr}}{3(1+\omega_0{}^2\tau_{sr}^2)} \tag{4.40}$$

* It has been pointed out[6] that equation (4.36) is somewhat misleading in that the interaction depends upon **Ω** rather than **J**. This follows from the nature of the origin of $\mathbf{H}_{sr}$ since the current associated with the circulating charges is clearly proportional to **Ω**. Hence

$$\mathbf{H}_{sr} \propto \boldsymbol{\Omega} \propto \mathbf{J}/I_0.$$

where $C_{\perp}$ and $C_{\parallel}$ are the components of the spin–rotation tensor perpendicular and parallel to the axis of symmetry. The spin–rotational correlation time τ_{sr} is implicitly defined by

$$\overline{J(J+1)(0)J(J+1)(t)} = \overline{J(J+1)(0)\,.\,J(J+1)(0)}\,.\,\exp(-t/\tau_{sr}) \quad (4.41)$$

or since $J = I_0\Omega$, by

$$\overline{\Omega(0)\Omega(0+t)} = \overline{\Omega(0)\Omega(0)}\,.\,\exp(-t/\tau_{sr}) \quad (4.42)$$

4.3 THE ROTATIONAL BEHAVIOUR OF MOLECULES IN LIQUIDS

The model upon which most theoretical discussions of the rotational behaviour of molecules in the liquid state are based is that of Brownian rotational diffusion. This was first used by Debye in his treatment of dielectric relaxation of dipolar molecules[8]. It is assumed that the molecule has spherical symmetry and that its rotational diffusion is isotropic. Deviations from these properties are still discussed in the more general framework of the basic model.

The appropriate diffusion equation is

$$\mathrm{d}P(\theta, \phi, t)/\mathrm{d}t = D_r \nabla_s^{\,2} P(\theta, \phi, t) \quad (4.43)$$

where $P(\theta, \phi, t)$ is the probability that a molecular axis, say the internuclear vector between two magnetic nuclei, makes angles θ, ϕ with an arbitrary fixed set of axes in the laboratory system at time t. D_r is the rotational diffusion coefficient and $\nabla_s^{\,2}$ is the Laplacian operator on the surface of a sphere. The solution of equation (4.43) is found to be an expansion in spherical harmonics and gives

$$P(\theta_0, \phi_0, \theta, \phi, t) = \sum_{l,m} Y_l^{m*}(\theta_0, \phi_0) Y_l^m(\theta, \phi) \exp[-tD_r l(l+1)] \quad (4.44)$$

in which $P(\theta_0, \phi_0, \theta, \phi, t)$ is the probability that the chosen vector has angular coordinates θ, ϕ at time t when it had an orientation θ_0, ϕ_0 at an arbitrary zero of time.

Hertz[9] has recently generalized equation (4.44) somewhat. It is supposed that the molecule resides in an equilibrium position for a time t' where $0 < t' < t_1$ and within this time interval $P(\theta_0, \phi_0, \theta, \phi, t)$ is determined by equation (4.44). At $t = t_1$ the molecule performs a finite rotational jump. After this the probability is given by

$$P(\theta_0, \phi_0, \theta, \phi, t_1) = \sum_{l,m} Y_l^{m*}(\theta_0, \phi_0) Y_l^m(\theta, \phi) \exp[-D_r l(l+1)(t_1+\xi) \quad (4.45)$$

$$= \sum_{l,m} Y_l^{m*}(\theta_0, \phi_0) Y_l^m(\theta, \phi) C_l \exp[-D_r l(l+1)t_1] \quad (4.46)$$

in which C_l is defined by

$$C_l = \exp[-D_r l(l+1)\xi]; \qquad \xi \geqslant 0 \quad (4.47)$$

The assumption implicit in this treatment is that the finite rotational jump at time t_1 has the same effect on the probability function as would a true infinitesimal step, i.e. Brownian, diffusion for the period $t = t_1 + \xi$. After the rotational jump genuine diffusive motion is maintained for a period τ after which another finite jump occurs. Hertz shows that the total effect is a probability function of the form

$$P(\theta_0, \phi_0, \theta, \phi, t) = \sum_{l,m} Y_l^{m*}(\theta_0, \phi_0) Y_l^m(\theta, \phi) \exp[-t\{l(l+1)D_r + (1-C_l)/\tau\}] \quad (4.48)$$

Defining a time constant τ_l by

$$1/\tau_l = D_r l(l+1) \quad (4.49)$$

gives

$$P(\theta_0, \phi_0, \theta, \phi, t) = \sum_{l,m} Y_l^{m*}(\theta_0 \phi_0) Y_l^m(\theta, \phi) \exp[-t\{1/\tau_l + (1-C_l)/\tau\}] \quad (4.50)$$

Clearly, the correlation time τ_c is given by

$$1/\tau_c = 1/\tau_l + (1-C_l)/\tau \quad (4.51)$$

which, since the position functions involved are spherical harmonics of second order, becomes

$$\frac{1}{\tau_c} = \frac{1}{\tau_2} + \frac{(1-C_2)}{\tau}$$

Dielectric relaxation involves the first-order function $\cos\theta(t)$ and the correlation function in this case decays with the characteristic time τ_r, the molecular reorientation time. Hertz derives the relationship between τ_c and τ_r[9]. From equations (4.47) and (4.49) it is clear that $C_2 = C_1{}^3$

and $\tau_2 = \tau_1/3$ whence

$$\frac{1}{\tau_r} = \frac{1}{3}\frac{1}{\tau_c} + \frac{1}{\tau}\left\{\frac{2}{3} - C_1\left(1 - \frac{1}{3}C_1{}^2\right)\right\} \tag{4.52}$$

In the limit $C_1 = 1$ the motion is true infinitesimal step diffusion and $\tau_r = 3\tau_c$. However, for $C_1 = 0$ and $\tau_1 \gg \tau$ then $\tau_r = \tau_c$. The conditions imply that the mode of rotation of the molecule is by large jumps, perhaps of the order of one radian. Powles[10] has reviewed this point and, with some exceptions, finds τ_r/τ_c to be nearly three for many liquids, giving support to the diffusion model.

Before considering this theory and its refinements in more detail it is interesting to consider the case where D_r becomes very large. For this situation the correlation functions are Gaussian[11].

$$f_0(t) = \exp(-\tau^{*2}) \tag{4.53a}$$

$$g_m(t) = \exp(-3\tau^{*2}) \tag{4.53b}$$

$$\tau^* = t(\mathbf{k}T/I_0)^{1/2} \tag{4.53c}$$

where $f_0(t)$ and $g_m(t)$ are the correlation functions for the first and second order spherical harmonics, $\mathbf{k}$ is Boltzmann's constant, T the absolute temperature and I_0 the moment of inertia of the molecule. For dipole–dipole relaxation in the limit of extreme narrowing the relaxation rate for $I = \frac{1}{2}$ is[11]

$$(T_1)^{-1} = \frac{3}{2}\gamma^4\hbar^2 r^{-6} \cdot \frac{1}{2}\frac{[\pi I_0]^{1/2}}{3\mathbf{k}T} \tag{4.54}$$

Comparison with equation (4.29) shows that the two types of correlation function lead to essentially the same result if τ_c in equation (4.29) is identified with $\frac{1}{2}[\pi I_0/3\mathbf{k}T]^{1/2}$. This may be expressed semi-quantitatively by considering the angular velocity of a classical free rotator. By equipartition

$$\tfrac{1}{2}I_0\overline{\Omega^2} = \tfrac{3}{2}\mathbf{k}T$$

which gives a root-mean-square angular velocity given by

$$[\overline{\Omega^2}]^{1/2} = \left(\frac{3\mathbf{k}T}{I_0}\right)^{1/2} \tag{4.55}$$

Identifying τ_c with the mean time for a molecule to rotate through one radian, then clearly

$$\tau_c \approx \left(\frac{I_0}{3\mathbf{k}T}\right)^{1/2} \tag{4.56}$$

which is sufficiently close to the required $\frac{1}{2}[\pi I_0/3\mathbf{k}T]^{1/2}$ to justify the somewhat crude picture given above.

It might be expected that if this form of correlation function obtained in the liquid state, then it would do so at temperatures close to the critical temperature. Powles[10] has collected a considerable amount of data of ν_c, where $\nu_c = (2\pi\tau_c)^{-1}$, as a function of temperature for a wide variety of liquids. This is summarized in Figure 4.5. There are several points of interest about the curves in Figure 4.5 but for the moment, two are pertinent to the present discussion. Firstly, the plots deviate without exception from an Arrhenius dependence of ν_c (or τ_c) upon temperature; although quite a large section of any individual curve obeys this type of law up to a little below the melting point, at this point the curve possesses a 'knee'. Secondly, and perhaps more remarkably, there seems to be an upper limit to the critical frequency of 10^{12} Hz, or a little less, for a wide range of liquids. Powles has given two possible semi-quantitative interpretations of this observation. The first uses activated-state rate theory from which

$$\nu_c \approx (\mathbf{k}T/\hbar)\exp(-\Delta H^{\ddagger}/\mathbf{k}T) \tag{4.57}$$

At high values of $\mathbf{k}T$, $\nu_c \rightarrow \mathbf{k}T/\hbar$ and the assumption of a critical temperature of about 250°C, $\nu_c \approx \mathbf{k}T/\hbar \approx 10^{13}$ Hz. τ_c in this limit is the uncertainty principle time for the thermal energy $\mathbf{k}T$. Alternatively, there is an anisotropic intermolecular interaction energy ΔE_{anis} which sets a limit to τ_c:

$$\Delta E_{\text{anis}}\tau_c \geqslant \hbar \tag{4.58}$$

giving $\Delta E_{\text{anis}} \geqslant 0{\cdot}8$ kcal mole^{-1} for $\tau_c \approx 10^{-13}$ s. Secondly, the molecule may be treated as a classical free rotator as in equation (4.55). Clearly

$$\nu_c \approx \frac{1}{2\pi}\left(\frac{3\mathbf{k}T}{I_0}\right)^{1/2}$$

which for $T \approx 250$°C and $I_0 \approx 10^{-38}$ g cm^2 gives $\nu_c \approx 10^{11.5}$ Hz. This is not sharply dependent upon the critical temperature nor on I_0 and

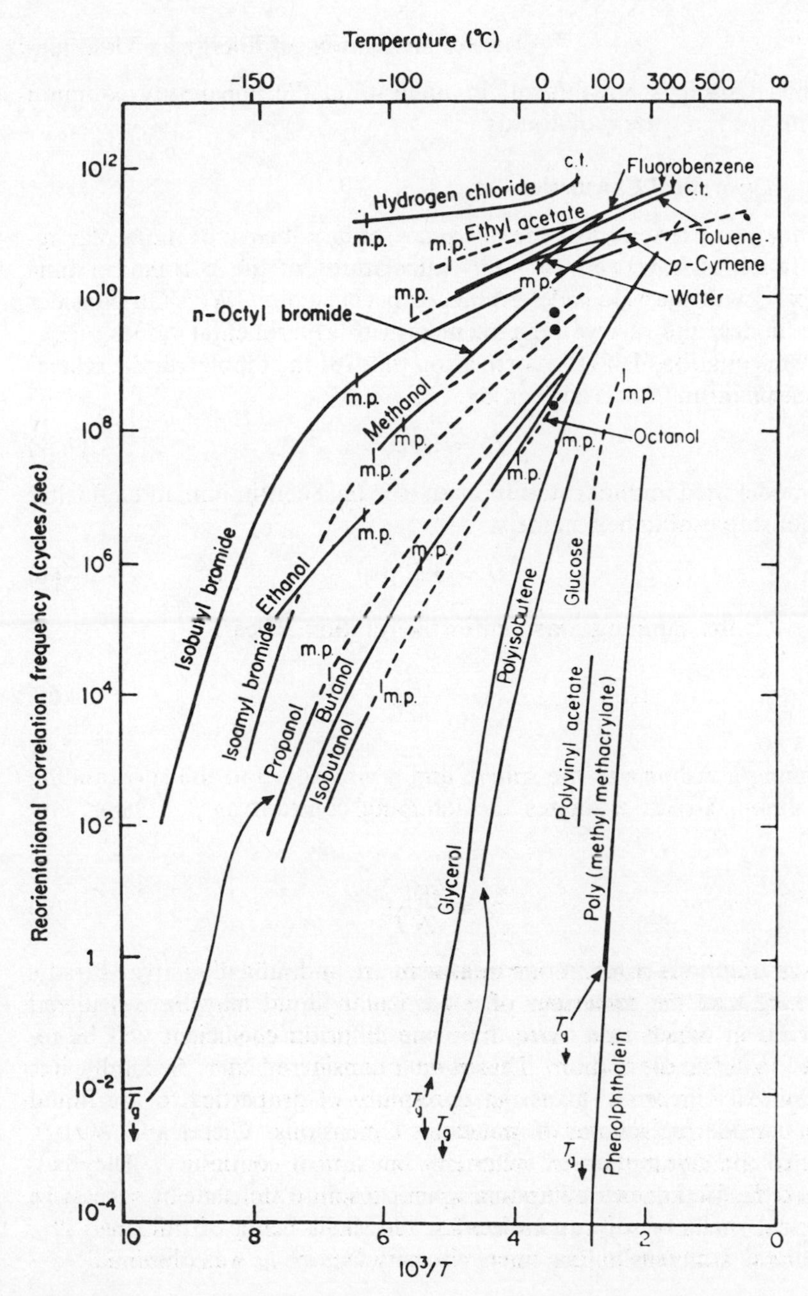

Figure 4.5 Temperature dependence of reorientational correlation frequencies for a variety of liquids. (From Reference 10 by courtesy of the author and publishers)

on this basis it is not difficult to understand the apparently common limit to ν_c for a variety of liquids.

4.3.1 Theoretical Calculations of τ_c

An important criterion of the adequacy of any theory of molecular reorientation is how well *ab initio* calculations of the correlation time compare with the values determined experimentally. We shall consider three models and review the agreement with experimental values of τ_c.

From equation (4.49) the correlation time for the dipole–dipole relaxation mechanism ($l = 2$) is given by

$$\tau_c = 1/6D_r \tag{4.59}$$

The model used implies rotation in an isotropic continuum and a further relationship is obtained, namely,

$$D_r = \mathbf{k}T/\beta \tag{4.60}$$

where β is the damping constant for the rotation. Clearly

$$\tau_c = \frac{\beta}{6\mathbf{k}T} \tag{4.61}$$

Assigning a radius r to the sphere and a viscosity η to the surrounding continuum, Stokes evaluates the damping constant as $\beta = 8\pi\eta r^3$ and thus

$$\tau_c = \frac{4\pi\eta r^3}{3\mathbf{k}T} \tag{4.62}$$

Two assumptions made in this treatment are undoubtedly naïve. Firstly, it is rare that the molecules of a particular liquid may be considered spherical in which case more than one diffusion coefficient will be required to define the motion. This point is considered later. Secondly, it is undoubtedly incorrect to assign continuity of properties to the liquid when considering spheres of molecular dimensions. Gierer and Wirtz[12] retained the assumption of sphericity but not of continuity. They calculated the friction exerted upon a spherical solute molecule by successive spherical shells of solvent molecules, the shells being of thickness $2r_L$. In general, a dimensionless 'microviscosity factor', f_r, was obtained

$$f_r = \{6r_L/r + 1/(1 + r_L/r)^3\}^{-1} \tag{4.63}$$

and β was shown to be

$$\beta = 8\pi\eta r^3 f_r = 8\pi\eta r^3\{6r_L/r + 1/(1+r_L/r)^3\}^{-1} \tag{4.64}$$

There are two important special cases of this model. Firstly, it is expected that for $r \gg r_L$ the discontinuous nature of the solvent will be unimportant in determining the damping constant for rotation of the large solute molecules and hence will not be reflected in τ_c. An equivalent statement is that the summation of the frictional effects of successive solvent shells may be expressed as an integral which should, in this limit, give Stokes' law. This is found to be the case. The second case is that of pure liquids for which $r = r_L$ and, to a reasonably good approximation, equation (4.64) reduces to

$$\beta \approx 8\pi\eta r^3/6 \tag{4.65}$$

giving a correlation time one-sixth of that expected from Stokes' law.

Hill has calculated β[13] using a theory due to Andrade. In this the cause of friction in a liquid is assumed to be the temporary union of molecules vibrating about their mean positions. The damping constant for rotation is given by

$$\beta = 6f_A \frac{I_{AB}I_B}{I_{AB}+I_B} \cdot \frac{m_A+m_B}{m_A m_B} \cdot \eta_{AB}\sigma_{AB} + 3(3-\sqrt{2})f_B \frac{I_{BB}I_B}{I_{BB}+I_B} \cdot \frac{2}{m_B} \cdot \eta_B\sigma_B \quad (4.66)^*$$

for a solution of A in B, the two components having mole fractions f_A and f_B and molecular masses m_A and m_B respectively. I_A and I_B are the moments of inertia of the molecules and I_{AB} and I_{BB} are the moments of inertia of type A and B molecules respectively about the centre of mass of a type B molecule during collision. Finally η_B is the viscosity of pure B, σ_B the mean distance between B molecules, η_{AB} is the mutual viscosity and σ_{AB} is the mean distance between A and B molecules.

An approximate value for β, and hence for the correlation time, may be found in the limit of infinite dilution of B in A. For this case the second term in equation (4.66) vanishes and the reduced moment of inertia is replaced by I_B. Further, $\eta_{AB}\sigma_{AB} \approx 2\eta\alpha$ where α is the average radius of

* The product $\eta_{AB}\sigma_{AB}$, in equation (4.66) is determined by measuring the viscosities η_m of mixtures of A and B using the relation

$$\eta_m\sigma_m = f_A{}^2\eta_A\sigma_A + f_B{}^2\eta_B\sigma_B + 2f_A f_B\eta_{AB}\sigma_{AB}$$

where σ_m are the mean intermolecular distances.

molecule A, and so

$$\tau_c = \frac{\beta}{6\mathbf{k}T} = \frac{2I_B\eta\alpha}{\mu\mathbf{k}T} \tag{4.67}$$

in which μ is the reduced mass $m_A m_B/(m_A + m_B)$. A knowledge of the value of $\eta_{AB}\sigma_{AB}$ is not required in this limit.

The Debye, microviscosity and Hill theories have been tested experimentally by Mitchell and Eisner[14,15] and, rather more completely, by Pritchard and Richards[16,17]. The first series of compounds studied by Pritchard and Richards were the aromatic hydrocarbons benzene, toluene, *p*-xylene, *o*-xylene and pyrene[16]. These are discussed below in the context of internal rotation. It is sufficient to note that depending on the choice of molecular dimensions the correlation times calculated on the basis of the Stokes relationship were at least four times the experimental values and often were an order of magnitude greater. However, the results did not allow a clear cut choice to be made between the microviscosity and the Hill theories.

A more severe test was found in the proton spin–lattice relaxation times of the halogenated methanes CH_3Br, CH_3I, CH_2Cl_2, CH_2Br_2, CH_2I_2, $CHCl_3$ and $CHBr_3$. In order to separate the intermolecular contributions to the relaxation rate* from the intramolecular effects, the halides were studied for a range of concentrations in carbon disulphide. This solvent has the advantage that, owing to the low abundance ($\sim 1\%$) of magnetic ^{13}C nuclei, its contribution to the relaxation rate is negligible. The theoretical values of $(T_1)^{-1}$ were calculated for the different models using the measured viscosities of the solutions together with known interatomic distances and van der Waals radii. The molecules were assumed to be spherical with radii equal to the mean molecular semiaxes. One other method of calculation was used. This involved measuring the density of the pure liquid at the temperature of the relaxation measurements (25°C). The molecules were again assumed to be spheres with a filling factor of 0·74, the value for hexagonal close-packing. The effective radius thus obtained was used to re-calculate the microviscosity factors.

* The intermolecular contribution has been neglected in this chapter since, as has been stated, the translational correlation time involved gives no information on molecular rotation. However, it is clearly necessary to take it into account when comparing experiment with theory. The general theory is to be found in Reference 4 and the specific expressions for the translational correlation times for the Stokes, microviscosity and Hill theories are in Reference 16.

Figure 4.6 compares the proton relaxation rates for CH_3I calculated using Hill theory, the simple microviscosity theory and the corrected microviscosity theory (hexagonal close-packing). Both the simple and corrected microviscosity theories closely reproduce the experimental relaxation rates. On the other hand, the Hill theory does not give the correct concentration dependence of $(T_1)^{-1}$. Pritchard and Richards

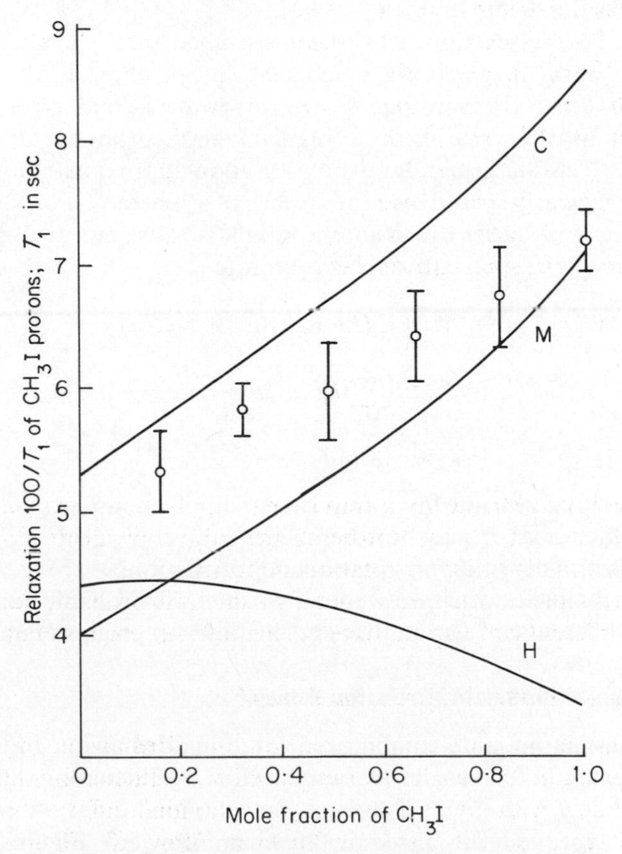

Figure 4.6 Relaxation rates of CH_3I protons in CS_2 solutions. ⫯, experimental values; H curve, predicted by Hill theory; M curve, predicted by simple microviscosity theory; C corrected microviscosity theory. (From Reference 17 by courtesy of the authors and of the Faraday Society)

ascribe this discrepancy to an incorrect prediction of the Hill theory, namely that the rotational correlation time will increase as the viscosity decreases. They point out that the Hill theory, whilst still inferior to the microviscosity approaches, gives better results for $CHBr_3$ than for CH_2Br_2 or CH_2Cl_2 and suggest that the 'stickiness' of the collisions increases with molecular polarizability. For the lighter molecules, however, transfer of angular momentum during a collision is not as complete as the theory implies.

In the above discussion it has been assumed that where a molecule contains several magnetically equivalent nuclei, these have the same correlation time. Hubbard has shown this to be a good approximation for nuclei at the vertices of a regular tetrahedron or an isosceles triangle[18,19]. More generally it is also a good approximation for nuclei which are regularly spaced over the surface of a sphere[20]. For less regular arrangements of nuclei the assumption is less satisfactory[21]. A general expression for the spin–lattice relaxation rate is

$$\begin{aligned}(T_1)^{-1} &= \tfrac{3}{2}\gamma_i^4\hbar^2 \sum_i I_i(I_i+1)[J_1(\omega_i)+J_2(2\omega_i)] \\ &\quad +\gamma_i^2\gamma_j^2\hbar^2 \sum_j I_j(I_j+1)[\tfrac{1}{12}J_0(\omega_i-\omega_j)+\tfrac{3}{2}J_1(\omega_i) \\ &\quad +\tfrac{3}{4}J_2(\omega_i+\omega_j)]\end{aligned} \tag{4.68}$$

This is the relaxation rate for a spin i due to dipolar couplings with several spins j, which may or may not be magnetically equivalent to i. The J's are the spectral densities for rotation and translation.

Clearly, for nuclei which are non-equivalent it will probably be necessary to assign different correlation times to the different groups of nuclei.

4.3.2 Spin–Rotational Correlation Times

Before considering more complicated rotational behaviour in the liquid phase, it is useful to consider two models for the fluctuation of the spin–rotational field with time. These are the rotational diffusion model and the 'pulse' type model discussed by Green and Powles[6]. Figure 4.7 shows the behaviour of $\mathbf{H}_{sr}$ with time for the two models. These are, of course, the infinitesimal step diffusion and finite rotational jump models discussed above for dipolar interactions.

Green and Powles derive a relationship between τ_{sr} and τ_c for these situations in the following way. For rotational diffusion it is assumed

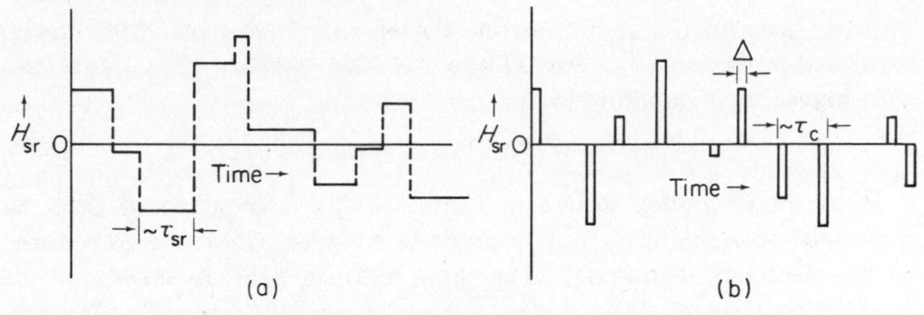

Figure 4.7 Spin–rotation magnetic field according to (a) a Brownian rotational diffusion model and (b) a pulse type model. (From Reference 6 by courtesy of the authors, The Institute of Physics and The Physical Society)

that at an arbitrary zero of time the molecule is rotating with angular velocity Ω. For a very short time t, that is, before the first collision, the molecule rotates through an angle $\psi = \Omega t$ and so $\psi^2 = \Omega^2 t^2$ for short times. The rotational motion is interrupted after a time which is to be identified with the spin–rotational correlation time, τ_{sr}. For the true infinitesimal step diffusion which is probably descriptive of many liquids at moderate temperatures, ψ will change only slightly in this time and after many intervals τ_{sr}, t^2 may be replaced by $\tau_{sr}t$ and so

$$\overline{\psi^2} = n\overline{\Omega^2}\tau_{sr}t \tag{4.69}$$

in which the factor n arises since reorientation is not restricted to be about a fixed axis. Hubbard has shown that $n = 6$. For appreciable reorientation $\overline{\psi^2} = 1$ and $t \approx \tau_c$, thus

$$6\overline{\Omega^2}\tau_{sr}\tau_c = 1 \tag{4.70}$$

The principle of equipartition gives

$$\tfrac{1}{2}I_0\overline{\Omega^2} = \tfrac{1}{2}\mathbf{k}T \tag{4.71}$$

and substituting for $\overline{\Omega^2}$ leads to the important relationship

$$\tau_{sr}\tau_c = \frac{I_0}{6\mathbf{k}T} \tag{4.72}$$

The spin–rotational correlation time *for this rotational model* may thus be expressed in terms of τ_c. For liquids, we may write $\omega_0{}^2\tau_{sr}^2 \ll 1$ and substituting for τ_{sr} in equation (4.40),

$$(T_1)_{sr}^{-1} = \tfrac{1}{3}\{\tfrac{1}{3}[2C_\perp{}^2 + C_\parallel{}^2]\}I_0{}^2\hbar^{-2}\tau_c^{-1} \tag{4.73}$$

In the pulse model, shown in Figure 4.7(b), it is proposed that the molecules reorientate by jumps in between which they are considered to be effectively stationary. The spin–rotation field as shown in the figure is only switched on during the jumps and if these are sufficiently large (~ 1 radian), the mean time interval between the pulses in $\mathbf{H}_{sr}$ is τ_c. The time-dependence of the correlation function for $\mathbf{H}_{sr}$ differs in the two cases shown in Figure 4.2. For the diffusion case it is exponential but for the pulse model, correlation falls linearly to zero from the beginning of a pulse ($t = 0$) to $t = \Delta$ and remains zero for $t > \Delta$. τ_{sr} is thus identified with $\Delta/2$. An implicit assumption of this model is that $\Delta \ll \tau_c$, i.e. $\tau_{sr} \ll \tau_c$. It follows that the mean isotropic fluctuating field $\overline{H_{sr}^2(t)}$ is given by $\overline{H_i{}^2}\,.\,\Delta/\tau_c$ where H_i is the value of the field during a jump. Referring to the general formula for T_1 (equation 4.38) the further assumption that $\omega_0{}^2\tau_{sr}^2 \ll 1$ gives

$$(T_1)_{sr}^{-1} = \frac{2\gamma^2\overline{H_i{}^2}\Delta}{3\tau_c}\cdot\frac{\Delta}{2} = \frac{1}{3}\gamma^2\overline{H_i{}^2}\frac{\Delta^2}{\tau_c} \tag{4.74}$$

To calculate $\overline{H_i{}^2}$ we note that the molecule only rotates significantly during the pulses. In identifying τ_c with the interval between collisions it is implied that during a pulse the molecule will rotate through approximately one radian. The angular velocity during this jump is thus Δ^{-1}. Treating the spin–rotational interaction constant as a scalar, c, then

$$H_i = cJ/\gamma = cI_0\Omega/\gamma\hbar = cI_0\Delta^{-1}/\gamma\hbar \tag{4.75}$$

and

$$\overline{H_i{}^2} = \left(\frac{cI_0}{\gamma\hbar}\right)^2\Delta^{-2} \tag{4.76}$$

Substitution of this value of $\overline{H_i{}^2}$ into equation (4.74) gives

$$(T_1)_{sr}^{-1} = \tfrac{1}{3}c^2I_0{}^2\hbar^{-2}\tau_c^{-1} \tag{4.77}$$

Accepting the limitations imposed by the assumptions made in deriving equations (4.73) and (4.77) it may be seen that the rotational diffusion model and the pulse model give the same result for $(T_1)_{sr}^{-1}$, at least semi-quantitatively. It has been suggested[6] that the relaxation rates due to spin–rotational interaction and also those with correlation time τ_c are relatively insensitive indicators of whether the Brownian diffusion model or the finite jump ('pulse') model pertains to a particular liquid. Methods which may be used to distinguish between the two types of behaviour are discussed later.

4.3.3 Temperature Dependence of τ_c and τ_{sr}

It was noted earlier that the correlation time τ_c deviated markedly from an Arrhenius-type dependence upon temperature when measured over a wide range of temperatures (Figure 4.5). In practice, however, it is

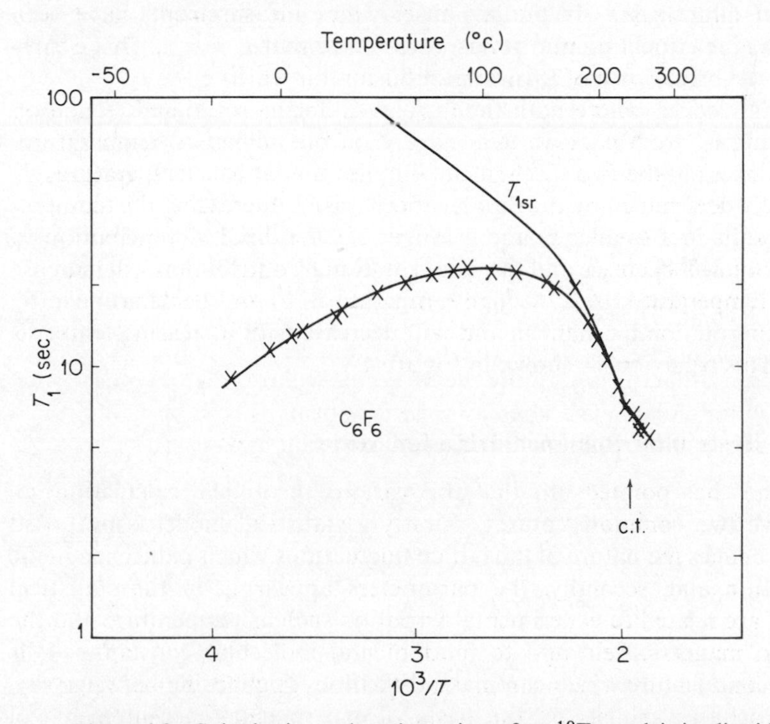

Figure 4.8 Spin–lattice relaxation time for ^{19}F nuclei in liquid perfluorobenzene. Also shown are calculated values for the spin–rotational contribution to T_1. (From Reference 6 by the courtesy of the authors, The Institute of Physics and The Physical Society)

frequently possible to fit measured values of τ_c to an equation of the form

$$\tau_c = \tau_0 \exp(E_a/\mathbf{k}T) \tag{4.78}$$

over a limited range of temperatures. In any event τ_c is expected to decrease sharply as the temperature is increased. Combining equation (4.78) with equation (4.72) shows that τ_{sr} increases with increasing temperature. This contrasting behaviour of the two correlation times may be qualitatively understood in the following way: an increase in τ_{sr} implies that a particular molecule spins for longer times at the same mean angular velocity as the temperature is increased. The angular position of the molecule changes more rapidly and so τ_c decreases as τ_{sr} increases. This argument depends for its validity on the condition $\tau_{sr} \ll \tau_c$. This is usually true in liquids over a considerable temperature range but is not true for dilute gases. In the few cases where measurements have been made at the critical point[10] it has been found that $\tau_c \approx \tau_{sr}$. This clearly shows the breakdown of Brownian diffusion for which $\tau_{sr} \ll \tau_c$.

For molecules where both the dipolar and spin–rotational relaxation mechanisms are operative the nature of the different temperature-dependences of the two mechanisms implies that at low temperatures T_1 is largely determined by the dipolar mechanism. Increasing the temperature results in a smaller τ_c and a longer τ_{sr}, the dipolar contribution to relaxation will decrease and the spin–rotational contribution will increase as the temperature rises. At high temperatures T_1 will be determined by the spin–rotation mechanism and will decrease with increasing temperature. This behaviour is shown in Figure 4.8.

4.3.4 Tests of the Rotational Diffusion Model

Waugh[22] has pointed out that the various theoretical calculations of T_1 have two common features. Firstly a statistical model is proposed which defines the nature of the lattice fluctuations which induce magnetic relaxation and, secondly, the parameters appearing in the statistical model are related to experimental variables such as temperature and the applied magnetic field and to fundamental molecular constants. It is this second feature which can make difficulties in choosing between, say, two stochastic models on the basis of experimental measurements of T_1 alone. This is illustrated by the Brownian diffusion model together with the use of equation (4.62) relating the correlation time to the bulk viscosity η, the molecular radius r and the temperature. It is well known that the simple Stokes law treatment leads to correlation times which are frequently an order of magnitude greater than the observed values. Despite the excellent results of Pritchard and Richards[17] using this theory with the 'microviscosity' modification, it would seem that the uncertainties involved in relating the microscopic behaviour under investigation to the macroscopic properties and known molecular constants are too great to allow of an unambiguous choice between two statistical models.

Two possibilities remain, the first being a comparison of τ_c with the dielectric relaxation time which is often taken to be the reorientation time τ_r of equation (4.52). In discussing that equation it was noted that for infinitesimal step diffusion $\tau_r = 3\tau_c$ whilst for reorientation by finite jumps $\tau_c \ll \tau_r < 3\tau_c$. The weakness in this method lies in identifying the macroscopic dielectric relaxation time with the molecular, or microscopic, reorientation time. In a discussion of the rôle of electric dipole–dipole interaction in dielectric relaxation, Scaife[23] has shown that the use of a correct cavity field has two effects. Firstly, if a single microscopic relaxation time is postulated this will be shifted by dipole–dipole coupling and, secondly, a distribution of observable macroscopic relaxation times results. More detailed calculations by Budó[24] and Zwanzig[25] show that electrostatic interactions between dipoles lead to a distribution of microscopic relaxation times. Empirically, however, it is found that for simple liquids the dielectric dispersion data may often be adequately represented either by a single relaxation time or a small number of discrete times. Except for highly associated liquids, possible continuous distributions of relaxation times reproduce the experimental data no better than single or discrete relaxation times. It may be, therefore, that a comparison of τ_r and τ_c will still give useful information about the statistical nature of the molecular rotation with especial reference to deviations from Brownian diffusion.

The second possibility of checking rotational models is an experiment proposed by Waugh[22]. It is supposed that a liquid is obtainable for which the only relaxation mechanisms are spin–rotational interactions and the chemical shift mechanism. The observed relaxation rate is given by the sum of the rates produced by the individual mechanisms and therefore

$$(T_1)^{-1} = (T_1)_{\mathrm{sr}}^{-1} + (T_1)_{\sigma}^{-1} \tag{4.79}$$

in which the subscripts sr and σ indicate the mechanism responsible for the associated term. The two contributions are readily evaluated separately by making use of the fact that the chemical shift term is proportional to H_0^2. Assuming that $\omega_0^2\tau_{sr}^2 \ll \omega_0^2\tau_c^2 \ll 1$ and that the spin–rotational coupling may be considered as a scalar, c, then the use of equations (4.31), (4.40) and (4.72) gives

$$(T_1)_{sr}(T_1)_\sigma = \frac{45}{2}\left[\frac{\hbar}{H_0(\sigma_\parallel - \sigma_\perp)cI_0}\right]^2 \tag{4.80}$$

If the shift anisotropy, the spin–rotation constant and the moment of inertia of the molecule are known then it is possible to evaluate $(T_1)_{sr}(T_1)_\sigma$ precisely. This value is independent of temperature, density and all other variables provided that the rotational motion of the molecules is described by infinitesimal step diffusion since it is only under this condition that the relationship between τ_{sr} and τ_c (equation 4.72) is valid. The values of c and $(\sigma_\parallel - \sigma_\perp)$ are often not known with any accuracy. This does not, however, detract from the value of the proposed experiment since the constancy of $(T_1)_{sr}(T_1)_\sigma$ under all conditions of temperature, pressure, etc., is still a rigorous test of the diffusion model. Compounds suitable for this experiment have been suggested by Waugh[22] but to the best of the present author's knowledge no extensive results have yet been reported.

4.3.5 Internal Rotations and Anisotropic Diffusion

So far, it has been assumed that the nuclei undergoing relaxation are in fixed positions with respect to a set of axes referred to the molecular framework and that therefore their relaxation behaviour is governed by a single correlation time. However, a situation commonly encountered is one in which two or more nuclei are present in a group which rotates about an axis fixed in the frame of the rest of the molecule. This axis undergoes rotational motion and it is clearly important to be able to relate the observed correlation time to the time constants governing the two rotational processes. Woessner[26] has treated this problem. With some change in notation, his result for dipolar relaxation of a spin pair is

$$(T_1)^{-1} = \tfrac{3}{2}\gamma^4\hbar^2 r^{-6} I(I+1)[\tfrac{1}{3}(3\cos^2\Delta - 1)^2\tau_{c1} + (\sin^2 2\Delta)\tau_{c2} + (\sin^4\Delta)\tau_{c3}] \tag{4.81}$$

in the limit $(\omega_0\tau_{c1})^2 \ll 1$. In this, Δ is the angle between the internuclear vector and the axis fixed in the molecular frame, the rotation of which is described by the time constant τ_{c1}. Two cases are of special interest, and in both it is assumed that the group which undergoes rotation about the fixed axis has three-fold symmetry. Firstly, infinitesimal step rotation about the internal axis leads to the following relationships between the various time constants in equation (4.81),

$$\tau_{c2} = [(1/\tau_{c1}) + (1/\tau_0)]^{-1}$$

$$\tau_{c3} = [(1/\tau_{c1}) + (4/\tau_0)]^{-1}$$

where τ_0 describes the internal rotation of the spin–spin vectors. Secondly, random finite jumps between adjacent equilibrium positions about the internal axis gives

$$\tau_{c3} = \tau_{c2}$$

if the rate of jumping is $(3\tau_0)^{-1}$.

Reference has been made to the work of Pritchard and Richards[16] on the spin–lattice relaxation behaviour of the protons in a series of aromatic hydrocarbons. Correlation times were calculated from relaxation rates extrapolated to the limit of infinite dilution of the hydrocarbons in CS_2 on the assumption of isotropic rotational diffusion of all protons. These values are given in Table 4.1. Hertz[9] has discussed the ratios $\tau_{\text{methyl}}/\tau_{\text{ring}}$, with special reference to toluene, in the light of Woessner's treatment of the effect on T_1 of internal rotation. He finds that after making a semi-quantitative estimate of the effect of anisotropic rotation of the fixed internal axis, the ratio $\tau_{\text{methyl}}/\tau_{\text{ring}} \approx 0{\cdot}75$ may be obtained from equation (4.81) if $\tau_{c1}/\tau_0 \approx \frac{1}{3}$. The interpretation of this is that after a time corresponding to one reorientation process of the whole molecule the methyl group makes one jump from one equilibrium position to another immediately adjacent.

Table 4.1

Rotational correlation times for Protons in various aromatic hydrocarbons at 25°C

Compound	$10^{11}\,\tau_c$ Ring	(seconds) Methyl
Benzene	0·12	—
Toluene	0·17	0·13
p-Xylene	0·24	0·18
o-Xylene	0·19	0·14
Pyrene	0·60	—

A further example of the application of n.m.r. in the study of anisotropic rotational motions in the liquid phase is to be found in the work done by Moniz and Gutowsky[27]. These authors measured the ^{14}N and proton spin–lattice relaxation times in liquid CH_3CN over a range of temperatures, the proton T_1 being determined at a resonance frequency of 20 MHz and that of ^{14}N at 3·74 MHz. The temperature dependences of these two relaxation times are shown in Figure 4.9. In the pure liquid it is to be expected that the proton relaxation will be caused by both inter- and intramolecular effects and these may be of comparable magnitudes. For ^{14}N, however, the dominant relaxation mechanism will be modulation of quadrupole coupling caused by molecular reorientation. This difference is reflected in the different activation energies found from plots of T_1 against $1/T$. Moniz and Gutowsky point out that CH_3CN is unusual in that rotations about the C—CN axis can contribute to the proton T_1 but not to the ^{14}N T_1. Evidently two correlation times are necessary

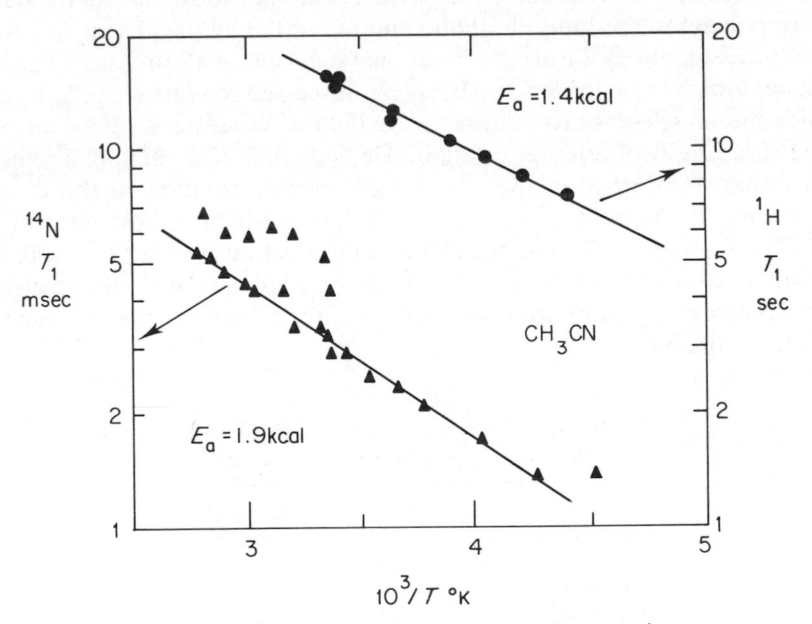

Figure 4.9 The temperature dependence of the ^{14}N and ^1H spin–lattice relaxation times observed in liquid CH_3CN. The proton T_1 was measured at a resonance frequency of 20 MHz and the ^{14}N at 3·74 MHz. (From Reference 27 by courtesy of the authors and the American Institute of Physics)

to describe the reorientation of this molecule. These may be denoted τ_{c1} and τ_{c3} for rotations about axes perpendicular and parallel to the C—CN bond respectively. The quantitative contribution of the reorientation of the CH_3 group about the C—CN bond to the relaxation time of the protons depends on the model chosen for the reorientation process. Since there is no barrier to internal rotation of this group it seems that a stochastic Brownian motion is much more plausible than a 'jump' motion. Using the further approximation[18] that the dipole–dipole interactions in the CH_3 group are completely separable, then the purely rotational contribution (τ_{c1}) to the dipolar relaxation of the proton is[27]

$$[T_{1\text{H}}]_{\text{rot}}^{-1} = \tfrac{3}{4}\gamma_{\text{H}}{}^{4}\hbar^{2}r^{-6}\tau_{c1} \tag{4.82}$$

Spin–lattice relaxation of ^{14}N is solely determined by τ_{c1}, i.e. by rotations about axes perpendicular to the C—CN axis and thus τ_{c1} is calculable. It was found that $[T_{1\text{H}}]_{\text{rot}}^{-1}$ calculated from equation (4.82) is about one-half of the observed value, the remaining contribution to the relaxation probability coming from the diffusion-modulated intermolecular interactions. The contributions were separated using the relationship

$$[T_1]_{\text{diff}}^{-1} = [T_1]_{\text{obs}}^{-1} - [T_1]_{\text{rot}}^{-1} \tag{4.83}$$

and it was found that a plot of ln $[T_1]_{\text{diff}}$ against $1/T$ gave a good straight line with slope which corresponded to an activation energy for self-diffusion of approximately 1 kcal mole^{-1}.

Moniz and Gutowsky conclude that their results clearly indicate the anisotropic nature of the rotational motion of CH_3CN molecules in the liquid phase. Their reasoning is as follows. If it is assumed that the rotation is isotropic and described by a single correlation time τ_c, then values for τ_c may be calculated from the ^{14}N measurements and hence values for $[T_1]_{\text{rot}}^{-1}$ are obtained. These, however, exceed the observed values by a factor of two and the authors claim that such a discrepancy is well outside the errors and approximations made in the analysis. Despite a recent criticism of this conclusion[9], it is difficult to find fault in the above discussion and the experiment demonstrates the considerable potentialities of relaxation studies of different nuclei in the same molecule.

4.4 RELAXATION IN GASES

An obvious and important distinction between the rotational motions of molecules in the liquid and gaseous phases is that in the latter J, the rotational quantum number, is a good quantum number whilst in liquids a particular molecule will undergo many collisions during a complete rotation. Previous sections have discussed diffusional and 'jump' models for this situation. It will become apparent that for gases it is frequently possible to ignore ΔJ transitions and that changes in m_J, whilst still infrequent in some cases, are sufficient to account for nuclear spin relaxation times in the gas phase.

4.4.1 Sources of Relaxation in Hydrogen Gas

Waugh and Johnson[28] have given a good qualitative discussion of the origins of nuclear relaxation in dilute gases with special reference to solutions of hydrogen in a variety of other gases and the relationship between the experimental results and anisotropic intermolecular potentials. These authors begin with the expression

$$(T_1)^{-1} = \gamma^2 H_{\text{eff}}^2 \frac{\tau_c}{1+\omega^2 \tau_c{}^2} \tag{4.84}$$

in which H_{eff} is a constant and is defined as the root mean squared amplitude of the local field fluctuation at the nucleus concerned. Figure 4.10(a) shows a possible schematic representation of the local field in a dilute gas.

Whilst the molecule is in free flight between collisions there will be an intramolecular field which will, however, remain constant and which cannot therefore induce nuclear transitions. Some collisions may change the internal quantum state of the molecule and in such cases the intramolecular field will change. These fluctuations constitute a relaxation mechanism. The mean free time between them will be defined as τ_1 and is shown in Figure 4.10. Further, it is necessary to consider the momentary perturbation during each collision. This lasts for a mean time τ_2. There are thus two local fields to be considered and these are assumed to be independent and hence separable as shown in Figure 4.10.

The peak intensity of the intermolecular field during a collision may be large but its mean squared intensity is reduced by a factor τ_2/τ_1. Waugh and Johnson have performed an order of magnitude calculation for relaxation by this impulsive field for transient dipolar interactions between nuclei on different molecules in a gas under standard conditions of temperature and pressure. The result is a relaxation time of the order of 10^6s. This is many orders of magnitude longer than the relaxation times recorded later in this section, and hence this relaxation mechanism

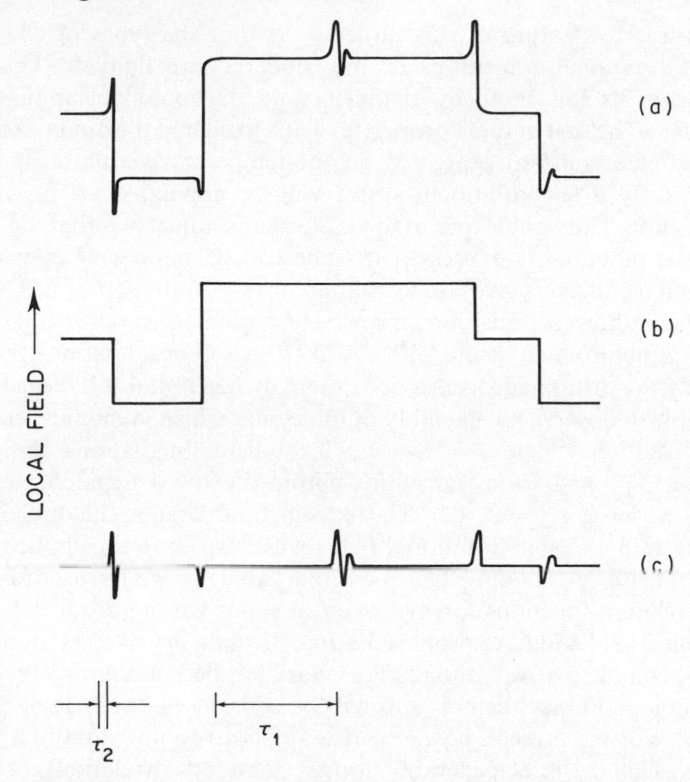

Figure 4.10 Local field fluctuations in a dilute polyatomic gas. (From Reference 28 by courtesy of the authors and the Faraday Society)

may be neglected. Clearly, the local field fluctuations responsible for the nuclear spin relaxations are those represented by Figure 4.10(b). For a dilute gas it is reasonable to assume that these fluctuations result from binary collisional transitions between rotational molecular quantum states with constant transition probabilities. Further, the correlation time τ_1 may be directly related to these transition probabilities.

The molecule which has received most attention, both experimental and theoretical, is the hydrogen molecule. In the rotational level $J = 1$, each proton in this molecule experiences a local field H' of 27 gauss arising from the spin–rotational interaction and a field H'' of 34 gauss arising from the dipole–dipole interaction with the other proton.

An attractive feature of this molecule is that the types of collision-induced rotational transitions it may undergo are limited. This fact arises from its low moment of inertia and its homonuclear diatomic symmetry. The first of these properties leads to a large rotational spacing, the rotational constant expressed on the temperature scale being 86°K. Clearly, only a few rotational states will be populated at, say, room temperature. The significance of the molecular symmetry is that the intermolecular potential is expressed in spherical harmonics of even order only so that collisions give rise to changes in J such that $\Delta J = 0, 2, 4, \ldots$. Even these allowed transitions have been estimated to occur only once in several hundred collisions[29,30]. Add to these considerations the fact that only the ortho hydrogen is detectable by n.m.r. and it is clear that it is possible to observe an assembly of molecules which are almost entirely in the rotational state $J = 1$. The local field fluctuations shown in Figure 4.10(b) arise from transitions within the $J = 1$ manifold, namely among the levels $m_J = 0, \pm 1$. These transitions change the direction of the spin–spin and spin–rotational fields with respect to the applied field.

Oppenheim and Bloom have, in a remarkable *tour de force*, considered the correlation functions for hydrogen in states varying from the dilute gas to the liquid and have proposed a statistical theory of relaxation valid over the complete density range. This work has been admirably reviewed by Deutch and Oppenheim[31] and will be considered briefly later in this chapter. For the present, however, it is sufficient to note that in a dilute gas in which τ_1 is considerably longer than τ_2 correlations between successive collisions may be ignored. In such a case it may be shown[4,32] that the spin–lattice relaxation time in the limit $\omega_0 \tau_c \ll 1$ is given by

$$(T_1)^{-1} = 2\gamma^2 J(J+1)\tau_c \left\{ \frac{1}{3} H'^2 + \frac{3H''^2}{(2J-1)(2J+3)} \right\} \tag{4.85}$$

if there are no selection rules on m_J. From considerations expressed above, τ_c is identified with τ_1. For $J = 1$, equation (4.85) reduces to

$$(T_1)^{-1} = 2{\cdot}7 \times 10^{12} \tau_c \tag{4.86}$$

Changes in m_J require a torque and this can only arise from an anisotropic intermolecular force acting during a collision, isotropic interactions being ineffective. It is thus possible from measurements of T_1 in gases to obtain information about this anisotropic interaction, information which is not readily obtainable by other methods.

The experimental results obtained by Waugh and Johnson are summarized in Figure 4.11. Extrapolations for a particular mixture to infinite dilution of H_2 clearly refer to interaction of H_2 with the other component alone. It should be mentioned that the fact that for all mixtures T_1 was found to be a linear function of the mole fractions is evidence for the validity of the assumption that the behaviour of the systems studied is governed by binary collisions of two or more distinct kinds, the collisions

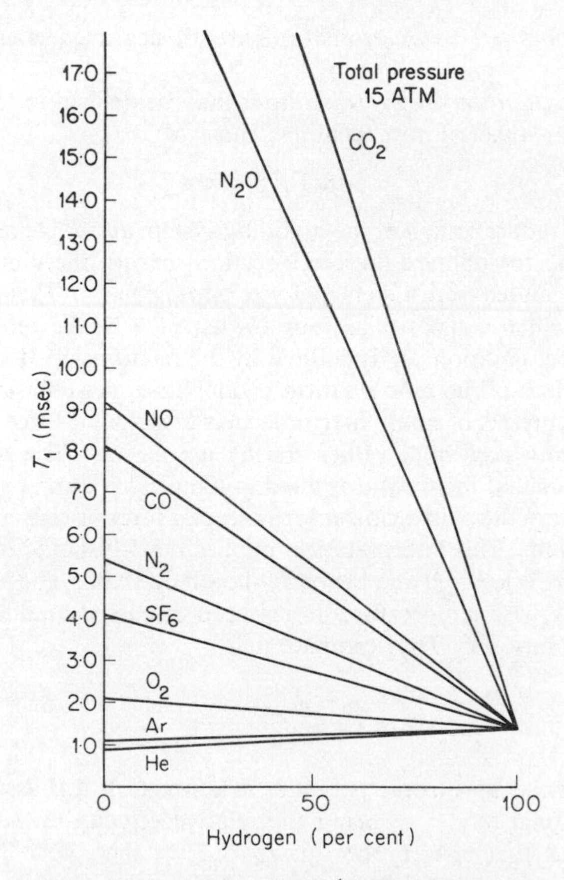

Figure 4.11 Dependence of ^{1}H spin–lattice relaxation time on composition in several gas mixtures. The total pressure was 15 atmospheres and the temperature 25°C. (From Reference 28 by courtesy of the authors and the Faraday Society)

being statistically independent. In other words, successive collisions are uncorrelated.

The measurements made by Waugh and Johnson were at a total pressure of 15 atmospheres and a temperature of 25°C. At this temperature there is some contribution from the $J = 3$ states. This was allowed for by modifying equation (4.86) to

$$(T_1)^{-1} = 3{\cdot}1 \times 10^{12}\tau_c \qquad (4.87)$$

It must be emphasized that ΔJ transitions are still neglected as a relaxation process.

A cross-section σ for the m_J transitions may be defined in terms of τ_c using the kinetic theory for an ideal gas, thus

$$\sigma = (\pi\mu \mathbf{k} T)^{1/2}/6P\tau_c \qquad (4.88)^*$$

where μ is the reduced mass of the colliding system and P is the pressure. It is interesting to compare the cross-sections σ with the kinetic cross-sections σ' calculated from Lennard-Jones parameters[33]. This is done in Table 4.2 in which is also to be found a list of W's, the reorientation probabilities per collision. W is defined by the relationship $W = \sigma/\sigma'$.

The low values of the reorientation probabilities per collision cannot be explained in terms of small intermolecular anisotropic forces. Indeed, these may attain very high values during a collision. The qualitative interpretation is that the torque applied to a molecule during a collision, which arises from the anisotropic intermolecular force, is only applied for a very short time. This interpretation implies that W should increase as the temperature is lowered and hence T_1 should decrease. The form of the variation of W with temperature T has been discussed qualitatively by Waugh and Johnson[28]. They estimate that

$$W \approx \frac{\tau_2{}^2\langle m_J|V_1|m'_J\rangle^2}{\hbar^2} \qquad (4.89)$$

in which V_1 is the anisotropic potential at contact. If it is assumed that τ_2 is proportional to v_r^{-1}, v_r being the relative velocity of the colliding molecules, and that v_r is proportional to $T^{1/2}$, then $W \propto T^{-1}$. This relationship has been tested for mixtures of H_2 with CO_2 over the range 273–373°K[28] and it was found that $W \propto T^{-1.1}$.

* The definition of σ in equation (4.88) assumes that the probabilities for $\Delta m_J = 0$, 1 and 2 are equal. Only the last two are included in σ.

Table 4.2

Relaxation times, cross-sections σ for m_J transitions, kinetic theory cross-section σ' and reorientation probabilities per collision W for mixtures of H_2 with various gases at 25°C and 15 atm pressure

Solvent gas	T_1(msec) $P = 15$ atm	σ (Å^2)	σ' (Å^2)	W
He	0·9	0·24	24	0·010
Ar	1·0	0·30	31	0·010
H_2	1·4	0·35	26	0·013
O_2	2·0	0·59	33	0·018
SF_6	4·0	1·3	55	0·024
CF_4	4·7	1·5	45	0·033
N_2	5·3	1·6	34	0·047
CO	7·7	2·3	34	0·068
NO	8·6	2·7	29	0·093
$CClF_3$	11·5	3·6	46	0·078
N_2O	26	8·1	44	0·18
CO_2	31	9·6	43	0·22
CHF_3	43	13	46	0·28

(The values of T_1 are for infinite dilution of H_2.)

4.4.2 The Oppenheim–Bloom Theory

In a series of papers[34–37], Oppenheim and Bloom, together with others, have made an extensive theoretical investigation of the relationship between nuclear spin relaxation times and intermolecular interactions. It has been noted above that this work has been well reviewed by Deutch and Oppenheim[31] and the theory will not be described in detail here. Two of the assumptions made by these authors are that anisotropic interactions are sufficiently weak that perturbation theory may be used to describe the collisions and that the translational motion of the molecules is determined solely by the isotropic interactions. The systems treated were molecular H_2 with different ortho and para concentrations; mixtures of H_2 with He, Ne, Ar, N_2, CO_2, O_2, CO, NO and N_2O; and normal D_2. Their analysis enabled them to determine the form of the anisotropic intermolecular potentials provided that the form of the isotropic part was known. These latter potentials were assumed to be of the Lennard-Jones form.

Whilst in general the anisotropic potentials may be represented by multipole expansions, it was found that the quadrupole–quadrupole

Table 4.3

Molecular quadrupole moments derived from nuclear spin relaxation times and by other methods. Units are 10^{-26} e.s.u.

Molecule	Spin–relaxation experiment	Other experiments
H_2	0·60	$0{\cdot}63^a$; $1{\cdot}0^b$
O_2	0·92	
N_2	1·70	$1{\cdot}64^c$; $1{\cdot}48^d$; $1{\cdot}5^e$; $1{\cdot}9^f$; $1{\cdot}0^b$
CO	2·0	$2{\cdot}8^f$
NO	2·05	
N_2O	4·25	
CO_2	4·85	$4{\cdot}1^g$; $4{\cdot}6^e$; $5{\cdot}9^b$

References and notes:

a. J. Harrick and N. F. Ramsey, *Phys. Rev.*, **88**, 288 (1952)—molecular beam techniques.

b. G. Birnbaum and A. A. Maryott, *J. Chem. Phys.*, **36**, 2032 (1962)—collision-induced microwave absorption.

c. J. van Kranendonk, *Physica*, **24**, 347 (1954)—collision-induced vibrational spectrum of H_2—N_2 mixtures.

d. J. D. Poll, *Phys. Rev. Letters*, **7**, 32 (1963)—induced absorption in the far infrared.

e. J. A. A. Ketelaar and R. P. H. Rettschnick, *Mol. Phys.*, **7**, 191 (1963)—pressure-induced rotational spectrum of N_2.

f. R. H. Orcutt, *J. Chem. Phys.*, **39**, 605 (1963)—second virial coefficients.

g. A. D. Buckingham and R. L. Disch, *Proc. Roy. Soc.* (*London*), **A273**, 275 (1963)—bi-refringence induced by an inhomogeneous electric field. This value is said to be in doubt to within $\pm 20\%$ because of lack of knowledge of the anisotropic polarizability of CO_2.

terms were dominant. Table 4.3 lists values of molecular quadrupole moments derived from spin–lattice relaxation times. Where possible these are compared with values obtained by other methods. In general the agreement is good and is usually within the combined experimental errors. It has been pointed out[37] that the values of the molecular quadru-pole moments obtained from nuclear spin relaxation methods must be considered as upper limits. Despite this reservation, it is clear that through the Oppenheim–Bloom theory spin relaxation measurements may now be considered an important tool in obtaining detailed informa-tion about the anisotropic interactions between molecules.

4.4.3 Gordon's Theory of Spin Relaxation in Gases

Gordon[38] has emphasized the unique properties of H_2 in the gas phase which permit the use of quantum-mechanical perturbation theory in the calculation of the reorientational probabilities. These properties are the comparative weakness of the intermolecular torques and the wide spacing of the rotational levels mentioned above. These conditions do not obtain in many other gases and perturbation calculations for these systems are limited in their usefulness. Gordon has presented a non-perturbative method of calculation which is valid for intermolecular torques of any size. Only binary collisions are considered and it is assumed that there is no correlation between successive collisions in either dilute gases or mixtures of dilute gases. Finally, the translational and rotational motions of the molecules are treated by the methods of classical mechanics. This limitation implies that the method is inapplicable to H_2 but should be quite accurate for heavier gases. The method is thus in a sense complementary to the work of Bloom and Oppenheim. The calculations performed by Gordon give expressions which relate the nuclear spin relaxation times to the rotational angular momentum transferred by collisions and it is possible to calculate spin relaxation times for given anisotropic potentials.

This theory has been used by Tward and Armstrong to calculate effective cross-sections for the collisional transfer of angular momentum in HCl, HBr and HI from measurements of nuclear spin relaxation times[39]. Further, using the ^{19}F resonance, these authors have made similar measurements on the spherical, non-polar molecules CF_4 and

Table 4.4
Probabilities for the transfer of angular momentum per collision for hydrogen chloride and iodide

Gas	Temp (°K)	W
HCl	228	3·32
	240	3·06
	263	2·88
	316	2·77
	338	2·41
HI	293	2·62
	303	2·43
	314	2·43
	337	2·40

Table 4.5

Probabilities for the transfer of angular momentum per collision for CF_4 and SiF_4

Temperature (°K)		W
244·4	CF_4	0·58
300		0·49
334		0·41
371·5		0·37
231·5	SiF_4	0·98
244		0·97
252		0·88
295		0·83
320·5		0·69
337·5		0·62

SiF_4[40]. In marked contrast to the systems studied by Waugh and Johnson (Table 4.2), it was found that for the hydrogen halides, the probabilities for the transfer of angular momentum per collision were greater than unity and relaxation occurred solely via the mechanism of spin–rotational interaction. Values for W have been calculated from Reference 39 and are shown in Table 4.4. Tward and Armstrong ascribe the large cross-sections to the long-range dipole–dipole interactions between these molecules. This conclusion is supported to some extent by the relaxation measurements made by these authors on the non-polar molecules CF_4 and SiF_4[40] where it was found that between one and three binary collisions were necessary to randomize the rotational angular momentum of these molecules. The temperature dependence of the spin–lattice relaxation times again indicated that relaxation was induced by spin–rotational interaction. The probabilities for transfer of angular momentum per collision at different temperatures are given in Table 4.5.

At a given temperature collisions between SiF_4 molecules are more effective in randomizing the rotational angular momentum than are collisions between CF_4 molecules.

It should be possible to obtain intermolecular potentials by fitting cross-sections obtained by the application of collision dynamics to experimental ones derived from spin–lattice relaxation measurements.

REFERENCES

1. R. Kubo and K. Tomita, *J. Phys. Soc. Japan*, **9**, 888 (1954).
2. A. G. Redfield, 'The theory of relaxation processes', in *Advances in Magnetic Resonance* (Ed. J. S. Waugh), Vol. 1, Academic Press, New York, 1965, pp. 28–29.
3. C. P. Slichter, *Principles of Magnetic Resonance*, Harper, New York, 1964, Appendix C.
4. A. Abragam, *The Principles of Nuclear Magnetism*, Oxford University Press, London, 1961, Chap. VIII.
5. H. M. McConnell and C. H. Holm, *J. Chem. Phys.*, **25**, 1289 (1956).
6. D. K. Green and J. G. Powles, *Proc. Phys. Soc. (London)*, **85**, 87 (1965).
7. Reference 3, p. 153.
8. P. Debye, *Polar Molecules*, Dover Publications, New York, 1945, Chap. 5.
9. H. G. Hertz, 'Microdynamic behaviour of liquids as studied by n.m.r. relaxation times', in *Progress in Nuclear Magnetic Resonance Spectroscopy* (Eds. J. W. Emsley, J. Feeney and L. H. Sutcliffe), Vol. 3, Pergamon Press, London, 1967, Chap. 5, pp. 169–170.
10. J. G. Powles, 'The relaxation of molecular orientation in liquids', in *Molecular Relaxation Processes*, Chem. Soc. Spec. Publ. 20, London, 1966.
11. W. B. Moniz, W. A. Steele and J. A. Dixon, *J. Chem. Phys.*, **38**, 2418 (1963).
12. A. Gierer and K. Wirtz, *Z. Naturforsch.*, **8A**, 532 (1953).
13. N. E. Hill, *Proc. Phys. Soc. (London)*, **67B**, 149 (1954).
14. R. W. Mitchell and M. Eisner, *J. Chem. Phys.*, **33**, 86 (1960).
15. R. W. Mitchell and M. Eisner, *J. Chem. Phys.*, **34**, 651 (1961).
16. A. M. Pritchard and R. E. Richards, *Trans. Faraday Soc.*, **62**, 1388 (1966).
17. A. M. Pritchard and R. E. Richards, *Trans. Faraday Soc.*, **62**, 2014 (1966).
18. P. S. Hubbard, *Phys. Rev.*, **109**, 1153 (1958).
19. P. S. Hubbard, *Phys. Rev.*, **128**, 650 (1962).
20. P. S. Hubbard, *Phys. Rev.*, **131**, 275 (1963).
21. H. Shimizu, *J. Chem. Phys.*, **37**, 765 (1962).
22. J. S. Waugh, 'Nuclear relaxation and molecular relaxation', in *Molecular Relaxation Processes*, Chem. Soc. Spec. Publ. 20, London, 1966.
23. B. K. P. Scaife, 'The rôle of dipole–dipole coupling in dielectric relaxation', in *Molecular Relaxation Processes*, Chem. Soc. Spec. Publ. 20, London, 1966.
24. A. Budó, *J. Chem. Phys.*, **17**, 686 (1949).
25. R. Zwanzig, *J. Chem. Phys.*, **38**, 2766 (1963).
26. D. E. Woessner, *J. Chem. Phys.*, **36**, 1 (1962).
27. W. B. Moniz and H. S. Gutowsky, *J. Chem. Phys.*, **38**, 1155 (1963).
28. J. S. Waugh and C. S. Johnson, Jr., *Discussions Faraday Soc.*, **34**, 191 (1962).
29. R. Brout, *J. Chem. Phys.*, **22**, 934 (1954).
30. I. Zartman, *J. Acoust. Soc. Am.*, **21**, 171 (1949).
31. J. M. Deutch and I. Oppenheim, 'Nuclear relaxation in hydrogen gas and liquid', in *Advances in Magnetic Resonance* (Ed. J. S. Waugh), Vol. 2, Academic Press, New York, 1966, pp. 225–262.
32. N. Bloembergen, E. M. Purcell and R. V. Pound, *Phys. Rev.*, **73**, 679 (1948).

33. J. O. Hirschfelder, C. F. Curtiss and R. B. Bird, *Molecular Theory of Gases and Liquids*, John Wiley, New York, 1954.
34. I. Oppenheim and M. Bloom, *Can. J. Phys.*, **39**, 845 (1961).
35. M. Bloom and I. Oppenheim, *Can. J. Phys.*, **41**, 1580 (1963).
36. I. Oppenheim, M. Bloom and H. C. Torrey, *Can. J. Phys.*, **42**, 70 (1964).
37. M. Bloom, I. Oppenheim, M. Lipsikas, C. G. Wade and C. F. Yarnell, *J. Chem. Phys.*, **43**, 1036 (1965).
38. R. G. Gordon, *J. Chem. Phys.*, **44**, 228 (1966).
39. E. Tward and R. L. Armstrong, *J. Chem. Phys.*, **47**, 4068 (1967).
40. R. L. Armstrong and E. Tward, *J. Chem. Phys.*, **48**, 332 (1968).

Appendix 1

Some Comments on Section 1.2.12

Although it does qualify as a rough survey, the description of molecular multipoles suffers from many limitations and inaccuracies. For example, it has been implicit in our presentation that internal couplings among the various contributors to the magnetic moment are stronger than the interactions of these components with the external field. When this assumption is invalid, as it often will be, a different but not necessarily more difficult analysis must be applied. Furthermore, in very many cases neither the LS or *jj* coupling scheme (or analogues thereof) provides an adequate description of the situation and so an 'intermediate coupling scheme' must be invoked. The literature associated with these issues is enormous but a few papers which are especially relevant to examples of current interest in kinetic theory are: (a) 'The Microwave Spectra of the Free Radicals OH and OD' (the intermediate coupling scheme), G. C. Dousmanis, T. M. Sanders, Jr. and C. H. Townes, *Phys. Rev.*, **100**, 1735 (1955); (b) 'On the Pressure Dependence of the Transport Properties of Dilute Polyatomic Gases' (includes a detailed discussion of the transition with field strength from one coupling mode to another), F. R. McCourt and H. Moraal, *Z. Naturforsch.*, **24a**, 1687 (1969); (c) 'The Senftleben Effects for a Gas of $^2\Sigma$ Molecules', H. Moraal and F. R. McCourt, *Chem. Phys. Letters*, **3**, 691 (1969); and (d) 'On the Senftleben Effect for a Dilute Gas of $^2\Pi$ Molecules', J. A. R. Coope, F. R. McCourt and H. Moraal, *Chem. Phys. Letters*, **4**, 84 (1969).

Appendix 2

The Thermal Conduction Senftleben Effect (Section 1.3.1)

Soon after the first report of Senftleben's experiments C. J. Gorter [*Naturwiss.*, **26**, 140 (1938)] devised a free-path theory to account for the observed phenomena. Although this theory, which was subsequently amended by F. Zernike and C. van Lier [*Physica*, **6**, 959 (1939)], is inadequate, it does correctly identify the basic mechanism and handles it with dispatch.

The collision cross-section of a rotating molecule was recognized to be dependent upon the angle between its direction of motion and its axis of rotation, that is, upon the helicity $\zeta \equiv \mathbf{L} \, . \, \mathbf{c}/Lc$. This cross-section was identified with the projection of the molecule's shape upon a plane perpendicular to the direction of its motion. Molecules with different helicities can be expected to contribute different amounts to the rates of transfer of energy and momentum. An imposed magnetic field will alter the directions of the molecular axes of rotation and so change the effective collision cross-sections. Indeed, one concludes almost immediately that precession of the spin will lead to an increase of the cross-section and so to diminution of the transport coefficients.

According to free-path theory the thermal conductivity is proportional to the average of $\lambda_{\mathrm{fp}} \cos^2 \gamma$ with λ_{fp} the free-path length and γ the angle between ∇T and the direction of molecular motion. Rotation causes the cross-section to vary so rapidly that we need only its average value. For example, in the case of a spherocylinder (sheath length, L, radius σ) the projection of this (helicity-dependent) averaged cross-section is found to be

$$A(\zeta) = \pi\sigma^2[1 + \delta^2|\cos \zeta| + \delta \frac{4}{\pi} E(\sin \zeta)] = \pi\bar{\sigma}^2[1 + \varepsilon P_2(\cos \zeta) + \cdots]$$

with

$$\delta = L/2\sigma, \qquad \varepsilon = (15\pi/64)(\delta + 4\delta^2/\pi)/(1+\pi\delta/2+\delta^2/2)$$

$$\approx (15\pi\delta/64) \quad \text{and} \quad \pi\bar{\sigma}^2 = \pi\sigma^2(1+\pi\delta/2+\delta^2/2)$$

Now the probability $f(x)$ for a molecule to move a distance x without collision is related to $p(\xi)$, the collision probability (per unit length of trajectory), by the familiar formula $f(x) = \exp\{-\int_0^x p(\xi)\,\mathrm{d}(\xi)\}$: the free path length is given by $\lambda_{\mathrm{fp}} = \int_0^\infty \mathrm{d}x\, f(x)$. We identify $p(\xi)$ with $nA(\zeta)$ and relate $\cos\zeta = \cos\theta_c \cos\theta_L + \sin\theta_c \sin\theta_L \cos(\varphi_L - \varphi_c)$ to the angular co-ordinates (θ_c, φ_c) and (θ_L, φ_L) of **c** and **L** respectively. Finally, $\varphi = \varphi_L - \varphi_c = \psi + \omega t \approx \psi + (\omega/\bar{c})\xi$ where $\omega = \mu H/|\mathbf{L}|$ is the angular velocity of precession and where $\bar{c} \approx \sqrt{3\mathbf{k}T/m}$ denotes the average molecular speed. It then was proved by Zernike and van Lier that for $\zeta = 0$ and $\zeta = \pi/2$ the average free path length

$$\bar{\lambda}_{\mathrm{fp}}(\zeta) = \int_{-1}^{1} \frac{1}{2}\mathrm{d}(\cos\theta_c) \int_{-1}^{1} \frac{1}{2}\mathrm{d}(\cos\theta_L) \int_0^{2\pi} \frac{\mathrm{d}\psi}{4\pi} \lambda_{\mathrm{fp}}(\zeta)\cos^2\gamma$$

assumes the values

$$\bar{\lambda}_{\mathrm{fp}}(0) = \lambda_{\mathrm{fp}}^0 \left[1 + \frac{4\varepsilon^2}{1575}\{11 + 18(1+\Theta^2)^{-1} + 6(1+4\Theta^2)^{-1}\}\right]$$

and

$$\bar{\lambda}_{\mathrm{fp}}\left(\frac{\pi}{2}\right) = \lambda_{\mathrm{fp}}^0 \left[1 + \frac{\varepsilon^2}{1575}\{5 + 12(1+\Theta^2)^{-1} + 18(1+4\Theta^2)^{-1}\}\right]$$

respectively. Here $\lambda_{\mathrm{fp}}^0 = 1/n\pi\bar{\sigma}^2$, $\Theta = \omega\tau$ and $\tau = \lambda_{\mathrm{fp}}^0/\bar{c}$.

From these results it follows immediately that

$$\Delta\lambda_{\parallel}/\lambda_0 = \frac{8\varepsilon^2}{525}\left\{3\frac{\Theta^2}{1+\Theta^2} + \frac{(2\Theta)^2}{1+(2\Theta)^2}\right\}$$

$$\Delta\lambda_{\perp}/\lambda_0 = \frac{8\varepsilon^2}{525}\left\{2\frac{\Theta^2}{1+\Theta^2} + 3\frac{(2\Theta)^2}{1+(2\Theta)^2}\right\}$$

and $(\Lambda\lambda_{\perp})_{\mathrm{sat}}/(\Delta_{\parallel})_{\mathrm{sat}} = 5/4$. The values of Θ at which $\Delta\lambda_{\parallel}$, $\Delta\lambda_{\perp}$ and $\Delta\lambda$ equal half their saturation values are given by $\Theta_{1/2}^{\parallel} = [\{3+(55)^{1/2}\}/16]^{1/2} \doteq 0{\cdot}80$, $\Theta_{1/2}^{\perp} = [\{-3+(391)^{1/2}\}/40]^{1/2} \doteq 0{\cdot}65$ and

$\Theta_{1/2} = [\{1+(145)^{1/2}\}/24]^{1/2} \doteq 0{\cdot}74$, respectively. The single and double frequencies, Θ and 2Θ, which appear in these formulas also occur in simplified theories based upon the Boltzmann or Waldmann–Snider equation but no such simple functional dependence emerges from the more rigorous theory.

To test these results we adopt the spherocylinder model and assume the molecules to be diamagnetic. Then, it can be established that $\Theta = \mathfrak{B}_D[(3\pi)^{1/2}(\sigma/d)^2]^{-1}$. In their studies Klein, Dahler, Hoffman and Cooper[67] used for N_2 the values $L = 0{\cdot}52$ Å, $\sigma = 1{\cdot}884$ Å and $d = 1{\cdot}098$ Å. With these $\varepsilon = 0{\cdot}0992$ and the Gorter–Zernike–van Lier theory predicts for $(\Delta\lambda/\lambda_0)_{sat} \times 10^3$, $(\Delta\lambda_\perp/\lambda_0)_{sat} \times 10^3$, $(\Delta\lambda_\parallel/\lambda_0)_{sat} \times 10^3$, $\mathfrak{B}_{1/2}$, $\mathfrak{B}_{1/2}^\perp$ and $\mathfrak{B}_{1/2}^\parallel$ the values 0·67, 0·75, 0·60, −6·66, −5·88 and −7·23. Klein, *et al.* obtained for these the values 8·04 (7·9), 9·67 (10·5), 6·40, −30·33 (−34·51), −25·18 and −39·15 where the numbers in parentheses refer to experimental observations.

Although we see that it fails by a large margin to account for the quantitative aspects of the thermal conduction Senftleben effect, this insightful theory does succeed excellently in predicting the correct functional forms of the field shifts, in accurately identifying $(\Delta\lambda_\parallel)_{sat}$ and $(\Delta\lambda_\perp)_{sat}$ as strongly dependent upon molecular non-sphericity, and in correlating $\mathfrak{B}_{1/2}^\parallel$ and $\mathfrak{B}_{1/2}^\perp$ with the magnitude of the molecular magnetic moment and with the 'elastic' cross-section $\pi\bar{\sigma}^2$.

The basic ideas imbodied in the Gorter theory have their analogues in the more complex theories based upon the Boltzmann and Waldmann–Snider equations. In particular, the non-central forces of interaction lead to a collisional coupling of linear and angular momentum which in the presence of a velocity or temperature gradient gives rise to an anisotropic distribution in spin space. (The effects of these distortions upon the values of the gas transport coefficient were considered previously.) The imposition of an external field partially destroys this anisotropy.

Appendix 3

Comments Added in Proof

During the few months which have elapsed since our preparation of this manuscript we have discovered that it contains a number of oversights and inadequacies. Within this same period of time there also have appeared several relevant technical publications.

(i) For an excellent bibliography and a brief but useful commentary on the theoretical and experimental aspects of inelastic scattering processes—with particular emphasis upon vibrational and rotation relaxation—see R. G. Gordon, W. Klemperer and J. I. Steinfeld, *Ann. Rev. Phys. Chem.*, H. Eyring, Ed. (Annual Reviews, Inc., Palo Alto, 1968).

(ii) In addition to the classical trajectory studies of Cross and Herschbach[41] and of van der Ree[44] we should also have cited the recent investigations of rotational relaxation by L. M. Raff [*J. Chem. Phys.*, **46**, 520 (1967); **47**, 1884 (1967)] and by G. C. Berend and S. W. Benson [*J. Chem. Phys.*, **47**, 4199 (1967)].

(iii) Carl Nyeland [*J. Chem. Phys.*, **46**, 63 (1967)] has used a classical perturbative method which appears to be less restrictive than that of Parker[43] to calculate the rotational relaxation time for rotors restricted to lie in the same plane.

(iv) A classical time-dependent perturbation theory, which closely resembles the familiar quantum theory, has been developed by Garrido [L. M. Garrido, *Proc. Phys. Soc.* (*London*), **76**, 33 (1960); *J. Math. Anal. Appl.*, **3**, 295 (1961); L. M. Garrido and F. Gascon, *Proc. Roy. Soc.* (*London*), **81**, 1115 (1963)]. This theory constitutes a significant advance beyond Parker's perturbation scheme and those used by Cross and Herschbach[41] and by Nyeland. It has been applied by F. J. Zeleznik [*J. Chem. Phys.*, **47**, 3410 (1967)] to the calculation of the relaxation time for a two-dimensional rotor.

L. Rowley and J. S. Dahler (unpublished) have applied this theory to elastic scattering and to the vibrational relaxation of 'breathing spheres' and are currently investigating its applicability to three-dimensional collisions of rotors.

(v) R. A. Marcus [*J. Chem. Phys.*, **52**, 4803 (1970)] has recently developed a perturbation method which differs somewhat from Garrido's.

(vi) The distorted-wave Born series has been applied by R. E. Roberts and J. Ross [*J. Chem. Phys.*, **52**, 5011 (1970)] to rotationally inelastic scattering of an atom by a diatomic molecule.

(vii) A. Tip [*Phys. Lett.*, **30A**, 147 (1969)] has derived a modification of the Waldmann–Snider equation which is applicable to polyatomic gases subject to high frequency disturbances.

(viii) For a brief summary of the current status of the Scott effect the reader is referred to the abstracts of the papers presented at the March 1970, Dallas meeting of the American Physical Society (Bulletin of the APS, Series II, Vol. 15, pp. 284–286).

(ix) G. A. Stevens [*Physica*, **44**, 401 (1969)] has developed an approximation technique for evaluating transport cross-sections (for rigid regular tetrahedral molecules) which treats the translational motions in the classical limit. This scheme appears to be computationally tractible and may turn out to be a valuable tool for the numerical estimation of polyatomic gas transport coefficients.

(x) A recent theory due to William H. Miller (to be published) shows how one can use exact solutions of the classical equations of motion to construct the corresponding classical approximation to the collision cross-sections. By coupling this theory with a perturbation technique it should be possible to generate useful approximations to many of the cross-sections of importance in the kinetic theory of polyatomic gases.

Author Index

Subject Index